Microprocessor Design

Microprocessor Design

A Practical Guide from Design Planning to Manufacturing

Grant McFarland

McGraw-Hill

New York Chicago San Francisco Lisbon London Madrid
Mexico City Milan New Delhi San Juan Seoul
Singapore Sydney Toronto

The McGraw·Hill Companies

CIP Data is on file with the library of Congress

Copyright © 2006 by The McGraw-Hill Publishing Companies, Inc. All rights reserved. Printed in the United States of America. Except as permitted under the United States Copyright Act of 1976, no part of this publication may be reproduced or distributed in any form or by any means, or stored in a data base or retrieval system, without the prior written permission of the publisher.

1 2 3 4 5 6 7 8 9 0   DOC/DOC   0 1 9 8 7 6

ISBN 0-07-145951-0

The sponsoring editor for this book was Wendy Rinaldi and the production supervisor was David Zielonka. It was set in Century Schoolbook by International Typesetting and Composition. The art director for the cover was Anthony Landi.

Printed and bound by RR Donnelley.

McGraw-Hill books are available at special quantity discounts to use as premiums and sales promotions, or for use in corporate training programs. For more information, please write to the Director of Special Sales, McGraw-Hill Professional, Two Penn Plaza, New York, NY 10121-2298. Or contact your local bookstore.

Information contained in this work has been obtained by The McGraw-Hill Companies, Inc. ("McGraw-Hill") from sources believed to be reliable. However, neither McGraw-Hill nor its authors guarantee the accuracy or completeness of any information published herein and neither McGraw-Hill nor its authors shall be responsible for any errors, omissions, or damages arising out of use of this information. This work is published with the understanding that McGraw-Hill and its authors are supplying information but are not attempting to render engineering or other professional services. If such services are required, the assistance of an appropriate professional should be sought.

To Elaine

Brief Contents

Chapter 1. The Evolution of the Microprocessor — 1

Chapter 2. Computer Components — 37

Chapter 3. Design Planning — 71

Chapter 4. Computer Architecture — 95

Chapter 5. Microarchitecture — 127

Chapter 6. Logic Design — 171

Chapter 7. Circuit Design — 199

Chapter 8. Layout — 239

Chapter 9. Semiconductor Manufacturing — 263

Chapter 10. Microprocessor Packaging — 303

Chapter 11. Silicon Debug and Test — 331

Glossary — 357

Index — 385

Contents

Preface xv
Acknowledgments xix

Chapter 1. The Evolution of the Microprocessor   1

    Overview   1
    Objectives   1
    Introduction   1
    The Transistor   3
    The Integrated Circuit   10
    The Microprocessor   14
    Moore's Law   17
        Transistor scaling   19
        Interconnect scaling   24
        Microprocessor scaling   27
        The future of Moore's law   30
    Conclusion   33
    Key Concepts and Terms   34
    Review Questions   34
    Bibliography   34

Chapter 2. Computer Components   37

    Overview   37
    Objectives   37
    Introduction   37
    Bus Standards   38
    Chipsets   41
    Processor Bus   44
    Main Memory   47
    Video Adapters (Graphics Cards)   51
    Storage Devices   53

x   Contents

Expansion Cards   55
Peripheral Bus   57
Motherboards   58
Basic Input Output System   62
Memory Hierarchy   63
Conclusion   69
Key Concepts and Terms   69
Review Questions   69
Bibliography   70

Chapter 3. Design Planning   71

Overview   71
Objectives   71
Introduction   71
Processor Roadmaps   74
Design Types and Design Time   79
Product Cost   85
Conclusion   91
Key Concepts and Terms   92
Review Questions   92
Bibliography   93

Chapter 4. Computer Architecture   95

Overview   95
Objectives   95
Introduction   95
Instructions   98
   Computation instructions   99
   Data transfer instructions   103
   Control flow instructions   111
Instruction Encoding   115
   CISC versus RISC   118
   RISC versus EPIC   120
   Recent x86 extensions   122
Conclusion   124
Key Concepts and Terms   125
Review Questions   125
Bibliography   126

Chapter 5. Microarchitecture   127

Overview   127
Objectives   127
Introduction   128
Pipelining   128
Designing for Performance   134
Measuring Performance   137
Microarchitectural Concepts   142

Contents   xi

Cache memory	143
Cache coherency	147
Branch prediction	149
Register renaming	152
Microinstructions and microcode	154
Reorder, retire, and replay	157
Life of an Instruction	160
Instruction prefetch	161
L2 cache read	162
Instruction decode	162
Branch prediction	162
Trace cache write	163
Microbranch prediction	163
Uop fetch and drive	163
Allocation	164
Register rename	165
Load instruction queue	165
Schedule and dispatch	165
Register file read	166
Execute and calculate flags	166
Retirement and drive	167
Conclusion	168
Key Concepts and Terms	168
Review Questions	168
Bibliography	169

Chapter 6. Logic Design   171

Overview	171
Objectives	171
Introduction	171
Hardware Description Language	173
Design automation	175
Pre-silicon validation	178
Logic Minimization	182
Combinational logic	182
Sequential logic	191
Conclusion	196
Key Concepts and Terms	197
Review Questions	197
Bibliography	197

Chapter 7. Circuit Design   199

Overview	199
Objectives	199
Introduction	199
MOSFET Behavior	200
CMOS Logic Gates	207
Transistor sizing	212

xii Contents

Sequentials	216
Circuit Checks	220
Timing	221
Noise	226
Power	231
Conclusion	235
Key Concepts and Terms	236
Review Questions	236
Bibliography	237

Chapter 8. Layout   239

Overview	239
Objectives	239
Introduction	239
Creating Layout	240
Layout Density	245
Layout Quality	253
Conclusion	259
Key Concepts and Terms	260
Review Questions	260
Bibliography	261

Chapter 9. Semiconductor Manufacturing   263

Overview	263
Objectives	263
Introduction	263
Wafer Fabrication	265
Layering	268
Doping	268
Deposition	272
Thermal oxidation	276
Planarization	278
Photolithography	279
Masks	280
Wavelength and lithography	282
Etch	286
Example CMOS Process Flow	289
Conclusion	300
Key Concepts and Terms	301
Review Questions	301
Bibliography	302

Chapter 10. Microprocessor Packaging   303

Overview	303
Objectives	303
Introduction	303

Package Hierarchy	304
Package Design Choices	308
Number and configuration of leads	309
Lead types	311
Substrate type	313
Die attach	318
Decoupling capacitors	319
Thermal resistance	320
Multichip modules	323
Example Assembly Flow	325
Conclusion	328
Key Concepts and Terms	329
Review Questions	329
Bibliography	330

Chapter 11. Silicon Debug and Test — 331

Overview	331
Objectives	331
Introduction	331
Design for Test Circuits	333
Post-Silicon Validation	338
Validation platforms and tests	339
A bug's life	341
Silicon Debug	344
Silicon Test	350
Conclusion	353
Key Concepts and Terms	354
Review Questions	355
Bibliography	355

Glossary   357
Index   385

Preface

Reading This Book

Microprocessor design isn't hard, but sometimes it seems that way. As processors have grown in complexity and processor design teams have grown in size, individual design engineers have become more specialized, focusing on only one part of the design process. Each step in the design flow has its own jargon; today it is not at all hard to be working on a processor design team and still not have a clear understanding of aspects of design that don't involve you personally. Likewise, most textbooks focus on one particular aspect of processor design, often leaving out information about what steps came before or what will happen afterward. The intent of this book is to provide an overall picture of the microprocessor design flow, from the initial planning of a processor through all the steps required to ship to customers.

Covering the entire design flow in a single book means that only the most important aspects of each step are covered. This book provides the key concepts of processor design and the vocabulary to enable the reader to learn more about each step. Students may use this book to gain a broad knowledge of design and to decide which area they are most interested in pursuing further. Engineers already working in design will find out how their specialty fits into the overall flow. Nonengineers who interact with design teams, such as managers, marketers, or customers, will learn the jargon and concepts being used by the engineers with whom they work.

The flow of the book follows the life of a microprocessor design. The first two chapters cover the concepts required before a design really begins. Chapter 1 discusses transistors and how their evolution drives processor design. Chapter 2 describes some of the other components with which the processor will communicate. Chapter 3 begins the processor design flow with planning, and the following chapters take the design through each of the needed steps to go all the way from an idea to shipping a finished product.

Chapter list

Chapter 1—The Evolution of the Microprocessor: Describes the development of the microprocessor and how transistor scaling has driven its evolution.

Chapter 2—Computer Components: Discusses computer components besides the microprocessor and the buses through which they interact with the processor.

Chapter 3—Design Planning: Explains the overall steps required to design a processor and some of the planning required to get started.

Chapter 4—Computer Architecture: Examines trade-offs in choosing an instruction set and how both instructions and data are encoded.

Chapter 5—Microarchitecture: Explains the operation of the different functional areas of a processor and how they determine performance.

Chapter 6—Logic Design: Discusses converting a microarchitectural design into the logic equations required to simulate processor behavior.

Chapter 7—Circuit Design: Shows logic design equations being converted into a transistor implementation.

Chapter 8—Layout: Demonstrates circuit designs being converted to layout drawings of the different layers of material required for fabrication.

Chapter 9—Semiconductor Manufacturing: Shows how integrated circuits are manufactured from layout.

Chapter 10—Microprocessor Packaging: Discusses how completed die are packaged for use and the trade-offs of different types of packages.

Chapter 11—Silicon Debug and Test: Explains how designs are checked for flaws and completed die are tested before shipping to customers.

The many specialized terms and acronyms of processor design are explained as they are introduced in the text, but for reference there is also a glossary at the end of the book. After reading this, microprocessor design won't seem that hard after all.

The Future of Processor Design

The rapid changes in the semiconductor industry make predicting the future of processor design difficult at best, but there are two critical questions designers must address in the coming years.

- How can design make best use of ever-increasing numbers of transistors?
- How can processors be designed to be more power efficient?

The first of these questions has remained the biggest question facing processor designers since the beginning of the industry. By the end of 2006, that is, when this book gets published, the highest transistor count processors on the market should include more than 1 billion devices. If the current rate of increase continues, a 10-billion device processor is likely before 2015 and a 100-billion device processor by 2025. What will these processors be like? The most recent answer for how to make use of more transistors is to put multiple processor cores onto a single die. Does this mean that a 10-billion transistor processor will merely be a combination of ten 1-billion transistor processors? This is certainly possible, but a 100-billion transistor processor will almost certainly not be a hundred core processor. At least today, most software problems cannot be divided into this many separate pieces. Perhaps new methods will be found, but it is likely that the number of cores in future processors will be limited more by software than by hardware.

If most software applications will only be able to make use of a very small number of cores, will each single core contain tens of billions of transistors? Design tools and methods of today are not up to creating the design for such a processor. We may be moving from a time when processors designs are no longer determined by the limits of fabrication, but instead by the limits of the design process itself. Perhaps the processor will absorb the functionality of other computer components, as has happened in the past. A microprocessor with several general-purpose cores as well as a graphics processor, memory controller, and even main memory itself, all built into a single die, could make use of a very large number of transistors indeed. Maybe this type of true "system-on-a-chip" will be the future. In the past, it has always been feared that the end of fabrication improvements was just a few years away. It is physically impossible for the shrinking of transistors to continue at its current pace forever, but every prediction so far of the end of scaling has been wrong. Today, the problems of new hardware and software design methodologies threaten to slow processor improvements before manufacturing limits.

The second critical question of power efficiency has received serious attention from the industry only recently. Early "low-power" processors were simply desktop designs operated at lower voltages and frequencies in order to save power. Only recently has the rapidly growing popularity of portable computing products led to the creation of a number of processor designs intended from conception as low power. Power efficiency has become even more important as high-performance desktops and server processors have reached the limits of cost-effective power delivery and cooling. Suddenly, 100-W server processors and 1-W embedded processors have started to be designed for low power.

The industry has used a number of circuit designs and some fabrication techniques to reduce power, but we have barely scratched the surface of designing power-efficient architectures or microarchitectures. How should things like design validation, test, and packaging change to support power reduction? How should the design of the transistors themselves be altered? These questions are made more complex by a growing need to focus not just on maximum power, but also on average power across a variety of applications. Performance per watt is rapidly becoming more important than simple performance, and our design methods have only just begun to take this into account.

It is not for this book to say what the answers to these questions will be. Instead, time will tell, but a future edition of this book will likely have some of the answers as well as a whole list of new questions. In the meantime, the methods described in this book are the foundation of these future designs. With an understanding of the overall design flow, the reader is ready to ask questions about today's methods, and asking questions is the first step toward finding answers.

Grant McFarland
February, 2006

Acknowledgments

I would like to thank George Alfs, Elaine La Joie, and Robert McFarland for their time and effort in reviewing the drafts of this book. I would also like to thank Mark Louis for getting me started teaching at Intel, and a special thanks to everyone who has ever asked a question in one of my classes. I have learned as much from my students as I have from any teacher.

Microprocessor Design

Chapter 1

The Evolution of the Microprocessor

Overview

This chapter describes the development of the transistor, how transistors evolved into integrated circuits and microprocessors, and how scaling has driven the evolution of microprocessors.

Objectives

Upon completion of this chapter, the reader will be able to:

1. Describe N-type and P-type semiconductors.
2. Understand the operation of diodes and transistors.
3. Understand the difference between different types of transistors.
4. Describe Moore's law and some of its implications.
5. Be aware of limits to scaling transistors and wires in integrated circuits.
6. Describe different schemes for scaling interconnects.
7. Explain the difference between "lead microprocessors" and "compactions."
8. List possibilities for future transistor scaling.

Introduction

Processors are the brains of computers. Other components allow a computer to store or retrieve data and to input or output data, but the processor performs computations and does something useful with the data.

It is the processor that determines what action will happen next within the computer and directs the overall operation. Processors in early computers were created out of many separate components, but as technology improved it became possible to integrate all of the components of a processor onto a single piece, or chip, of silicon. These integrated circuits are called *microprocessors*.

Today microprocessors are everywhere. Supercomputers are designed to perform calculations using hundreds or thousands of microprocessors. Even personal computers that have a single central processor use other processors to control the display, network communication, disk drives, and other functions. In addition, thousands of products we don't think of as computers make use of microprocessors. Cars, stereos, cell phones, microwaves, and washing machines all contain microprocessors. This book focuses on the design of largest and most complex microprocessors, which are used as the central processing units of computers, but what makes processors so ubiquitous is their ability to provide many different functions.

Some computer chips are designed to perform a single very specific function, but microprocessors are built to run programs. By designing the processor to be able to execute many different instructions in any order, the processor can be programmed to perform whatever function is needed at the moment. The possible uses of the processor are limited only by the imagination of the programmer. This flexibility is one of the keys to the microprocessor's success. Another is the steady improvement of performance.

Over the last 30 years, as manufacturing technologies have improved, the performance of microprocessors has doubled roughly every 2 years.[1] For most products, built to perform a particular function, this amount of improvement would be unnecessary. Microwave ovens are an improvement on conventional ovens mainly because they cook food more quickly, but what if instead of heating food in a few minutes, they could be improved even more to only take a few seconds? There would probably be a demand for this, but what about further improvements so that it took only tenths of a second, or even just hundredths of a second.

At some point, further improvements in performance of a single task become meaningless because the task being performed is fast enough. However, the flexibility of processors allows them to constantly make use of more performance by being programmed to perform new tasks. All a processor can do is run software, but improved performance makes new software practical. Tasks that would have taken an unreasonable amount of time suddenly become possible.

---

[1]Moore, "No Exponential is Forever," 20.

If I never changed the software on my computer, it is likely that at some point it would become fast enough. Spell checking an entire large document in a few seconds is a useful feature, but the capability to do it a 100 times in a few seconds is overkill. What drives the need for performance is new functionality. People will sometimes say they need to buy a new computer because their old one has become too slow. This is of course only a matter of perception. Their computer has the exact same speed as the day they bought it. What has changed to make it appear slower is the software. As the performance of computers improves, new software is written to perform new tasks that require higher performance, so that installing the latest software on a computer that is a few years old makes it appear very slow indeed.

Being designed to run programs allows microprocessors to perform many different functions, and rapid improvements in performance are constantly allowing for new functions to be found. Continuing demand for new applications funds manufacturing improvements, which make possible these performance gains.

Despite all the different functions a microprocessor performs, in the end it is only a collection of transistors and wires. The job of microprocessor design is ultimately deciding how to connect transistors to be able to quickly execute the commands that run programs. As the number of transistors on a processor has grown from thousands to millions that job has become steadily more complicated, but a microprocessor is still just a collection of transistors connected to operate as the brain of a computer. The story of the first microprocessor is therefore also the story of the invention of the transistor and the integrated circuit.

The Transistor

In 1940, many experiments were performed with semiconductor crystals to try and create better diodes. Diodes allow electricity to flow in one direction but not the other and are required for radio and radar receivers. Vacuum tube diodes could be used but did not work well at the high frequencies required by accurate radar. Instead, crystals of semiconductors were used. How these crystals worked was very poorly understood at the time, but it was known that a metal needle touching the surface of some crystals could form a diode. These cat whisker diodes could operate at high frequencies but were extremely temperamental. The crystals had many defects and impurities and so the user was required to try different points on the crystal at random until finding one that worked well. Any vibration could throw the whole apparatus off. This was an application in need of a sturdy and reliable electrical switch.

At AT&T® Bell Laboratories, Russell Ohl was working with a silicon crystal when he noticed a very curious phenomenon. The crystal produced an electric current when exposed to light. This type of effect had been observed in other crystals but never to this degree, and it had never been well understood. Upon examining the crystal more closely Ohl discovered a crack that had formed in the crystal as it was made. This crack had caused the impurities in the crystal to be distributed unevenly between the two sides.

One side had impurities with **electrons** that were free to move through the crystal. He called this side the **N-type** silicon because it had negative charge carriers. The other side had impurities that produced spaces that electrons could occupy but were empty. The spaces or **holes** could move through the crystal as one electron after another moved to fill the hole, like moving one car after another into an empty parking space and causing the empty space to move. The holes acted as positive charge carriers and so Ohl called this **P-type** silicon. The junction formed by these two types of silicon allowed electricity to flow in only one direction, which meant that the energy added by light could produce a current in only one direction. The single-piece, solid-state diode had been discovered.

Today we have a much better understanding of why some impurities produce N-type and others P-type semiconductors, and the operation of a **junction diode**. Everything is made of atoms and since all atoms contain electrons, anything can conduct electricity. Anyone who has seen a lightning storm has seen proof that with a sufficiently large electric field, even air can conduct large amounts of electricity. Materials are classified by how easily they carry electricity. Materials like copper that conduct easily are called *conductors*, and materials like glass that do not conduct easily are called *insulators*. Some materials, such as silicon, normally do not conduct easily, but very small amounts of impurities cause them to become good conductors. These materials are called *semiconductors*. The reason for this behavior is electron energy bands and band gaps (Fig. 1-1).

Quantum mechanics tells us that electrons can occupy only a finite number of discrete energy levels. In any noncrystalline material the

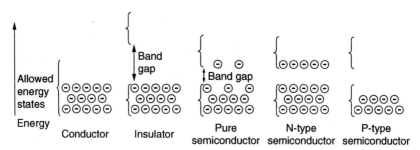

Figure 1-1 Energy bands.

spacing of atoms is nonuniform so the available quantum states vary through the material, but in a crystal the spacing of atoms is uniform so that continuous bands of allowed energy states are created. In a conductor the electrons do not completely fill these bands so that any energy at all will allow them to move through the crystal.

In an insulator the lowest energy band, the **valence band**, is completely filled with electrons and there is a large energy gap to the next band, the **conduction band**, which is completely empty. No current flows, because there are no open spaces for electrons in the valence band to move to, and they do not have enough energy to reach the conduction band. The conduction band has open spaces but has no electrons to carry current.

Semiconductors have a very small band gap between the valence and conduction band. This means that at room temperature a small number of electrons from the valence band will have enough energy to reach the conduction band. These electrons in the conduction band and the holes they leave behind in the valence band are now free to move under the influence of an electric field and carry current.

In a pure semiconductor there are very few free carriers, but their number is greatly increased by adding impurities. On the periodic table of elements, shown in Fig. 1-2, silicon is a column 4 element which means it has 4 outer electrons that can bond to neighboring atoms. If a

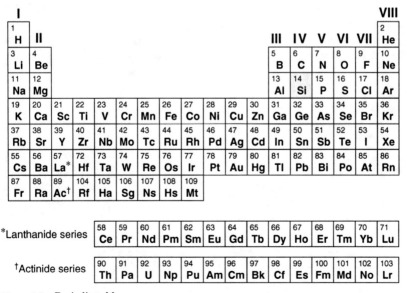

Figure 1-2   Periodic table.

column 5 element such as arsenic or phosphorous is added to a silicon crystal, these atoms will bond with the 4 neighboring silicon atoms leaving one electron left over. This electron is free to move in the conduction band and acts as a negative carrier. Therefore, adding column 5 elements creates N-type silicon. Column 3 elements, such as boron, have 3 outer electrons so they naturally create holes in the valence band, which act as positive charge carriers. Therefore, adding column 3 elements creates P-type silicon. Understanding N-type and P-type silicon allows us to understand Ohl's P-N junction diode.

Figure 1-3 shows a junction diode. The P-type silicon has positive carriers, and the N-type silicon has negative carriers. When a positive voltage is applied to the P-side of the diode and a negative voltage to the N-side, the positive voltage repels the positive carriers and the negative voltage repels the negative carriers. At the junction the holes and electrons recombine when the electrons drop from the conduction band to fill the holes in the valence band. Meanwhile, the voltage source is pushing more free electrons into the N-side and creating more holes in the P-side by removing electrons. As the process continues electric current flows. If the voltages are reversed, the negative voltage attracts the holes and the positive voltage attracts the electrons. The free carriers move away from each other and no current flows. Because a junction diode is made of a single piece of silicon, it is much more reliable than earlier cat whisker diodes and operates at much higher frequency than vacuum tube diodes.

Ohl's junction diode showed the potential of semiconductors to revolutionize electronics, but what was needed was a semiconductor device that could act as a switch or amplifier like a vacuum tube. In 1945, William Shockley was put in charge of semiconductor research at Bell Labs with the goal of creating a solid-state amplifier. Shockley's idea was

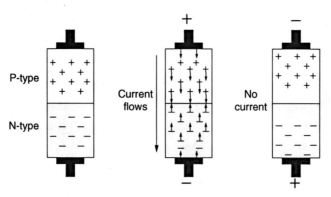

Figure 1-3 P-N junction diode.

to use a metal plate placed near the surface of a semiconductor, but not touching, to control the flow current within the semiconductor. He believed the electric field from the plate could create a channel of charge carriers within the semiconductor, allowing it to conduct dramatically more current than possible when the field was not present. The idea seemed sound, but when Shockley built an apparatus to test his theory, it didn't work. Even with 1000 V on the metal plate, he could not measure any amplification at all.

Shockley gave this puzzle to two of the researchers on his team. John Bardeen was one of very few men in the world at the time with a thorough grasp of the physics of semiconductors, and Walter Brattain was an experienced experimentalist with a reputation as a man who could build anything. Together these two began experiments to determine why Shockley's "field-effect" device did not work. They came to believe that the problem was charges trapped at the surface of the semiconductor. If the semiconductor naturally formed a slight excess of electrons at its surface and these charges were unable to move under the influence of an electric field, they would prevent an electric field from penetrating the surface and creating amplification. Bardeen and Brattain began looking for a way to neutralize these surface charges. They tried placing the semiconductor in various liquids and got some improvement but never with enough amplification at a high enough frequency to be useful.

Bardeen and Brattain had a piece of germanium (another column 4 semiconductor) prepared with a thin oxide layer on top. They hoped the oxide would somehow neutralize the surface charges. Brattain carefully cleaned the crystal and began testing. At first his results made no sense at all. The device responded as if the oxide layer wasn't even there. Then to his horror, Brattain realized that in cleaning the crystal he had accidentally removed the oxide layer. Germanium oxide is soluble in water. Without the oxide layer what he had created was a number of cat whisker diodes on the same piece of semiconductor. Frustrated, he continued experimenting with the device anyway and was surprised to find some amplification between the diodes. Perhaps neutralizing the surface charges wasn't necessary at all.

They decided to try two cat whisker diodes touching the same semiconductor crystal at almost the same point. In December 1947, they wrapped a piece of gold foil around a triangular piece of plastic and then used a razor blade to make a cut in the gold at the point of the triangle. By touching the point of the triangle to a piece of germanium crystal, they created two cat whisker diodes. They discovered that electrons emitted into the semiconductor by one diode were collected by the other. The voltage on the crystal base could increase or suppress this effect allowing their device to amplify the signal at the base. Because a control could vary the resistance between emitter and collector, it was later decided to call the device a transfer-resistor or transistor.

It was immediately apparent that the transistor would be an extremely important innovation and Bell Labs began the process of applying for a patent immediately. Shockley felt that Bardeen and Brattain's success had come as a result of his field-effect concept and that Bell Labs should file a broad patent to include his field-effect ideas and the newly created transistor. Furthermore, as leader of the team, Shockley felt that his name alone should appear on the patent. Shockley's harmonious research team quickly became polarized over the question of who should receive credit for the new invention. Bell Lab attorneys might have given Shockley what he wanted, but in reviewing the patents already on record they found something disturbing.

An obscure physics professor named Julius Lilienfeld had been granted two patents in 1930 and 1933 for the idea of controlling the current in a semiconductor by means of an electric field. Although there was no record of Lilienfeld ever trying to build his device and surface charges would almost certainly have prevented it from working if he did, the idea was clearly the same as Shockley's. A patent on a field-effect transistor would almost certainly be overturned. The attorneys at Bell Labs decided to write a more narrow patent application focusing just on Bardeen and Brattian's point-contact transistor, which worked on different principles. Ultimately patent 2524035, "Three-Electrode Circuit Element Utilizing Semiconductive Materials" lists only Bardeen and Brattain as inventors.

Shockley felt he had been robbed of proper credit and began working on his own to devise a new and better transistor. The point-contact transistor shared all the same problems as cat whisker diodes. Because it required two metal lines to just touch the surface of the semiconductor it was extremely temperamental and hard to consistently reproduce. Shockley reasoned that if a transistor could be formed out of two cat whisker diodes sharing one terminal, perhaps he could make a transistor out of two junction diodes sharing one terminal.

He began working out the theory of a junction transistor but told no one at Bell Labs until February 1948, when experiments with the point-contact transistor suggested that charge carriers could travel through the bulk of semiconductors and not just at the surface as Bardeen and Brattain had been assuming. Knowing this was proof that his concept for a transistor could work, Shockley now described his ideas to the group. It was obvious to Bardeen and Brattain that Shockley had been thinking about this for some time and intentionally kept them in the dark. In June 1948, despite not yet having a working prototype, Bell Labs applied for a patent of the junction transistor listing only Shockley's name. Bardeen and Brattain would never work effectively with Shockley again.

Under Shockley's direction, Morgan Sparks built the first working junction transistor in 1949; by 1950, Sparks and Gordon Teal had vastly improved their techniques. For the previous 2 years, Teal had been advocating making semiconductor devices out of single crystals. The "crystals" used at the time were really just amalgams of many crystals all with different orientations and many noncrystalline defects in between. Teal built an apparatus to grow single crystals semiconductors without these defects using the ideas of the Polish scientist J. Czochralski.

Using Czochralski's method, a very small seed crystal is dipped into a container of molten semiconductor and very slowly pulled out. As the crystal is pulled out, atoms from the melt freeze onto its surface, gradually growing the crystal. Each atom tends to freeze into proper place in the crystal lattice producing a large defect-free crystal. Also important is that impurities in the melt tend to stay in the melt. The crystal drawn out is purer than the starting material. By repeatedly drawing crystals and then melting them down to be drawn again, Teal achieved purity levels vastly better than anything Shockley had worked with thus far. In later years, the cylindrical ingots pulled from the melt would be sliced into the round silicon wafers used today. Using these new techniques, Sparks and Teal created the junction transistor Shockley had imagined.

Rather than two diodes side by side, Shockley imagined a sandwich of three semiconductor layers alternating N-type, P-type, and N-type as shown in Fig. 1-4. The emitter at the bottom injects electrons into the base in the center. These electrons diffuse across the base to be captured by the collector at the top. The voltage on the base controls the injection of electrons by the emitter. Because this device contains two P-N

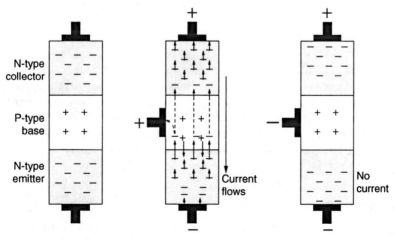

Figure 1-4 Junction transistor.

junctions, this device is known as the **Bipolar Junction Transistor (BJT)**. Sparks and Teal were able to create this structure by adding dopants to the melt as they pulled the crystal. Pulling the crystal more slowly created thinner layers and improved performance.

As Shockley used the lab's team of experimentalists to pursue his own theories of how to make a better transistor, he actively worked to exclude John Bardeen from further work on semiconductors. Shockley was going to make sure he was the one to make any further breakthroughs. In 1951, Bardeen, frustrated and prevented from contributing to the development of the devices he helped create, left Bell Labs to become a professor at the University of Illinois. In the same year, AT&T began licensing the rights to manufacture both point contact and junction transistors, and the first few products using transistors began to appear. However, there was a serious limitation.

The first commercially available transistors were all made of germanium. They tended to be very sensitive to temperature and at temperatures above 75°C they didn't work at all. Packed tightly with other electronics these temperatures would not be hard to achieve. The military in particular mandated components that would function properly in a wide range of conditions, and military applications were the most lucrative for early sales of transistors. The problem was that the band gap of germanium was too small. At room temperature only the carriers intentionally added by dopants could jump the gap and conduct electricity, but at higher temperatures many more electrons had enough energy so that germanium stopped behaving like a semiconductor and began behaving like a conductor. The solution was to replace germanium with silicon.

Silicon's band gap is almost twice as large as germanium's, so it is far less sensitive to temperature and has a much higher maximum operating temperature. In 1952, Gordon Teal left Bell Labs to join Texas Instruments® (TI) with the goal of creating a silicon transistor. Unfortunately silicon's melting point is also much higher than germanium's, and it reacts more easily with oxygen and other contaminants. Creating pure crystal silicon was far more difficult, but as the man who had pioneered growing high-quality crystals at Bell Labs, Teal was probably better suited than any man in the world to overcome these problems. In 1954, TI produced the first silicon junction transistor. In the same year, TI began manufacturing components for the first commercially available transistor radio, still using germanium. The radio sold out wherever it was offered, and the transistor revolution had truly begun.

The Integrated Circuit

Bill Shockley probably understood the enormous potential of the transistor better than any man in the world at the time, but he felt he wasn't being given sufficient authority at Bell Labs. More than once he had watched other men promoted over him. Also, Shockley personally would not receive any money from the transistor patents he had helped AT&T secure. He decided to leave and found his own company to make semiconductor products.

In the fall of 1955, Shockley secured financially backing from California businessman Arnold Beckman and chose Palo Alto, California, as the location for his new company. At the time Palo Alto was notable only as the home of Stanford University. Shockley had received strong encouragement from Frederick Terman, the Stanford Dean of Engineering, to locate nearby. The university's engineering school would be an ideal place to recruit new employees, but perhaps Shockley was ultimately persuaded by more personal reasons. He had grown up in Palo Alto, and his mother still lived there. Over the next 20 years, Shockley's decision would cause this sleepy valley of orchards to be transformed into the famous "Silicon Valley."

Perhaps not surprisingly, Shockley had little luck hiring any of his former colleagues from Bell Labs, and so he turned to a younger generation. He recruited some of the brightest engineers, physicists, and chemists from around the country, and in February 1956, Shockley Semiconductor was founded. That same year, Shockley, Bardeen, and Brattain were together awarded the Nobel Prize in physics for their invention of the transistor. Shockley's brilliance as a scientist had received the highest recognition, and yet it was his management style that would doom his company.

Feeling that his subordinates at Bell Labs had tried to steal the credit for his ideas, Shockley was determined that it would not happen at his own company. All ideas and all development direction were to come from him. The talented men he hired grew restless under his heavy-handed management. In 1957, eight of Shockley's recruits, led by Robert Noyce and Gordon Moore, approached Beckman to ask that Shockley be removed from management and allowed to act only as a technical consultant. Beckman considered this seriously for a month before deciding to leave Shockley in charge. On September 18, 1957, the group that would become known in Silicon Valley as the "traitorous eight" resigned. Shockley Semiconductor continued operating another 11 years but never turned a profit. Bill Shockley never again had a significant influence on the semiconductor industry he had helped to start.

The day after the "eight" resigned they founded a new company, Fairchild Semiconductor, in which Fairchild Camera and Instruments provided the financial backing. The company was becoming involved in components for missiles and satellites, and it was clear that transistors would play an important part. Fairchild Semiconductor made rapid progress and by 1959 was profitably selling silicon junction transistors.

Silicon transistors were cheap enough and reliable enough to allow designs using hundreds or thousands of them. At that point the biggest roadblock to using more transistors became making all the connections between them. At this time hundreds of transistors could be made on a single 1-inch silicon wafer. The wafer was then cut up into individual transistors, and leads were soldered by hand onto each one. After being sold to customers, the packaged transistors would then be soldered together, again by hand, to form the needed circuit. It wouldn't take long for the market for transistors to become limited by the difficulty of assembling them into useful products. The first person to conceive of a solution to this bottleneck was Jack Kilby at Texas Instruments.

Kilby had already been working with transistors for a number of years when he started work at Texas Instruments in May 1958. At that time everyone at TI took time off work in July, but Kilby hadn't been with the company long enough to qualify for vacation. He spent his time alone in the lab thinking about how to make it easier to assemble complicated circuits from transistors. The idea occurred to him that perhaps all the components of the circuit could be integrated into a single piece of semiconductor. By September Kilby had constructed the first **Integrated Circuit (IC)**. All the computer chips of today are realizations of this simple but revolutionary idea.

A few months later at Fairchild Semiconductor, Bob Noyce had the same idea for an integrated circuit, but he carried the idea even further. Constructing the components together still left the problem of making the needed connections by hand. Noyce imagined simultaneously making not only transistors but the wires connecting them as well. Silicon naturally forms silicon dioxide (the main ingredient in glass) when exposed to air and heat, and silicon dioxide is an excellent insulator. By growing a layer of silicon dioxide on top of a silicon chip, the components could be isolated from wires deposited on top. Acid could be used to cut holes in the insulator where connections needed to be made. By enabling the fabrication of both transistors and wires as a single, solid structure, Noyce made Kilby's idea practical. All modern integrated circuits are made in this fashion.

Fairchild Semiconductor and Texas Instruments both filed for patents. Kilby had the idea first and created the first working integrated circuit. Noyce's idea of including the wiring on the chip made real products possible. The legal battle of who invented the integrated circuit would continue for 11 years after Kilby first had his insight. However, Noyce and Kilby, both men of fair and generous natures (and from whom William Shockley could have learned a great deal), consistently acknowledged the importance of the other's contribution.

For either company there was the chance of obtaining sole rights to one of the most lucrative ideas in history, or being cut out of the business altogether. The courts could easily decide for either side. Rather than pursue total victory at the risk of total destruction, in 1966, both companies agreed to cross license their patents to each other. They also agreed to sell licenses for both patents for a small percentage of the profits, to anyone else who wanted to make integrated circuits. When later court cases found first in favor of Kilby and then in favor of Noyce, it no longer mattered, and today they are considered coinventors of the integrated circuit.

In 1960, Bell Labs made another key advance, creating the first successful field-effect transistor. This was the same transistor that Shockley had originally set out to build 15 years earlier and Lilienfeld had first proposed 15 years before that, but no one had been able to make it work. The key turned out to be the same silicon dioxide layer that Noyce had proposed using. A group headed by John Atalla showed that by carefully cleaning the silicon surface and then growing an oxide layer the surface states could be eliminated. The electric field from a metal wire deposited on top of the oxide could penetrate the silicon and turn on and off a flow of current. Because it was a vertical stack of metal, oxide, and semiconductor, this device was named the **Metal Oxide Semiconductor Field-Effect Transistor (MOSFET).**

Figure 1-5 shows a MOSFET that is off and one that is on. Each consists of metal input and output wires that touch the silicon. Where they touch the silicon, impurities have been added to allow the silicon to conduct. These regions of dopants added to the silicon are the transistor's source and drain. The source will provide the electrical charges that will be removed by the drain. In between the source and drain, a third conducting wire passes very close to but does not touch the silicon. This wire is the gate of the transistor, and we can imagine it swinging open or

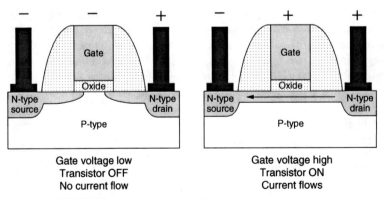

Figure 1-5 MOSFET transistor cross sections.

closed to turn the transistor on or off. The voltage of the gate wire produces an electric field which determines whether current can flow from the source to the drain or not. In the left figure, a low voltage at the gate repels the negative charges in the source and drain and keeps them separated. In the right figure, a high voltage at the gate attracts the negative charges and pulls them together into a channel connecting the source and drain. Where bipolar junction transistors require a constant current into the base to remain switched on, MOSFETs require only a voltage to be held at the gate. This allows MOSFETs to use less power than equivalent BJT circuits. Almost all transistors made today are MOSFETs connected together in integrated circuits.

The Microprocessor

The integrated circuit was not an immediate commercial success. By 1960 the computer had gone from a laboratory device to big business with thousands in operation worldwide and more than half a billion dollars in sales in 1960 alone.[2] International Business Machines (IBM®) had become the leading computer manufacturer and had just begun shipping its first all-transistorized computer. These machines still bore little resemblance to the computers of today. Costing millions these "mainframe" computers filled rooms and required teams of operators to man them. Integrated circuits would reduce the cost of assembling these computers but not nearly enough to offset their high prices compared to discrete transistors. Without a large market the volume production that would bring integrated circuit costs down couldn't happen. Then, in 1961, President Kennedy challenged the United States to put a man on the moon before the end of the decade. To do this would require extremely compact and light computers, and cost was not a limitation. For the next 3 years, the newly created space agency, NASA, and the U.S. Defense Department purchased every integrated circuit made and demand soared.

The key to making integrated circuits cost effective enough for the general market place was incorporating more transistors into each chip. The size of early MOSFETs was limited by the problem of making the gate cross exactly between the source and drain. Adding dopants to form the source and drain regions requires very high temperatures that would melt a metal gate wire. This forced the metal gates to be formed after the source and drain, and ensuring the gates were properly aligned was a difficult problem. In 1967, Fedrico Faggin at Fairchild Semiconductor experimented with making the gate wires out of silicon. Because the silicon was deposited on top of an oxide layer, it was not a single crystal

---

[2]Einstein and Franklin, "Computer Manufacturing," 10.

but a jumble of many small crystals called **polycrystalline silicon**, polysilicon, or just poly. By forming polysilicon gates before adding dopants, the gate itself would determine where the dopants would enter the silicon crystal. The result was a self-aligned MOSFET. The resistance of polysilicon is much higher than a metal conductor, but with heavy doping it is low enough to be useful. MOSFETs are still made with poly gates today.

The computers of the 1960s stored their data and instructions in "core" memory. These memories were constructed of grids of wires with metal donuts threaded onto each intersection point. By applying current to one vertical and one horizontal wire a specific donut or "core" could be magnetized in one direction or the other to store a single bit of information. Core memory was reliable but difficult to assemble and operated slowly compared to the transistors performing computations. A memory made out of transistors was possible but would require thousands of transistors to provide enough storage to be useful. Assembling this by hand wasn't practical, but the transistors and connections needed would be a simple pattern repeated many times, making semiconductor memory a perfect market for the early integrated circuit business.

In 1968, Bob Noyce and Gordon Moore left Fairchild Semiconductor to start their own company focused on building products from integrated circuits. They named their company Intel® (from INTegrated ELectronics). In 1969, Intel began shipping the first commercial integrated circuit using MOSFETs, a 256-bit memory chip called the 1101. The 1101 memory chip did not sell well, but Intel was able to rapidly shrink the size of the new silicon gate MOSFETs and add more transistors to their designs. One year later Intel offered the 1103 with 1024 bits of memory, and this rapidly became a standard component in the computers of the day.

Although focused on memory chips, Intel received a contract to design a set of chips for a desktop calculator to be built by the Japanese company Busicom. At that time, calculators were either mechanical or used hard-wired logic circuits to do the required calculations. Ted Hoff was asked to design the chips for the calculator and came to the conclusion that creating a general purpose processing chip that would read instructions from a memory chip could reduce the number of logic chips required. Stan Mazor detailed how the chips would work together and after much convincing Busicom agreed to accept Intel's design. There would be four chips altogether: one chip controlling input and output functions, a memory chip to hold data, another to hold instructions, and a central processing unit that would eventually become the world's first microprocessor.

The computer processors that powered the mainframe computers of the day were assembled from thousands of discrete transistors and logic chips.

This was the first serious proposal to put all the logic of a computer processor onto a single chip. However, Hoff had no experience with MOSFETs and did not know how to make his design a reality. The memory chips Intel was making at the time were logically very simple with the same basic memory cell circuit repeated over and over. Hoff's design would require much more complicated logic and circuit design than any integrated circuit yet attempted. For months no progress was made as Intel struggled to find someone who could implement Hoff's idea.

In April 1970, Intel hired Faggin, the inventor of the silicon gate MOSFET, away from Fairchild. On Faggin's second day at Intel, Masatoshi Shima, the engineering representative from Busicom, arrived from Japan to review the design. Faggin had nothing to show him but the same plans Shima had already reviewed half a year earlier. Shima was furious, and Faggin finished his second day at a new job already 6 months behind schedule. Faggin began working at a furious pace with Shima helping to validate the design, and amazingly by February 1971 they had all four chips working. The chips processed data 4 bits at a time and so were named the 4000 series. The fourth chip of the series was the first microprocessor, the Intel 4004 (Fig. 1-6).

The 4004 contained 2300 transistors and ran at a clock speed of 740 kHz, executing on average about 60,000 instructions per second.[3] This gave it the same processing power as early computers that had filled entire rooms, but on a chip that was only 24 mm². It was an incredible engineering achievement, but at the time it was not at all clear that it had a commercial future. The 4004 might match the performance of the fastest computer in the world in the late 1940s, but the mainframe computers of 1971 were hundreds of times faster. Intel began shipping the 4000 series to Busicom in March 1971, but the calculator market had become intensely competitive and Busicom was unenthusiastic about the high cost of the 4000 series. To make matters worse, Intel's contract with Busicom specified Intel could not sell the chips to anyone else. Hoff, Faggin, and Mazor pleaded with Intel's management to secure the right to sell to other customers. Bob Noyce offered Busicom a reduced price for the 4000 series if they would change the contract, and desperate to cut costs in order to stay in business Busicom agreed. By the end of 1971, Intel was marketing the 4004 as a general purpose microprocessor. Busicom ultimately sold about 100,000 of the series 4000 calculators before going out of business in 1974. Intel would go on to become the leading manufacturer in what was for 2003—a $27 billion a year market for microprocessors. The incredible improvements in microprocessor performance and growth of the

---

[3]Real, "Revolution in Progress," 12.

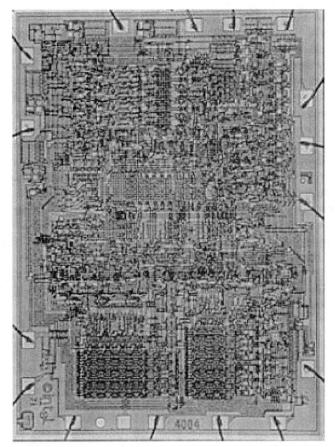

Figure 1-6 First microprocessor, Intel 4004. (*Courtesy: Intel Corporation.*)

semiconductor industry since 1971 have been made possible by steady year after year improvements in the manufacturing of transistors.

Moore's Law

Since the creation of the first integrated circuit, the primary driving force for the entire semiconductor industry has been process scaling. Process scaling is shrinking the physical size of the transistors and the wires interconnecting them, allowing more devices to be placed on each chip, which allows more complex functions to be implemented. In 1975, Gordon Moore observed that shrinking transistor dimensions were allowing the number of transistors on a die to double roughly every 18 months.[4] This trend has come to be known as **Moore's law**. For microprocessors, the trend has been closer to a doubling every 2 years, but amazingly this exponential increase has continued now for 30 years and seems likely to continue through the foreseeable future (Fig. 1-7).

The 4004 used transistors with a feature size of 10 microns (μm). This means that the distance from the source of the transistor to the drain was approximately 10 μm. A human hair is around 100 μm across. In 2003, transistors were being mass produced with a feature size of only 0.13 μm. Smaller transistors not only allow for more logic gates, but also allow the individual logic gates to switch more quickly. This has provided for even greater improvements in performance by allowing faster clock rates. Perhaps even more importantly, shrinking the size of a computer chip reduces its manufacturing cost. The cost is determined by the cost to process a wafer, and the smaller the chip, the more that are made from each wafer. The importance of transistor scaling to the semiconductor industry is almost impossible to overstate. Making transistors smaller allows for chips that provide more performance, and therefore sell for more money, to be made at a lower cost. This is the fundamental driving force of the semiconductor industry.

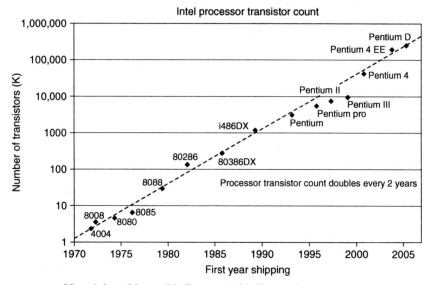

Figure 1-7 Moore's law. (Moore, "No Exponential is Forever," 10.)

---

[4]Moore, "Progress in Digital Integrated Electronics."

Transistor scaling

The reason smaller transistors switch faster is that although they draw less current, they also have less capacitance. Less charge has to be moved to switch their gates on and off. The delay of switching a gate (T_{DELAY}) is determined by the capacitance of the gate (C_{GATE}), the total voltage swing (V_{dd}), and the drain to source current (I_{DS}) drawn by the transistor causing the gate to switch.

$$T_{DELAY} \propto C_{GATE} \times \frac{V_{dd}}{I_{DS}}$$

Higher capacitance or higher voltage requires more charge to be drawn out of the gate to switch the transistor, and therefore more current to switch in the same amount of time. The capacitance of the gate increases linearly with the width (W) and length (L) of the gate and decreases linearly with the thickness of the **gate oxide** (T_{OX}).

$$T_{DELAY} \propto C_{GATE} \times \frac{V_{dd}}{I_{DS}} \propto \frac{W \times L}{T_{OX}} \times \frac{V_{dd}}{I_{DS}}$$

The current drawn by a MOSFET increases with the device width (W), since there is a wider path for charges to flow, and decreases with the device length (L), since the charges have farther to travel from source to drain. Reducing the gate oxide thickness (T_{OX}) increases current, since pushing the gate physically closer to the silicon channel allows its electric field to better penetrate the semiconductor and draw more charges into the channel (Fig. 1-8).

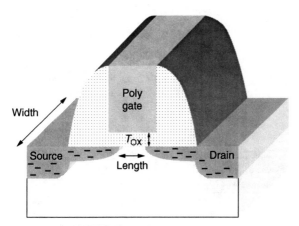

Figure 1-8 MOSFET dimensions.

To draw any current at all, the gate voltage must be greater than a certain minimum voltage called the **threshold voltage** (V_T). This voltage is determined by both the gate oxide thickness and the concentration of dopant atoms added to the channel. Current from the drain to source increases quadratically after the threshold voltage is crossed. The current of MOSFETs is discussed in more detail in Chap. 7.

$$I_{DS} \propto \frac{W}{L \times T_{OX}} \times (V_{dd} - V_T)^2$$

Putting together these equations for delay and current we find:

$$T_{DELAY} \propto L^2 \times \frac{V_{dd}}{(V_{dd} - V_T)^2}$$

Decreasing device lengths, increasing voltage, or decreasing threshold voltage reduces the delay of a MOSFET. Of these methods decreasing the device length is the most effective, and this is what the semiconductor industry has focused on the most. There are different ways to measure **channel length**, and so when comparing one process to another, it is important to be clear on which measurement is being compared. Channel length is measured by three different values as shown in Fig. 1-9.

The drawn gate length (L_{DRAWN}) is the width of the gate wire as drawn on the mask used to create the transistors. This is how wide the wire will be when it begins processing. The etching process reduces the width of the actual wire to less than what was drawn on the mask. The manufacturing of MOSFETs is discussed in detail in Chap. 9. The width of the gate wire

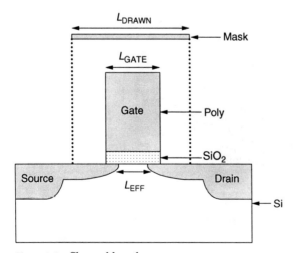

Figure 1-9 Channel length.

at the end of processing is the actual gate length (L_{GATE}). Also, the source and drain regions within the silicon typically reach some distance underneath the gate. This makes the effective separation between source and drain in the silicon less than the final gate length. This distance is called the *effective channel length* (L_{EFF}). It is this effective distance that is the most important to transistor performance, but because it is under the gate and inside the silicon, it can not be measured directly. L_{EFF} is only estimated by electrical measurements. Therefore, L_{GATE} is the value most commonly used to compare difference processes.

Gate oxide thickness is also measured in more than one way as shown in Fig. 1-10. The actual distance from the bottom of the gate to the top of the silicon is the physical gate oxide thickness (T_{OX-P}). For older processes this was the only relevant measurement, but as the oxide thickness has been reduced, the thickness of the layer of charge on both sides of the oxide has become significant. The electrical oxide thickness (T_{OX-E}) includes the distance to the center of the sheets of charge above and below the gate oxide. It is this thickness that determines how much current a transistor will produce and hence its performance. One of the limits to future scaling is that increasingly large reductions in the physical oxide thickness are required to get the same effective reduction in the electrical oxide thickness.

While scaling channel length alone is the most effective way to reduce delays, the increase in leakage current prevents it from being practical. As the source and drain become physically closer together, they become more difficult to electrically isolate from one another. In deep submicron MOSFETs there may be significant current flow from the drain to the source even when the gate voltage is below the threshold voltage. This is called *subthreshold leakage*. It means that even transistors that should be off still conduct a small amount of current like a leaky faucet. This current may be hundreds or thousands of times smaller than the current when the transistor is on, but for a die with millions of transistors this leakage current can rapidly become a problem. The most common solution for this is reducing the oxide thickness.

Moving the gate terminal physically closer to the channel gives the gate more control and limits subthreshold leakage. However, this

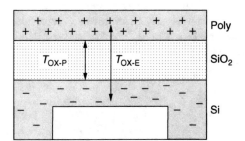

Figure 1-10 Gate oxide thickness.

reduces the long-term reliability of the transistors. Any material will conduct electricity if a sufficient electrical field is applied. In the case of insulators this is called *dielectric breakdown* and physically melts the material. At extremely high electric fields the electrons, which bind the molecules of the material together, are torn free and suddenly large amounts of current begin to flow. The gate oxides of working MOSFETs accumulate defects over time that gradually lower the field at which the transistor will fail. These defects can also reduce the switching speed of the transistors.[5] These phenomena are particularly worrisome to semiconductor manufacturers because they can cause a new product to begin failing after it has already been shipping for months or years.

The accumulation of defects in the gate oxide is in part due to "hot" electron effects. Normally the electrons in the channel do not have enough energy to enter the gate oxide. Its band gap is far too large for any significant number of electrons to have enough energy to surmount at normal operating temperatures. Electrons in the channel drift from source to drain due to the lateral electric field in the channel. Their average drift velocity is determined by how strong the electric field is and how often the electrons collide with the atoms of the semiconductor crystal. Typically the drift velocity is only a tiny fraction of the random thermal velocity of the electrons, but at very high lateral fields some electrons may get accelerated to velocities much higher than they would usually have at the operating temperature. It is as if these electrons are at a much higher temperature than the rest, and they may have enough energy to enter the gate oxide. They may travel through and create a current at the gate, or they may become trapped in the oxide creating a defect. If a series of defects happens to line up on a path from the gate to the channel, gate oxide breakdown occurs. Thus the reliability of the transistors is a limit to how much their dimensions can be scaled. In addition, as gate oxides are scaled below 5 nm, gate tunneling current becomes significant.

One implication of quantum mechanics is that the position of an electron is not precisely defined. This means that with a sufficiently thin oxide layer, electrons will occasionally appear on the opposite side of the insulator. If there is an electric field, the electron will then be pulled away and unable to get back. The current this phenomenon creates through the insulator is called a *tunneling current*. It does not damage the layer as occurs with hot electrons because the electron does not travel through the oxide in the classical sense, but this does cause unwanted leakage current through the gate of any ON device. The typical solution for both dielectric breakdown and gate tunneling current is to reduce the supply voltage.

---

[5]Chen, "Dynamic NBTI of p-MOS Transistors."

Scaling the supply voltage by the same amount as the channel length and oxide thickness keeps all the electrical fields in the device constant. This concept is called *constant field scaling* and was proposed by Robert Dennard in 1974.[6] Constant field scaling is an easy way to address problems such as subthreshold leakage and dielectric breakdown, but a higher supply voltage provides for better performance. As a result, the industry has scaled voltages as slowly as possible, allowing fields in the channel and the oxide to increase significantly with each device generation. This has required many process adjustments to tolerate the higher fields. The concentration of dopants in the source, drain, and channel is precisely controlled to create a three-dimensional profile that minimizes subthreshold leakage and hot electron effects. Still, even the very gradual scaling of supply voltages increases delay and hurts performance. This penalty increases dramatically when the supply voltage becomes less than about three times the threshold voltage.

It is possible to design integrated circuits that operate with supply voltages less than the threshold voltages of the devices. These designs operate using only subthreshold leakage currents and as a result are incredibly power efficient. However, because the currents being used are orders of magnitude smaller than full ON currents, the delays involved are orders of magnitude larger. This is a good trade-off for a chip to go into a digital watch but not acceptable for a desktop computer. To maintain reasonable performance a processor must use a supply voltage several times larger than the threshold voltage. To gain performance at lower supply voltages the channel doping can be reduced to lower the threshold voltage.

Lowering the threshold voltage immediately provides for more on current but increases subthreshold current much more rapidly. The rate at which subthreshold currents increase with reduced threshold voltage is called the *subthreshold slope* and a typical value is 100 mV/decade. This means a 100-mV drop in threshold will increase subthreshold leakage by a factor of 10. The need to maintain several orders of magnitude difference between the on and off current of a device therefore limits how much the threshold voltage can be reduced. Because the increase in subthreshold current was the first problem encountered when scaling the channel length, we have come full circle to the original problem. In the end there is no easy solution and process engineers are continuing to look for new materials and structures that will allow them to reduce delay while controlling leakage currents and reliability (Fig. 1-11).

---

[6]Dennard, "MOSFETs with Small Dimensions."

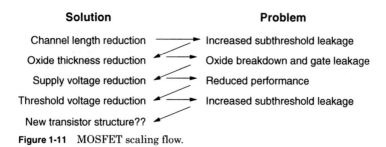

Figure 1-11 MOSFET scaling flow.

Interconnect scaling

Fitting more transistors onto a die requires not only shrinking the transistors but also shrinking the wires that interconnect them. To connect millions of transistors modern microprocessors may use seven or more separate layers of wires. These interconnects contribute to the delay of the overall circuit. They add capacitive load to the transistor outputs, and their resistance means that voltages take time to travel their length. The capacitance of a wire is the sum of its capacitance to wires on either side and to wires above and below (see Fig. 1-12).

Fringing fields make the wire capacitance a complex function, but for cases where the wire width (W_{INT}) is equal to the wire spacing (W_{SP}) and

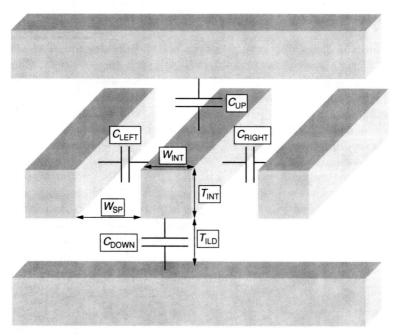

Figure 1-12 Wire capacitance.

thickness (T_{INT}) is equal to the vertical spacing of wires (T_{ILD}), capacitance per length (C_L) is approximated by the following equation.[7]

$$C_L = C_{LEFT} + C_{RIGHT} + C_{UP} + C_{DOWN} \approx 2K\varepsilon_0 \left(\frac{T_{INT}}{W_{SP}} + \frac{W_{INT}}{T_{ILD}} \right)$$

Wire capacitance is kept to a minimum by using small wires and wide spaces, but this reduces the total number of wires that can fit in a given area and leads to high wire resistance. The delay for a voltage signal to travel a length of wire (L_{WIRE}) is the product of the resistance of the wire and the capacitance of the wire, the RC delay. The wire resistance per length (R_L) is determined by the width and thickness of the wire as well as the resistivity (ρ) of the material.

$$R_L = \frac{\rho}{W_{INT} \times T_{INT}}$$

Engineers have tried three basic methods of scaling interconnects in order to balance the need for low capacitance and low resistance. These are ideal scaling, quasi-ideal scaling, and constant-R scaling.[8] For a wire whose length is being scaled by a value S less than 1, each scheme scales the other dimensions of the wire in different ways, as shown in Table 1-1.

---

[7]Bohr, "Interconnect Scaling," 106.
[8]Bakoglu, "Circuits and Interconnections," 197.

TABLE 1-1 Scaling of Interconnects

Description	Parameter	Ideal scaling	Quasi-ideal scaling	Constant-R scaling
Length	L_{WIRE}	S	S	S
Width	W_{INT}	S	S	$\sqrt{S}$
Spacing	W_{SP}	S	S	$\sqrt{S}$
Thickness	T_{INT}	S	$\sqrt{S}$	$\sqrt{S}$
Interlevel dielectric	T_{ILD}	S	$\sqrt{S}$	$\sqrt{S}$
Aspect ratio	T_{INT}/W_{INT}	1	$1/\sqrt{S}$	1
Wire capacitance per length	C_L	1	$1/S^{0.1}$	1
Wire capacitance	$C_L L_{WIRE}$	S	$S^{0.9}$	S
Wire resistance per length	R_L	$1/S^2$	$1/S^{1.5}$	$1/S$
Wire resistance	$R_L L_{WIRE}$	$1/S$	$1/S^{0.5}$	1
RC delay	$R_L C_L L_{WIRE}^2$	1	$S^{0.4}$	S

Ideal scaling reduces all the vertical and horizontal dimensions by the same amount. This keeps the capacitance per length constant but greatly increases the resistance per length. In the end the reduction in wire capacitance is offset by the increase in wire resistance, and the wire delay remains constant. Scaling interconnects this way would mean that as transistors grew faster, processor frequency would quickly become limited by the interconnect delay.

To make interconnect delay scale with the transistor delay, constant-R scaling can be used. By scaling the vertical and horizontal dimensions of the wire less than its length, the total resistance of the wire is kept constant. Because the capacitance is reduced at the same rate as ideal scaling, the overall RC delay scales with the wire length. The downside of constant-R scaling is that if S is also scaling the device dimensions, then the area required for wires is not decreasing as quickly as the device area. The size of a chip will be rapidly determined not by the number of transistors but by the number of wires.

To allow for maximum scaling of die area while mitigating the increase in wire resistance, most manufactures use quasi-ideal scaling. In this scheme horizontal dimensions are scaled with wire length, but vertical dimensions are scaled more slowly. The capacitance per length increases only slightly and the increase in resistance is not as much as ideal scaling. Overall the RC delay will decrease although not as much constant-R scaling. The biggest disadvantage of quasi-ideal scaling is that it increases the aspect ratio of the wires, the ratio of thickness to width. This scaling has rapidly led to wires in modern processors that are twice as tall as they are wide, but manufacturing wires with ever-greater aspect ratios is difficult. To help in continuing to reduce interconnect delays, manufactures have turned to new materials.

In 2000, some semiconductor manufacturers switched from using aluminum wires, which had been used since the very first integrated circuits, to copper wires. The resistivity of copper is less than aluminum providing lower resistance wires. Copper had not been used previously because it diffuses very easily through silicon and silicon dioxide. Copper atoms from the wires could quickly spread throughout a chip acting as defects in the silicon and ruining the transistor behavior. To prevent this, manufacturers coat all sides of the copper wires with materials that act as diffusion barriers. This reduces the cross section of the wire that is actually copper but prevents contamination.

Wire capacitances have been reduced through the use of low-K dielectrics. Not only the dimensions of the wires determine wire capacitance but also by the permittivity or K value of the insulator surrounding the wires. The best capacitance would be achieved if there were simply air or vacuum between wires giving a K equal to 1, but of course this would provide no physical support. Silicon dioxide is traditionally used, but this has a *K* value of 4. New materials are being tried to reduce *K* to 3 or even 2, but these materials tend to be very soft and porous. When heated by high electrical currents the metal wires tend to flex and stretch and soft dielectrics do little to prevent this. Future interlevel dielectrics must provide reduced capacitance without sacrificing reliability.

One of the common sources of interconnect failures is called *electromigration*. In wires with very high current densities, atoms tend to be pushed along the length of the wire in the direction of the flow of electrons, like rocks being pushed along a fast moving stream. This phenomenon happens more quickly at narrow spots in the wire where the current density is highest. This leads these spots to become more and more narrow, accelerating the process. Eventually a break in the wire is created. Rigid interlevel dielectrics slow this process by preventing the wires from growing in size elsewhere, but the circuit design must make sure not to exceed the current carrying capacity of any one wire.

Despite using new conductor materials and new insulator materials, improvements in the delay of interconnects have continued to trail behind improvements in transistor delay. One of the ways in which microprocessors designs try to compensate for this is by adding more wiring layers. The lowest levels are produced with the smallest dimensions. This allows for a very large number of interconnections. The highest levels are produced with large widths, spaces, and thickness. This allows them to have much less delay at the cost of allowing fewer wires in the same area.

The different wiring layers connect transistors on a chip the way roads connect houses in a city. The only interconnect layer that actually connects to a transistor is the first layer deposited, usually called the metal 1 or M1 layer. These are the suburban streets of a city. Because they are narrow, traveling on them is slow, but typically they are very short. To travel longer distances, wider high speed levels must be used. The top layer wires would be the freeways of the chip. They are used to travel long distances quickly, but they must connect through all the lower slower levels before reaching a specific destination.

There is no real limit to the number of wiring levels that can be added, but each level adds to the cost of processing the wafer. In the end the design of the microprocessor itself will have to continue to evolve to allow for the greater importance of interconnect delays.

Microprocessor scaling

Because of the importance of process scaling to processor design, all microprocessor designs can be broken down into two basic categories: lead designs and compactions. **Lead designs** are fundamentally new designs. They typically add new features that require more transistors and therefore a larger die size. **Compactions** change completed designs to make them work on new fabrication processes. This allows for higher frequency, lower power, and smaller dies. Figure 1-13 shows to scale die photos of different Intel lead and compaction designs.

Each new lead design offers increased performance from added functionality but uses a bigger die size than a compaction in the same generation. It is the improvements in frequency and reductions in cost that come from compacting the design onto future process generations that make the new designs profitable. We can use Intel manufacturing processes of the last 10 years to show the typical process scaling from one generation to the next (Table 1-2).

On average the semiconductor industry has begun a new generation of fabrication process every 2 to 3 years. Each generation reduces horizontal dimensions about 30 percent compared to the previous generation. It would be possible to produce new generations more often if a smaller shrink factor was used, but a smaller improvement in performance might not justify the expense of new equipment. A larger shrink factor could provide more performance improvement but would require a longer time between generations. The company attempting the larger shrink factor would be at a disadvantage when competitors had advanced to a new process before them.

The process generations have come to be referred to by their "technology node." In older generations this name indicated the MOSFET

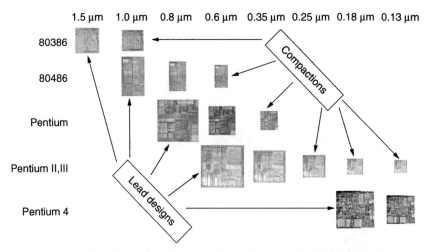

Figure 1-13 Intel lead designs and compactions. (*Courtesy: Intel Corporation.*)

TABLE 1-2 Intel Manufacturing Processes[9,10,11,12]

Year	Technology node (nm)	L_{GATE} (nm)	T_{OX-P} (nm)	V_{dd} (V)	Metal layers
1993	500	500	8.0	3.3	4
1995	350	350	5.2	2.5	4
1997	250	200	3.1	1.8	5
1999	180	130	2.0	1.6	6
2001	130	70	1.4	1.4	6
2003	90	50	1.2	1.2	7

[9]Walden, "90nm and Beyond"
[10]Chau, "Silicon Nano-Transistors."
[11]Bohr, "Silicon Trends and Limits."
[12]Bhandarkar, "Billion Transistor Processor Chips."

gate length of the process (L_{GATE}), but more recently some manufactures have scaled their gate lengths more aggressively than others. This means that today two different 90-nm processes may not have the same device or interconnect dimensions, and it may be that neither has any important dimension that is actually 90-nm. The technology node has become merely a name describing the order of manufacturing generations and the typical 30 percent scaling of dimensions. The important historical trends in microprocessor fabrication demonstrated by Table 1-2 and quasi-ideal interconnect scaling are shown in Table 1-3.

Although it is going from one process generation to the next that gradually moves the semiconductor industry forward, manufacturers do not stand still for the 2 years between process generations. Small incremental improvements are constantly being made to the process that allow for part of the steady improvement in processor frequency. As a result, a compaction microprocessor design may first ship at about the

TABLE 1-3 Microprocessor Fabrication Historical Trends

New generation every 2 years
35% reduction in gate length
30% reduction in gate oxide thickness
15% reduction in voltage
30% reduction in interconnect horizontal dimensions
15% reduction in interconnect vertical dimensions
Add 1 metal layer every other generation

same frequency as the previous generation, which has been gradually improving since its launch.

The motivation for the new compaction is not only the immediate reduction in cost due to a smaller die size, but the potential that it will be able to eventually scale to frequencies beyond what the previous generation could reach. As an example the 180-nm generation Intel Pentium® 4 began at a maximum frequency of 1.5 GHz and scaled to 2.0 GHz. The 130-nm Pentium 4 started at 2.0 GHz and scaled to 3.4 GHz. The 90-nm Pentium 4 started at 3.2 GHz. Each new technology generation is planned to start when the previous generation can no longer be easily improved.

The future of Moore's law

In recent years, the exponential increase with time of almost any aspect of the semiconductor industry has been referred to as Moore's law. Indeed, things like microprocessor frequency, computer performance, the cost of a semiconductor fabrication plant, or the size of a microprocessor design team have all increased exponentially. No exponential trend can continue forever, and this simple fact has led to predictions of the end of Moore's law for decades. All these predictions have turned out to be wrong. For 30 years, there have always been seemingly insurmountable problems about 10 years in the future. Perhaps one of the most important lessons of Moore's law is that when billions of dollars in profits are on the line, incredibly difficult problems can be overcome.

Moore's law is of course not a "law" but merely a trend that has been true in the past. If it is to remain true in the future, it will be because the industry finds it profitable to continue to solve "insurmountable" problems and force Moore's law to come true. There have already been a number of new fabrication technologies proposed or put into use that will help continue Moore's law through 2015.

Multiple threshold voltages. Increasing the threshold voltage dramatically reduces subthreshold leakage. Unfortunately this also reduces the on current of the device and slows switching. By applying different amounts of dopant to the channels of different transistors, devices with different threshold voltages are made on the same die. When speed is required, low V_T devices, which are fast but high power, are used. In circuits that do not limit the frequency of the processor, slower, more power-efficient, high V_T devices are used to reduce overall leakage power. This technique is already in use in the Intel 90-nm fabrication generation.[13]

---

[13]Ghani et al., "90nm Logic Technology."

Silicon on insulator (SOI). SOI transistors, as shown in Fig. 1-14, build MOSFETs out of a thin layer of silicon sitting on top of an insulator. This layer of insulation reduces the capacitance of the source and drain regions, improving speed and reducing power. However, creating defect-free crystalline silicon on top of an insulator is difficult. One way to accomplish this is called *silicon implanted with oxygen* (SIMOX). In this method oxygen atoms are ionized and accelerated at a silicon wafer so that they become embedded beneath the surface. Heating the wafer then causes silicon dioxide to form and damage to the crystal structure of the surface to be repaired.

Another way of creating an SOI wafer is to start with two separate wafers. An oxide layer is grown on the surface of one and then this wafer is implanted with hydrogen ions to weaken the wafer just beneath the oxide layer. The wafer is then turned upside down and bonded to a second wafer. The layer of damage caused by the hydrogen acts as a perforation, allowing most of the top wafer to be cut away. Etching then reduces the thickness of the remaining silicon further, leaving just a thin layer of crystal silicon on top. These are known as *bonded etched back silicon on insulator* (BESOI) wafers. SOI is already in use in the Advanced Micro Devices (AMD®) 90-nm fabrication generation.[14]

Strained silicon. The ability of charge carriers to move through silicon is improved by placing the crystal lattice under strain. Electrons in the conduction band are not attached to any particular atom and travel more easily when the atoms of the crystal are pulled apart to create more space between them. Depositing silicon nitride on top of the source and drain regions tends to compress these areas. This pulls the atoms in the channel farther apart and improves electron mobility. Holes in the valence band are attached to a particular atom and travel more easily

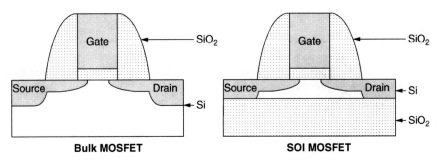

Figure 1-14 Bulk and SOI MOSFETs. (Thompson, "MOS Scaling.")

---

[14]"AMD and 90nm Manufacturing."

when the atoms of the crystal are pushed together. Depositing germanium atoms, which are larger than silicon atoms, into the source and drain tends to expand these areas. This pushes the atoms in the channel closer together and improves hole mobility. Strained silicon is already in use in the Intel 90-nm fabrication generation.[15]

High-K Gate Dielectric. Gate oxide layers thinner than 1 nm are only a few molecules thick and would have very large gate leakage currents. Replacing the silicon dioxide, which is currently used in gate oxides, with a higher permittivity material strengthens the electric field reaching the channel. This allows for thicker gate oxides to provide the same control of the channel at dramatically lower gate leakage currents.

Improved interconnects. Improvements in interconnect capacitance are possible through further reductions in the permittivity of interlevel dielectrics. However, improvements in resistance are probably not possible. Quasi-ideal interconnect scaling will rapidly reach aspect ratios over 2, beyond which fabrication and cross talk noise with neighboring wires become serious problems. The only element with less resistivity than copper is silver, but it offers only a 10 percent improvement and is very susceptible to electromigration. So, it seems unlikely that any practical replacement for copper will be found, and yet at dimensions below about 0.2 µm the resistivity of copper wires rapidly increases.[16]

The density of free electrons and the average distance a free electron travels before colliding with an atom determine the resistivity of a bulk conductor. In wires whose dimensions approach the mean free path length, the number of collisions is increased by the boundaries of the wire itself. The poor scaling of interconnect delays may have to be compensated for by scaling the upper levels of metal more slowly and adding new metal layers more rapidly to continue to provide enough

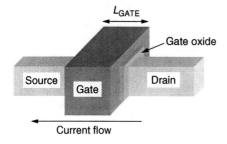

Figure 1-15 Triple gate transistor.

---

[15]Ghani et al., "Nanotechnology Featuring Strained-Silicon."

[16]David, "Silicon Research at Intel," 21.

TABLE 1-4 Microprocessor Fabrication Projection (2005–2015)

New generation every 2–3 years
30% reduction in gate length
30% increase in gate capacitance through high-K materials
15% reduction in voltage
30% reduction in interconnect horizontal and vertical dimensions for lower metal layers
15% reduction in interconnect horizontal and vertical dimensions for upper metal layers
Add 1 metal layer every generation

connections.[17] Improving the scaling of interconnects is currently the greatest challenge to the continuation of Moore's law.

Double/Triple Gate. Another way to provide the gate more control over the channel is to wrap the gate wire around two or three sides of a raised strip of silicon. In a triple gate device the channel is like a tunnel with the gate forming both sides and the roof (Fig. 1-15). This allows strong electric fields from the gate to penetrate the silicon and increases on current while reducing leakage currents.

These ideas allow at least an educated guess as to what the scaling of devices may look like over the next 10 years (Table 1-4).

Conclusion

Picturing the scaling of devices beyond 2015 becomes difficult. There is no reason why all the ideas discussed already could not be combined, creating a triple high-K gate strained silicon-on-insulator MOSFET. If this does happen, a high priority will have to be finding a better name. Although these combinations would provide further improvement, at current scaling rates the gate length of a 2030 transistor would be only 0.5 nm (about two silicon atoms across). It's not clear what a transistor at these dimensions would look like or how it would operate. As always, our predictions for semiconductor technology can only see about 10 years into the future.

Nanotechnology start-ups have trumpeted the possibility of single molecule structures, but these high hopes have had no real impact on the semiconductor industry of today. While there is the chance that carbon tubules or other single molecule structures will be used in everyday semiconductor products someday, it is highly unlikely that a technological leap will suddenly make this commonplace. As exciting as it is to think about structures one-hundredth the size of today's devices, of more immediate value is how to make devices two-thirds the size. Moore's law will continue, but it will continue through the steady evolution that has brought us so far already.

---

[17]Bohr, "Interconnect Scaling," 111.

Key Concepts and Terms

Bipolar junction transistor (BJT)
Channel length
Conduction and valence bands
Electrons and holes
Gate oxide
Integrated circuit (IC)
Junction diode
Lead design and compaction
Metal oxide semiconductor field-effect transistor (MOSFET)
Moore's law
N-type, P-type
Polycrystalline silicon (Poly)
Silicon on insulator (SOI)
Threshold voltage

Review Questions

1. What column elements are used as P-type or N-type dopants and why?
2. Describe the valence and conduction bands of a conductor, an insulator, and a semiconductor.
3. How is a BJT different than a MOSFET?
4. Why is poly commonly used for MOSFET gates?
5. What are some of the trade-offs in choosing channel length, gate oxide thickness, and threshold voltage?
6. How do microprocessor fabrication processes typically scale over time?
7. In addition to simple scaling what are other fabrication technologies to improve transistor performance?
8. Why is Moore's law so important to the semiconductor industry?
9. [Lab] Measure the current versus voltage of a junction diode. Is the current in reverse bias always zero?
10. [Discussion] What are some of the limits to Moore's law and ways of addressing these limits? How long will Moore's law continue? How might electronics continue to evolve after Moore's law?

Bibliography

"AMD and 90nm Manufacturing: Paving the Way for Tomorrow, Today." http://www.amd.com/us-en/Processors/ComputingSolutions.

Bakoglu, H., *Circuits, Interconnections, and Packaging for VLSI,* Reading, MA: Addison-Wesley, 1990.

Bhandarkar, Dileep. "Billion Transistor Processor Chips in the Mainstream Enterprise Platforms of the Future." *Ninth International Symposium on High Performance Computer Architecture,* Anaheim, CA: 2003.

Bohr, Mark et al. "A High Performance 0.25 µm Logic Technology Optimized for 1.8 V Operation." *International Electronic Devices Meeting,* San Francisco, CA: 1996, pp. 847–850. [Describes Intel's 250-nm process generation.]

Bohr, Mark. "Interconnect Scaling—The Real Limiter to High Performance ULSI." *Solid State Technology,* September 1996, pp. 105–111.

Bohr, Mark. "Silicon Trends and Limits for Advanced Microprocessors." *Communications of the ACM*, March 1998, pp. 80–87.
Ceruzzi, Paul. *A History of Modern Computing*. 2d ed., Cambridge, MA: The MIT Press, 2003.
Chau, Robert. "Silicon Nano-Transistors and Breaking the 10nm Physical Gate Length Barrier." *61st Device Research Conference*, Salt Lake City, UT: June 2003.
Chen, Gang et al. "Dynamic NBTI of p-MOS Transistors and Its Impact on MOSFET Scaling." *IEEE Electron Device Letters*, San Francisco, CA: 2002.
David, Ken. "Silicon Research at Intel." *Global Strategic Forum*, Washington DC: March 2004.
Dennard, Robert et al. "Design of Ion-Implanted MOSFETs with Very Small Dimensions." *IEEE Journal of Solid Static Circuits*, 1974, pp. 256–268.
Einstein, Marcus and James Franklin, "Computer Manufacturing Enters a New Era of Growth." *Monthly Labor Review*, September 1986, pp. 9–16.
Flatow, Ira et al. "Transistorized." ScienCentral Inc. and Twin Cities Public Television, 1998, http://www.pbs.org/transistor/index.html. [A fun and extremely accessible public television special on the invention of the transistor. Based mainly on the book *Crystal Fire*.]
Flynn, Michael and Patrick Hung, "Microprocessor Design Issues: Thoughts on the Road Ahead," *IEEE Computer Society*, Los Alamitos, CA: 2005. [A great discussion of the impact of Moore's law on design.]
Ghani, Tahir. et al. "A 90nm High Volume Manufacturing Logic Technology Featuring Novel 45nm Gate Length Strained Silicon CMOS Transistors." *International Electronic Devices Meeting*, Washington DC: 2003. [Describes Intel's 90-nm process generation.]
Ghani, Tahir. et al. "100nm Gate Length High Performance/Low Power CMOS Transistor Structure." *International Electronic Devices Meeting*, Washington DC: 1999, pp. 415–418. [Describes Intel's 180nm process generation.]
McFarland, Grant. "CMOS Technology Scaling and Its Impact on Cache Delay." Stanford University PhD Dissertation, 1995. [My own PhD dissertation did a fair job of predicting the characteristics of the 90-nm process generation, which was reached 10 years later.]
Moore, Gordon. "Cramming More Components onto Integrated Circuits." *Electronics*, April 1965, pp. 114–117. [The original paper showing Moore's law. Predicts a doubling of the number of transistors on a die every year from 1959 to 1975. Moore's 1975 paper revised this to a doubling every 18 months.]
Moore, Gordon. "Progress in Digital Integrated Electronics." *International Electronic Devices Meeting*, December 1975, pp. 11–13. [The second and most often quoted formulation of Moore's law.]
Moore, Gordon. "No Exponential is Forever... But We Can Delay Forever." *International Solid State Circuits Conference*, San Francisco, CA: February 2003. [A great update on Moore's law by the creator himself.]
Muller, Richard and Theodore Kamins. *Device Electronics for Integrated Circuits*. 2d ed., New York: John Wiley & Sons, 1986. Real, Mimi. *A Revolution in Progress: A History of Intel to Date*. Santa Clara, CA: Intel Corporation, 1984.
Reid, T. R. *The Chip*, Random House Trade Paperbacks, 2001. [Tells the story of the invention of the integrated circuit by Jack Kilby and Bob Noyce.]
Riordan, Michael and Lillian Hoddeson. *Crystal Fire*, New York: W. W. Norton & Company, 1997. [Tells the story of the invention of the transistor.]
Roberts, Bruce, Alain Harrus, and Robert Jackson. "Interconnect Metallization for Future Device Generations." *Solid State Technology*, February 1995, pp. 69–78.
Thompson, Scott et al. "A Logic Nanotechnology Featuring Strained-Silicon." *IEEE Electron Device Letters*, April 2004, pp. 191–193. [A description of the strained silicon technique used in Intel's 90-nm process generation.]
Thompson, Scott et al. "MOS Scaling: Transistor Challenges for the 21st Century." *Intel Technology Journal*, Q3 1998, pp. 1–19.
Walden, Josh. "90nm and Beyond: Moore's Law and More." *Intel Developer Forum*, San Jose, CA: April 2003.
Walker, Robert et al. "Silicon Genesis: An Oral History of Semiconductor Technology." http://silicongenesis.stanford.edu. [An incredible collection of interviews with many of the founders of silicon valley including Gordon Moore, Ted Hoff, Fedrico Faggin, Jerry Sanders, and many others.]

Chapter

2

Computer Components

Overview

This chapter discusses different computer components including buses, the chipset, main memory, graphics and expansion cards, and the motherboard; BIOS; the memory hierarchy; and how all these interact with the microprocessor.

Objectives

Upon completion of this chapter, the reader will be able to:

1. Understand how the processor, chipset, and motherboard work together.
2. Understand the importance of bus standards and their characteristics.
3. Be aware of the differences between common bus standards.
4. Describe the advantages and options when using a chipset.
5. Describe the operation of synchronous DRAM.
6. Describe the operation of a video adapter.
7. Explain the purpose of BIOS.
8. Calculate how memory hierarchy improves performance.

Introduction

A microprocessor can't do anything by itself. What makes a processor useful is the ability to input instructions and data and to output results, but to do this a processor must work together with other components.

Before beginning to design a processor, we must consider what other components are needed to create a finished product and how these components will communicate with the processor. There must be a main memory store that will hold instructions and data as well as results while the computer is running. Permanent storage will require a hard drive or other nonvolatile memory. Getting data into the system requires input devices like a keyboard, mouse, disk drives, or other peripherals. Getting results out of the system requires output devices like a monitor, audio output, or printer.

The list of available components is always changing, so most processors rely on a **chipset** of two or more separate computer chips to manage communications between the processor and other components. Different chipsets can allow the same processor to work with very different components to make a very different product. The **motherboard** is the circuit board that physically connects the components. Much of the performance difference between computers is a result of differences in processors, but without the right chipset or motherboard, the processor may become starved for data and performance limited by other computer components.

The chipset and motherboard are crucial to performance and are typically the only components designed specifically for a particular processor or family of processors. All the other components are designed independently of the processor as long as they communicate by one of the bus standards supported by the chipset and motherboard. For this reason, this chapter leaves out many details about the implementation of the components. Hard drives, CD drives, computer printers, and other peripherals are complex systems in their own right (many of which use their own processors), but from the perspective of the main processor all that matters is what bus standards are used to communicate.

Bus Standards

Most computer components are concerned with storing data or moving that data into or out of the microprocessor. The movement of data within the computer is accomplished by a series of buses. A **bus** is simply a collection of wires connecting two or more chips. Two chips must support the same bus standard to communicate successfully. Bus standards include both physical and electrical specifications.

The physical specification includes how many wires are in the bus, the maximum length of the wires, and the physical connections to the bus. Using more physical wires makes it possible to transmit more data in parallel but also makes the bus more expensive. Current bus standards use as few as 1 and as many as 128 wires to transmit data. In addition to wires for data, each bus standard may include additional wires to carry control signals, power supply, or to act as shields from electrical noise. Allowing physically long wires makes it easier to connect peripherals, especially ones that might be outside the computer case, but ultimately long wires mean long latency and reduced performance. Some buses are point-to-point buses connecting exactly two chips. These are sometimes called *ports* rather than buses. Other buses are designed to be multidrop, meaning that more than two chips communicate over the same set of wires. Allowing multiple chips to share one physical bus greatly reduces the number of separate buses required by the system, but greatly complicates the signaling on those buses.

The electrical specifications describe the type of data to be sent over each wire, the voltage to be used, how signals are to be transmitted over the wires, as well as protocols for bus arbitration. Some bus standards are single ended, meaning a single bit of information is read from a single wire by comparing its voltage to a reference voltage. Any voltage above the reference is read as a 1, and any voltage below the reference is read as a 0.

Other buses use differential signaling where a single bit of information is read from two wires by comparing their voltages. Whichever of the two wires has the higher voltage determines whether the bit is read as a 1 or a 0. Differential buses allow faster switching because they are less vulnerable to electrical noise. If interference changes the voltage of a single-ended signal, it may be read as the wrong value. Interference does not affect differential signals as long as each pair of wires is affected equally, since all that matters is the difference between the two wires, not their absolute voltages.

For point-to-point bus standards that only allow transmission of data in one direction, there is only one chip that will ever drive signals onto a particular wire. For standards that allow transmission in both directions or multidrop buses, there are multiple chips that might need to transmit on the same wire. In these cases, there must be some way of determining, which is allowed to use the bus next. This protocol is called *bus arbitration.*

Arbitration schemes can treat all users of the bus equally or give some higher priority access than others. Efficient arbitration protocols are critical to performance since any time spent deciding who will transmit data next is time that no one is transmitting. The problem is greatly simplified and performance improved by having only one transmitter on each wire, but this requires a great many more wires to allow all the needed communication.

All modern computer buses are synchronous buses that use a clock signal to synchronize the transmission of data over the bus. Chips transmitting or receiving data from the bus use the clock signal to determine when to send or capture data. Many standards allow one transfer of data every clock cycle; others allow a transfer only every other cycle, or sometimes two or even four transfers in a single cycle. Buses allowing two transfers per cycle are called *double-pumped*, and buses allowing four transfers per cycle are called *quad-pumped*. More transfers per cycle allows for better performance, but makes sending and capturing data at the proper time much more difficult.

The most important measure of the performance of a bus standard is its bandwidth. This is specified as the number of data transfers per second or as the number of bytes of data transmitted per second. Increasing bandwidth usually means either supporting a wider bus with more physical wires, increasing the bus clock rate, or allowing more transfers per cycle.

When we buy a computer it is often marketed as having a particular frequency, a 3-GHz PC, for example. The clock frequency advertised is typically that of the microprocessor, arguably the most important, but by no means the only clock signal inside your computer. Because each bus standard will specify its own clock frequency, a single computer can easily have 10 or more separate clock signals.

The processor clock frequency helps determine how quickly the processor performs calculations, but the clock signal used internally by the processor is typically higher frequency than any of the bus clocks. The frequency of the different bus clocks will help determine how quickly data moves between the different computer components. It is possible for a computer with a slower processor clock to outperform a computer with a faster processor clock if it uses higher performance buses.

There is no perfect bus standard. Trade-offs must be made between performance, cost, and complexity in choosing all the physical and electrical standards; the type of components being connected will have a large impact on which trade-offs make the most sense. As a result, there are literally dozens of bus standards and more appearing all the time. Each one faces the same dilemma that very few manufacturers will commit to building hardware supporting a new bus standard without significant demand, but demand is never significant until after some hardware support is already available.

Despite these difficulties, the appearance of new types of components and the demand for more performance from existing components steadily drive the industry to support new bus standards. However, anticipating which standards will ultimately be successful is extremely difficult, and it would add significant complexity and risk to the microprocessor design to try and support all these standards directly. This has led to the creation of chipsets that support the different bus standards of the computer, so that the processor doesn't have to.

Chipsets

The chipset provides a vital layer of abstraction for the processor. Instead of the processor having to keep up with the latest hard drive standards, graphics cards, or DRAM, it can be designed to interface only with the chipset. The chipset then has the responsibility of understanding all the different bus standards to be used by all the computer components. The chipset acts as a bridge between the different bus standards; modern chipsets typically contain two chips called the **Northbridge** and **Southbridge**.

The Northbridge communicates with the processor and the components requiring the highest bandwidth connections. Because this often includes main memory, the Northbridge is sometimes called the Memory Controller Hub (**MCH**). The connections of a Northbridge typically used with the Pentium® 4 or Athlon® XP are shown in Fig. 2-1.

In this configuration, the processor communicates only with the Northbridge and possibly another processor in a multiprocessor system. This makes bus logic on the processor as simple as possible and allows the most flexibility in what components are used with the processor. A single processor design can be sold for use with multiple different types of memory as long as chipsets are available to support each type.

Sometimes the Northbridge includes a built-in graphics controller as well as providing a bus to an optional graphics card. This type of Northbridge is called a Graphics Memory Controller Hub (**GMCH**). Including a graphics controller in the Northbridge reduces costs by avoiding the need to install a separate card, but it reduces performance

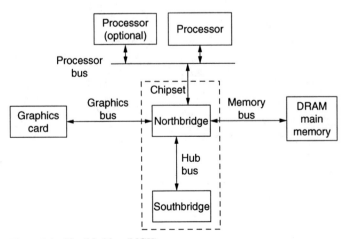

Figure 2-1 Northbridge (MCH).

by requiring the system's main memory to be used to store video images rather than dedicated memory on the graphics card.

Performance can be improved with the loss of some flexibility by providing a separate connection from the processor directly to memory. The Athlon 64 uses this configuration. Building a memory controller directly into the processor die reduces the overall latency of memory accesses. All other traffic is routed through a separate bus that connects to the Northbridge chip. Because it now interacts directly only with the graphics card, this type of Northbridge is sometimes called a *graphics tunnel* (Fig. 2-2).

Whereas a direct bus from processor to memory improves performance, the processor die itself now determines which memory standards will be supported. New memory types will require a redesign of the processor rather than simply a new chipset. In addition, the two separate buses to the processor will increase the total number of package pins needed.

Another tactic for improving performance is increasing the total memory bandwidth by interleaving memory. By providing two separate bus interfaces to two groups of memory modules, one module can be reading out data while another is receiving a new address. The total memory store is divided among the separate modules and the Northbridge combines the data from both memory channels to send to the processor. One disadvantage of memory interleaving is a more expensive Northbridge chip to handle the multiple connections. Another downside is that new memory modules must be added in matching pairs to keep the number of modules on each channel equal.

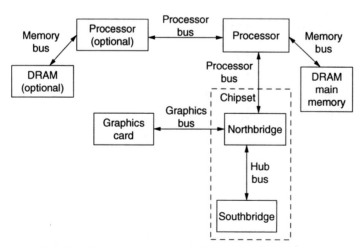

Figure 2-2 Northbridge (graphics tunnel).

Communication with all lower-performance components is routed through an *Input output Controller Hub* (**ICH**), also known as the *Southbridge chip*. The Southbridge typically controls communication between the processor and every peripheral except the graphics card and main memory (Fig. 2-3). The expansion bus supports circuit boards plugged directly into the motherboard. Peripheral buses support devices external to the computer case. Usually a separate storage bus supports access to hard drives and optical storage drives. To provide low-performance "legacy" standards such as the keyboard, serial port, and parallel port, many chipsets use a separate chip called the **super I/O** *chip*.

The main reason for dividing the functions of the processor, Northbridge, Southbridge, and super I/O chips among separate chips is flexibility. It allows different combinations to provide different functionality. Multiple different Northbridge designs can allow a single processor to work with different types of graphics and memory. Each Northbridge may be compatible with multiple Southbridge chips to provide even more combinations. All of these combinations might still use the same super I/O design to provide legacy standard support.

In recent years, transistor budgets for microprocessors have increased to the point where the functionality of the chipset could easily be incorporated into the processor. This idea is often referred to as *system-on-a-chip*, since it provides a single chip ready to interact with all the

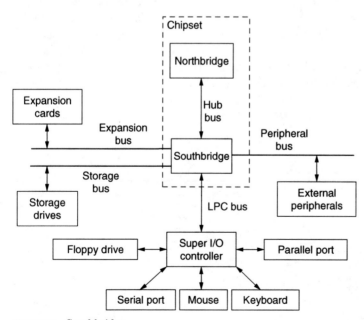

Figure 2-3 Southbridge.

common computer components. This is attractive because it requires less physical space than a separate processor and chipset and packaging costs are reduced. However, it makes the processor design dependent upon the different bus standards it supports.

Supporting multiple standards requires duplicate hardware for each standard built into the processor or supporting different versions of the processor design. Because the microprocessor is much more expensive to design, validate, and manufacture, it is often more efficient to place these functions which depend upon constantly improving bus standards on separate chips. As new bus standards become widely used, chipsets are quickly developed to support them without affecting the design of the microprocessor. For portable and handheld products where physical space is at a very high premium, it may be worth giving up the flexibility of a separate chipset in order to reduce the number of chips on the motherboard, but for desktop computers it seems likely that a separate chipset is here to stay.

Processor Bus

The processor bus controls how the microprocessor communicates with the outside world. It is sometimes called the *Front-Side Bus* (FSB). Early Pentium III and Athlon processors had high-speed cache memory chips built into the processor package. Communication with these chips was through a *back-side bus*, making the connection to the outside world the front-side bus. More recent processors incorporate their cache memory directly into the processor die, but the term front-side bus persists. Some recent processor bus standards are listed in Table 2-1.

The Athlon XP enables two data transfers per bus clock whereas the Pentium 4 enables four. For both processors, the number in the name of the bus standard refers to the number of millions of transfers per second. Because both processors perform more than one transfer per

TABLE 2-1 Processor Bus Standards

Processor bus	Bus width (b)	Processor bus clock (MHz)	Transfers per clock	Transfers per second (MT/s)	Max data bandwidth (MB/s)
Athlon XP FSB200	64	100	2	200	1600
Athlon XP FSB266	64	133	2	266	2133
Athlon XP FSB333	64	167	2	333	2667
Athlon XP FSB400	64	200	2	400	3200
Pentium 4 FSB400	64	100	4	400	3200
Pentium 4 FSB533	64	133	4	533	4267
Pentium 4 FSB800	64	200	4	800	6400

clock, neither FSB400 bus uses a 400-MHz clock, even though both are commonly referred to as "400-MHz" buses. From a performance perspective this makes perfect sense. The data buses for both processors have the same width (64 bits), so the data bandwidth at 400 MT/s is the same regardless of the frequency of the bus clock. Both FSB400 standards provide a maximum of 3.2 GB/s data bandwidth. Where the true bus clock frequency makes a difference is in determining the processor frequency.

Multiplying the frequency of the bus clock by a value set by the manufacturer generates the processor clock. This value is known as the *bus multiplier* or *bus ratio*. The allowable bus ratios and the processor bus clock frequency determine what processor frequencies are possible. Table 2-2 shows some of these possible clock frequencies for the Athlon XP and Pentium 4 for various bus speeds.

The Athlon XP allows for half bus ratios, so for a 200-MHz bus clock, the smallest possible increment in processor frequency is 100 MHz. The Pentium 4 allows only integer bus ratios, so for a 200-MHz bus clock the smallest possible increment is 200 MHz. As processor bus ratios get very high, performance can become more and more limited by communication through the processor bus. This is why improvements in bus frequency are also required to steadily improve computer performance.

Of course, to run at a particular frequency the processor must not only have the appropriate bus ratio, but also the slowest circuit path on the processor must be faster than the chosen frequency. Before processors are sold, their manufacturers test them to find the highest bus ratio they

TABLE 2-2 Processor Frequency versus Bus Ratio

	Athlon XP freq (GHz)		
Bus ratio	FSB267 (133 MHz)	FSB333 (167 MHz)	FSB400 (200 MHz)
10.0	1.33	1.67	2.00
10.5	1.40	1.75	2.10
11.0	1.47	1.83	2.20
11.5	1.53	1.92	2.30
	Pentium 4 freq (GHz)		
	FSB400 (100 MHz)	FSB533 (133 MHz)	FSB800 (200 MHz)
14	1.40	1.87	2.80
15	1.50	2.00	3.00
16	1.60	2.13	3.20
17	1.70	2.27	3.40

can successfully run. Changes to the design or the manufacturing process can improve the average processor frequency, but there is always some manufacturing variation.

Like a sheet of cookies in which the cookies in the center are overdone and those on the edge underdone, processors with identical designs that have been through the same manufacturing process will not all run at the same maximum frequency. An Athlon XP being sold to use FSB400 might first be tested at 2.3 GHz. If that test fails, the same test would be repeated at 2.2 GHz, then 2.1 GHz, and so on until a passing frequency is found, and the chip is sold at that speed. If the minimum frequency for sale fails, then the chip is discarded. The percentages of chips passing at each frequency are known as the *frequency bin splits*, and each manufacturer works hard to increase bin splits in the top frequency bins since these parts have the highest performance and are sold at the highest prices.

To get top bin frequency without paying top bin prices, some users overclock their processors. This means running the processor at a higher frequency than the manufacturer has specified. In part, this is possible because the manufacturer's tests tend to be conservative. In testing for frequency, they may assume a low-quality motherboard and poor cooling and guarantee that even with continuous operation on the worst case application the processor will still function correctly for 10 years. A system with a very good motherboard and enhanced cooling may be able to achieve higher frequencies than the processor specification.

Another reason some processors can be significantly overclocked is down binning. From month to month the demand for processors from different frequency bins may not match exactly what is produced by the fab. If more high-frequency processors are produced than can be sold, it may be time to drop prices, but in the meantime rather than stockpile processors as inventory, some high-frequency parts may be sold at lower frequency bins. Ultimately a 2-GHz frequency rating only guarantees the processor will function at 2 GHz, not that it might not be able to go faster.

There is more profit selling a part that could run at 2.4 GHz at its full speed rating, but selling it for less money is better than not at all. Serious overclockers may buy several parts from the lowest frequency bin and test each one for its maximum frequency hoping to find a very high-frequency part that was down binned. After identifying the best one they sell the others.

Most processors are sold with the bus ratio permanently fixed. Therefore, to overclock the processor requires increasing the processor bus clock frequency. Because the processor derives its own internal clock from the bus clock, at a fixed bus ratio increasing the bus clock will increase the processor clock by the same percentage. Some motherboards allow the user to tune the processor bus clock specifically for this purpose. Overclockers increase the processor bus frequency until their computer fails then decrease it a notch.

One potential problem is that the other bus clocks on the motherboard are typically derived from the processor bus frequency. This means increasing the processor bus frequency can increase the frequency of not only the processor but of all the other components as well. The frequency limiter could easily be some component besides the processor. Some motherboards have the capability of adjusting the ratios between the various bus clocks to allow the other buses to stay near their nominal frequency as the processor bus is overclocked.

Processor overclocking is no more illegal than working on your own car, and there are plenty of amateur auto mechanics who have been able to improve the performance of their car by making a few modifications. However, it is important to remember that overclocking will invalidate a processor's warranty. If a personally installed custom muffler system causes a car to break down, it's unlikely the dealer who sold the car would agree to fix it.

Overclocking reduces the lifetime of the processor. Like driving a car with the RPM in the red zone all the time, overclocked processors are under more strain than the manufacturer deemed safe and they will tend to wear out sooner. Of course, most people replace their computers long before the components are worn out anyway, and the promise and maybe more importantly the challenge of getting the most out of their computer will continue to make overclocking a rewarding hobby for some.

Main Memory

The main memory store of computers today is always based on a particular type of memory circuit, *Dynamic Random Access Memory* (**DRAM**). Because this has been true since the late 1970s, the terms main memory and DRAM have become effectively interchangeable. DRAM chips provide efficient storage because they use only one transistor to store each bit of information.

The transistor controls access to a capacitor that is used to hold an electric charge. To write a bit of information, the transistor is turned on and charge is either added to or drained from the capacitor. To read, the transistor is turned on again and the charge on the capacitor is detected as a change in voltage on the output of the transistor. A gigabit DRAM chip has a billion transistors and capacitors storing information.

Over time the DRAM manufacturing process has focused on creating capacitors that will store more charge while taking up less die area. This had led to creating capacitors by etching deep trenches into the surface of the silicon, allowing a large capacitor to take up very little area at the surface of the die. Unfortunately the capacitors are not perfect. Charge tends to leak out over time, and all data would be lost in less than a second. This is why DRAM is called a *dynamic memory*; the charge in all the capacitors must be refreshed about every 15 ms.

Cache memories are implemented using only transistors as *Static Random Access Memory* (**SRAM**). SRAM is a static memory because it will hold its value as long as power is supplied. This requires using six transistors for each memory bit instead of only one. As a result, SRAM memories require more die area per bit and therefore cost more per bit. However, they provide faster access and do not require the special DRAM processing steps used to create the DRAM cell capacitors. The manufacturing of DRAMs has diverged from that of microprocessors; all processors contain SRAM memories, as they normally do not use DRAM cells.

Early DRAM chips were asynchronous, meaning there was no shared timing signal between the memory and the processor. Later, synchronous DRAM (**SDRAM**) designs used shared clocking signals to provide higher bandwidth data transfer. All DRAM standards currently being manufactured use some type of clocking signal. SDRAM also takes advantage of memory accesses typically appearing in bursts of sequential addresses.

The memory bus clock frequency is set to allow the SDRAM chips to perform one data transfer every bus clock, but only if the transfers are from sequential addresses. This operation is known as burst mode and it determines the maximum data bandwidth possible. When accessing nonsequential locations, there are added latencies. Different DRAM innovations have focused on improving both the maximum data bandwidth and the average access latency.

DRAM chips contain grids of memory cells arranged into rows and columns. To request a specific piece of data, first the row address is supplied and then a column address is supplied. The row access strobe (RAS) and column access strobe (CAS) signals tell the DRAM whether the current address being supplied is for a row or column. Early DRAM designs required a new row address and column address be given for every access, but very often the data being accessed was multiple columns on the same row. Current DRAM designs take advantage of this by allowing multiple accesses to the same memory row to be made without the latency of driving a new row address.

After a new row is accessed, there is a delay before a column address can be driven. This is the RAS to CAS delay (T_{RCD}). After the column address is supplied, there is a latency until the first piece of data is supplied, the CAS latency (T_{CL}). After the CAS latency, data arrives every clock cycle from sequential locations. Before a new row can be accessed, the current row must be precharged (T_{RP}) to leave it ready for future accesses. In addition to the bus frequency, these three latencies are used to describe the performance of an SDRAM. They are commonly specified in the format "$T_{CL} - T_{RCD} - T_{RP}$." Typical values for each of these would be 2 or 3 cycles. Thus, Fig. 2-4 shows the operation of a "2-2-3" SDRAM.

Computer Components

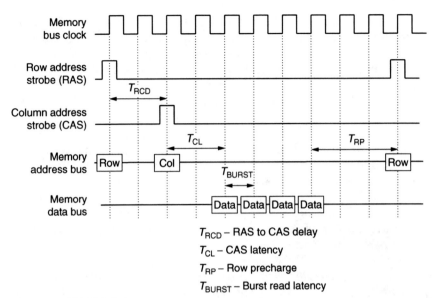

Figure 2-4 SDRAM timing.

Average latency is improved by dividing DRAM into banks where one bank precharges while another is being accessed. This means the worst-case latency would occur when accessing a different row in the same bank. In this case, the old row must be precharged, then a new row address given, and then a new column address given. The overall latency would be $T_{RP} + T_{RCD} + T_{CL}$.

Banking reduces the average latency because an access to a new row in a different bank no longer requires a precharge delay. When accessing one bank, the other banks are precharged while waiting to be used. So an access to a different bank has latency, $T_{RCD} + T_{CL}$. Accessing a different column in an already open row has only latency T_{CL}, and sequential locations after that column address are driven every cycle. These latencies are summarized in Table 2-3.

TABLE 2-3 SDRAM Latencies

Type of access	Latency	Typical bus cycles
New row in same bank	$T_{RP} + T_{RCD} + T_{CL}$	6–9
New row in different bank	$T_{RCD} + T_{CL}$	4–6
New column in same row	T_{CL}	2–3
Sequential address in open row	T_{Burst}	1

T_{RP} Row precharge delay
T_{RCD} RAS to CAS delay
T_{CL} CAS latency

The double data rate SDRAM (DDR SDRAM) standard provides more bandwidth by supplying two pieces of data per memory bus clock in burst mode instead of just one. This concept has been extended by the DDR2 standard that operates in the same fashion as DDR but uses differential signaling to achieve higher frequencies. By transmitting data as a voltage difference between two wires, the signals are less susceptible to noise and can be switched more rapidly. The downside is that two package pins and two wires are used to transmit a single bit of data.

Rambus DRAM (RDRAM) achieves even higher frequencies by placing more constraints on the routing of the memory bus and by limiting the number of bits in the bus. The more bits being driven in parallel, the more difficult it is to make sure they all arrive at the same moment. As a result, many bus standards are shifting toward smaller numbers of bits driven at higher frequencies. Some typical memory bus standards are shown in Table 2-4.

To make different DRAM standards easier to identify, early SDRAM standards were named "PC#" where the number stood for the bus frequency, but the advantage of DDR is in increased bandwidth at the same frequency, so the PC number was used to represent total data bandwidth instead. Because of the confusion this causes, DDR and DDR2 memory are often also named by the number of data transfers per second.

Just as with processor buses, transfers per cycle and clocks per cycle are often confused, and this leads to DDR266 being described as 266-MHz memory even though its clock is really only half that speed. As if things weren't confusing enough, the early RDRAM standards used the PC number to represent transfers per cycle, while later wider RDRAM bus standards have changed to being labeled by total bandwidth like DDR memory.

TABLE 2-4 Memory Bus Standards

Memory bus	Bus width (b)	Memory bus clock (MHz)	Transfers per clock	Transfers per second (MT/s)	Max data bandwidth (MB/s)
PC100 SDRAM	64	100	1	100	800
PC133 SDRAM	64	133	1	133	1066
PC2100 DDR (DDR266)	64	133	2	266	2133
PC2700 DDR (DDR333)	64	167	2	333	2667
PC2-3200 DDR (DDR2-400)	64	200	2	400	3200
PC2-4300 DDR2 (DDR2-533)	64	267	2	533	4267
PC800 RDRAM	16	400	2	800	1600
PC1066 RDRAM	16	533	2	1066	2133
PC3200 RDRAM	32	400	2	800	3200
PC4200 RDRAM	32	533	2	1066	4267

Suffice it to say that one must be very careful in buying DRAM to make sure to get the appropriate type for your computer. Ideally, the memory bus standard will support the same maximum bandwidth as the processor bus. This allows the processor to consume data at its maximum rate without wasting money on memory that is faster than your processor can use.

Video Adapters (Graphics Cards)

Most output devices consume data at a glacial pace compared with the processor's ability to produce it. The most important exception is the video adapter and display. A single high-resolution color image can contain 7 MB of data and at a typical computer monitor refresh rate of 72 Hz, the display could output data at more than 500 MB/s. If multiple frames are to be combined or processed into one, even higher data rates could be needed. Because of the need for high data bandwidth, the video adapter that drives the computer monitor typically has a dedicated high-speed connection to the Northbridge of the chipset.

Early video adapters simply translated the digital color images produced by the computer to the analog voltage signals that control the monitor. The image to be displayed is assembled in a dedicated region of memory called the *frame buffer*. The amount of memory required for the frame buffer depends on the resolution to be displayed and the number of bits used to represent the color of each pixel.

Typical resolutions range anywhere from 640×480 up to 1600×1200, and color is specified with 16, 24, or 32 bits. A display of 1600×1200 with 32-bit color requires a 7.3 MB frame buffer ($7.3 = 1600 \times 1200 \times 32/2^{20}$). The Random Access Memory Digital-to-Analog Converter (**RAMDAC**) continuously scans the frame buffer and converts the binary color of each pixel to three analog voltage signals that drive the red, green, and blue monitor controls.

Double buffering allocates two frame buffers, so that while one frame is being displayed, the next is being constructed. The RAMDAC alternates between the two buffers, so that one is always being read and one is always being written. To help generate 3D effects a z-buffer may also be used. This is a block of memory containing the effective depth (or z-value) of each pixel in the frame buffer. The z-buffer is used to determine what part of each new polygon should be drawn because it is in front of the other polygons already drawn.

Texture maps are also stored in memory to be used to color surfaces in 3D images. Rather than trying to draw the coarse surface of a brick wall, the computer renders a flat surface and then paints the image with a brick texture map. The sky in a 3D game would typically not be modeled as a vast open space with 3D clouds moving through it; instead it would be treated as a flat ceiling painted with a "sky" texture map.

Storing and processing all this data could rapidly use up the computer's main memory space and processing power. To prevent this all modern video adapters are also graphics accelerators, meaning they contain dedicated graphics memory and a graphics processor. The memory used is the same DRAM chips used for main memory or slight variations. Graphics accelerators commonly come with between 1 and 32 MB of memory built in.

The *Graphics Processor Unit* (GPU) can off-load work from the *Central Processing Unit* (CPU) by performing many of the tasks used in creating 2D or 3D images. To display a circle without a graphics processor, the CPU might create a bitmap containing the desired color of each pixel and then copy it into the frame buffer. With a graphics processor, the CPU might issue a command to the graphics processor asking for a circle with a specific color, size, and location. The graphics processor would then perform the task of deciding the correct color for each pixel.

Modern graphics processors also specialize in the operations required to create realistic 3D images. These include shading, lighting, reflections, transparency, distance fogging, and many others. Because they contain specialized hardware, the GPUs perform these functions much more quickly than a general-purpose microprocessor. As a result, for many of the latest 3D games the performance of the graphics accelerator is more important than that of the CPU.

The most common bus interfaces between the video adapter and the Northbridge are the Accelerated Graphics Port (AGP) standards. The most recent standards, PCI Express, began to be used in 2004. These graphics bus standards are shown in Table 2-5.

Some chipsets contain integrated graphics controllers. This means the Northbridge chips include a graphics processor and video adapter, so that a separate video adapter card is not required. The graphics performance of these built-in controllers is typically less than the latest separate video cards. Lacking separate graphics memory, these integrated controllers must use main memory for frame buffers and display information. Still,

TABLE 2-5 Graphics Bus Standards[1]

Graphics bus	Bus width (b)	Memory bus clock (MHz)	Transfers per clock	Transfers per second (MT/s)	Max data bandwidth (MB/s)
AGP	32	66	1	66	267
AGP × 2	32	66	2	133	533
AGP × 4	32	66	4	266	1067
AGP × 8	32	66	8	533	2133
PCI Express × 16	16	2000	1	2000	4000

[1]Mueller, *Upgrading and Repairing PCs.*

for systems that are mainly used for 2D applications, the graphics provided by these integrated solutions is often more than sufficient, and the cost savings are significant.

Storage Devices

Because hard drives are universally used by computers as primary storage, Southbridge chips of most chipsets have a bus specifically intended for use with hard drives. Hard drives store binary data as magnetic dots on metal platters that are spun at high speeds to allow the drive head to read or to change the magnetic orientation of the dots passing beneath.

Hard drives have their own version of Moore's law based not on shrinking transistors but on shrinking the size of the magnetic dots used to store data. Incredibly they have maintained the same kind of exponential trend of increasing densities over the same time period using fundamentally different technologies from computer chip manufacturing. By steadily decreasing the area required for a single magnetic dot, the hard drive industry has provided steadily more capacity at lower cost. This trend of rapidly increasing storage capacity has been critical in making use of the rapidly increasing processing capacity of microprocessors. More tightly packed data and higher spin rates have also increased the maximum data transfer bandwidth drives support. This has created the need for higher bandwidth storage bus standards shown in Table 2-6.

The most common storage bus standard is Advanced Technology Attachment (ATA). It was used with the first hard drives to include

TABLE 2-6 Storage Bus Standards[2]

Storage bus	Bus width (b)	Memory bus clock (MHz)	Transfers per clock	Transfers per second (MT/s)	Max data bandwidth (MB/s)
IDE (ATA-1)	16	8.3	0.5	4	8.3
EIDE (ATA-2)	16	8.3	1	8	16.6
Ultra-ATA/33 (UDMA-33)	16	8.3	2	16	33
Ultra-ATA/66 (UDMA-66)	16	16	2	33	66
Ultra-ATA/100 (UDMA-100)	16	25	2	50	100
SCSI	8	5	1	5	5
SCSI-Fast	8	10	1	10	10
SCSI-Ultra	8	20	1	20	20
SCSI-Ultra2	8	40	1	40	40
SCSI-Ultra3 (Ultra160)	16	40	2	80	160
SCSI-Ultra4 (Ultra320)	16	80	2	160	320
SATA-150	1	1500	0.8	1200	150
SATA-300	1	3000	0.8	2400	300
SATA-600	1	6000	0.8	4800	600

[2]Ibid.

built-in controllers, so the earliest version of ATA is usually referred to by the name Integrated Drive Electronics (IDE). Later increases in bandwidth were called Enhanced IDE (EIDE) and Ultra-ATA. The most common alternative to ATA is Small Computer System Interface (SCSI pronounced "scuzzy"). More commonly used in high performance PC servers than desktops, SCSI drives are also often used with Macintosh computers. Increasing the performance of the fastest ATA or SCSI bus standards becomes difficult because of the need to synchronize all the data bits on the bus and the electromagnetic interference between the different signals.

Beginning in 2004, a competing solution is Serial ATA (SATA), which transmits data only a single bit at a time but at vastly higher clock frequencies, allowing higher overall bandwidth. To help keep sender and receiver synchronized at such high frequencies the data is encoded to guarantee at least a single voltage transition for every 5 bits. This means that in the worst case only 8 of every 10 bits transmitted represent real data. The SATA standard is physically and electrically completely different from the original ATA standards, but it is designed to be software compatible.

Although most commonly used with hard drives, any of these standards can also be used with high-density floppy drives, tape drives, or optical CD or DVD drives. Floppy disks and tape drives store data magnetically just as hard drives do but use flexible media. This limits the data density but makes them much more affordable as removable media. Tapes store vastly more than disks by allowing the media to wrap upon itself, at the cost of only being able to efficiently access the data serially.

Optical drives store information as pits in a reflective surface that are read with a laser. As the disc spins beneath a laser beam, the reflection flashes on and off and is read by a photodetector like a naval signal light. CDs and DVDs use the same mechanism, with DVDs using smaller, more tightly packed pits. This density requires DVDs to use a shorter-wavelength laser light to accurately read the smaller pits.

A variety of writable optical formats are now available. The CD-R and DVD-R standards allow a disc to be written only once by heating a dye in the disc with a high-intensity laser to make the needed nonreflective dots. The CD-RW and DVD-RW standards allow discs to be rewritten by using a phase change media. A high-intensity laser pulse heats a spot on the disc that is then either allowed to rapidly cool or is repeatedly heated at lower intensity causing the spot to cool gradually. The phase change media will freeze into a highly reflective or a nonreflective form depending on the rate it cools. Magneto-optic (MO) discs store information magnetically but read it optically. Spots on the disc reflect light with a different polarization depending on the direction of the magnetic field. This field is very stable and can't be changed at room temperature, but heating the spot with a laser allows the field to be changed and the drive to be written.

All of these storage media have very different physical mechanisms for storing information. Shared bus standards and hardware device drivers allow the chipset to interact with them without needing the details of their operation, and the chipset allows the processor to be oblivious to even the bus standards being used.

Expansion Cards

To allow computers to be customized more easily, almost all motherboards include expansion slots that allow new circuit boards to be plugged directly into the motherboard. These expansion cards provide higher performance than features already built into the motherboard, or add entirely new functionality. The connection from the expansion cards to the chipset is called the *expansion bus* or sometimes the *input/output (I/O) bus*.

In the original IBM PC, all communication internal to the system box occurred over the expansion bus that was connected directly to the processor and memory, and ran at the same clock frequency as the processor. There were no separate processor, memory, or graphics buses. In these systems, the expansion bus was simply "The Bus," and the original design was called Industry Standard Architecture (ISA). Some mainstream expansion bus standards are shown in Table 2-7.

The original ISA standard transmitted data 8 bits at a time at a frequency of 4.77 MHz. This matched the data bus width and clock frequency of the

TABLE 2-7 Expansion Bus Standards[3]

Expansion bus	Bus width (b)	Memory bus clock (MHz)	Transfers per clock	Transfers per second (MT/s)	Max data bandwidth (MB/s)
ISA (PC/XT)	8	4.77	0.5	2	2.4
ISA (AT)	16	8.3	0.5	4	8.3
MCA	32	5	1	5	20
EISA	32	8	1	8	32
PCI	32	33	1	33	133
PCI 66 MHz	32	66	1	66	267
PCI 66 MHz/64 bits	64	66	1	66	533
PCI-X	64	133	1	133	1067
PCI Express × 1	1	2000	1	2000	250
PCI Express × 4	4	2000	1	2000	1000
PCI Express × 8	8	2000	1	2000	2000

[3]Ibid.

Intel 8088 processors used in the first IBM PC. Released in 1984, the IBM AT used the Intel 286 processor. The ISA bus was expanded to match the 16-bit data bus width of that processor and its higher clock frequency. This 16-bit version was also backward compatible with 8-bit cards and became enormously popular. IBM did not try to control the ISA standard and dozens of companies built IBM PC clones and ISA expansion cards for PCs. Both 8- and 16-bit ISA cards were still widely used into the late 1990s.

With the release of the Intel 386, which transferred data 32 bits at a time, it made sense that "The Bus" needed to change again. In 1987, IBM proposed a 32-bit-wide standard called Micro Channel Architecture (MCA), but made it clear that any company wishing to build MCA components or computers would have to pay licensing fees to IBM. Also, the MCA bus would not allow the use of ISA cards. This was a chance for IBM to regain control of the PC standard it had created and time for companies that had grown rich making ISA components to pay IBM its due.

Instead, a group of seven companies led by Compaq, the largest PC clone manufacturer at the time, created a separate 32-bit bus standard called Extended ISA (EISA). EISA would be backward compatible with older 8 and 16-bit ISA cards, and most importantly no licensing fees would be charged. As a result, the MCA standard was doomed and never appeared outside of IBM's own PS/2® line. EISA never became popular either, but the message was clear: the PC standard was now bigger than any one company, even the original creator, IBM.

The Peripheral Component Interconnect (PCI) standard was proposed in 1992 and has now replaced ISA. PCI offers high bandwidth but perhaps more importantly supports Plug-n-Play (PnP) functionality. ISA cards required the user to set switches on each card to determine which interrupt line the card would use as well as other system resources. If two cards tried to use the same resource, the card might not function, and in some cases the computer wouldn't be able to boot successfully. The PCI standard includes protocols that allow the system to poll for new devices on the expansion bus each time the system is started and dynamically assign resources to avoid conflicts. Updates to the PCI standard have allowed for steadily more bandwidth.

Starting in 2004, systems began appearing using PCI-Express, which cuts the number of data lines but vastly increases frequencies. PCI-Express is software compatible with PCI and expected to gradually replace it. The standard allows for bus widths of 1, 4, 8, or 16 bits to allow for varying levels of performance. Eventually PCI-Express may replace other buses in the system. Already some systems are replacing the AGP graphics bus with 16-bit-wide PCI-Express.

As users continue to put computers to new uses, there will always be a need for a high-performance expansion bus.

Peripheral Bus

For devices that cannot be placed conveniently inside the computer case and attached to the expansion bus, peripheral bus standards allow external components to communicate with the system.

The original IBM PC was equipped with a single bidirectional bus that transmitted a single bit of data at a time and therefore was called the *serial port* (Table 2-8). In addition, a unidirectional 8-bit-wide bus became known as the *parallel port*; it was primarily used for connecting to printers. Twenty years later, most PCs are still equipped with these ports, and they are only very gradually being dropped from new systems.

In 1986, Apple computer developed a dramatically higher-performance peripheral bus, which they called **FireWire**. This was standardized in 1995 as IEEE standard #1394. FireWire was a huge leap forward. Like the SATA and PCI-Express standards that would come years later, FireWire provided high bandwidth by transmitting data only a single bit at a time but at high frequencies. This let it use a very small physical connector, which was important for small electronic peripherals. FireWire supported Plug-n-Play capability and was also hot swappable, meaning it did not require a computer to be reset in order to find a new device. Finally, FireWire devices could be daisy chained allowing any FireWire device to provide more FireWire ports. FireWire became ubiquitous among digital video cameras and recorders.

Meanwhile, a group of seven companies lead by Intel released their own peripheral standard in 1996, Universal Serial Bus **(USB)**. USB is in many ways similar to FireWire. It transmits data serially, supports Plug-n-Play, is hot swappable, and allows daisy chaining. However, the original USB standard was intended to be used with low-performance, low-cost peripherals and only allowed 3 percent of the maximum bandwidth of FireWire.

TABLE 2-8 Peripheral Bus Standards[4]

Peripheral bus	Bus width (b)	Memory bus clock (MHz)	Transfers per clock	Transfers per second (MT/s)	Max data bandwidth (MB/s)
Serial port (RS-232)	1	0.1152	0.1	0.01	0.001
Parallel port (IEEE-1284)	8	8.3	0.17	1.4	1.4
FireWire (IEEE-1394a) S400	1	400	1	400	50
USB 1.1	1	12	1	12	1.5
USB 2.0	1	480	1	480	60
FireWire (IEEE-1394b) S800	1	800	1	800	100

[4]Ibid.

In 1998, Intel began negotiations with Apple to begin including FireWire support in Intel chipsets. FireWire would be used to support high-performance peripherals, and USB would support low-performance devices. Apple asked for a $1 licensing fee per FireWire connection, and the Intel chipset that was to support FireWire was never sold.[5] Instead, Intel and others began working on a higher-performance version of USB. The result was the release of USB 2.0 in 2000. USB 2.0 retains all the features of the original standard, is backward compatible, and increases the maximum bandwidth possible to greater than FireWire at the time. Standard with Intel chipsets, USB 2.0 is supported by most PCs sold after 2002.

Both USB and FireWire are flexible enough and low cost enough to be used by dozens of different devices. External hard drives and optical drives, digital cameras, scanners, printers, personal digital assistants, and many others use one or both of these standards. Apple has continued to promote FireWire by updating the standard (IEEE-1394b) to allow double the bandwidth and by dropping the need to pay license fees. In 2005, it remains to be seen if USB or FireWire will eventually replace the other. For now, it seems more likely that both standards will be supported for some years to come, perhaps until some new as yet unformed standard replaces them both.

Motherboards

The motherboard is the circuit board that connects the processor, chipset, and other computer components, as shown in Fig. 2-5. It physically implements the buses that tie these components together and provides all their physical connectors to the outside world.

The chipset used is the most important choice in the design of a motherboard. This determines the available bus standards and therefore the type of processor, main memory, graphics cards, storage devices, expansions cards, and peripherals the motherboard will support.

For each chip to be used on the motherboard, a decision must be made whether to solder the chip directly to the board or provide a socket that it can be plugged into. Sockets are more expensive but leave open the possibility of replacing or upgrading chips later. Microprocessors and DRAM are the most expensive required components, and therefore are typically provided with sockets. This allows a single motherboard design to be used with different processor designs and speeds, provided they are available in a compatible package. Slots for memory modules also allow the speed and total amount of main memory to be customized.

---

[5]Davis, "Apple Licensing FireWire for a Fee."

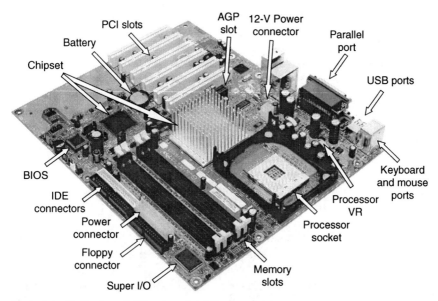

Figure 2-5 Motherboard. (*Courtesy: Intel Corporation.*)

The chipset determines the types of expansion slots available, and the physical size (or form factor) of the board limits how many are provided. Some common form factors are shown in Table 2-9.

By far the most common form factor for motherboards is the Advanced Technology Extended (ATX) standard. ATX motherboards come in four different sizes, with the main difference being that the smaller boards offer fewer expansion slots. All the ATX sizes are compatible, meaning that they use the same power supply connectors and place mounting holes in the same places. This means a PC case and power supply designed for any of the ATX sizes can be used with that size or any of the smaller ATX standards.

TABLE 2-9 Motherboard Form Factors

Form factor	Width (in)	Depth (in)	Max number expansion slots
ATX	12	9.6	7
Mini-ATX	11.2	8.2	6
Micro-ATX	9.6	9.6	4
Flex-ATX	9.0	7.5	3
BTX	12.8	10.5	7
Micro-BTX	10.4	10.5	4
Pico-BTX	8	10.5	1

In 2004, motherboards using the Balanced Technology Extended (BTX) standard began appearing. This new standard is incompatible with ATX and requires new cases although it does use the same power supply connectors. The biggest change with the BTX standard is rearranging the placement of the components on the board to allow for improved cooling. When the ATX standard first came into use, the cooling of the components on the motherboard was not a serious consideration. As processor power increased, large heavy heat sinks with dedicated fans became required.

More recently, chipsets and graphics cards have begun requiring their own heat sinks and fans. The performance possible from these components can be limited by the system's ability to cool them, and adding more fans or running the fans at higher speed may quickly create an unacceptable level of noise.

The BTX standard lines up the processor, chipset, and graphics card, so air drawn in from a single fan at the front of the system travels in a straight path over all these components and out the back of the system. This allows fewer total fans and slower fan speeds, making BTX systems quieter than ATX systems providing the same level of cooling. Like ATX, the different BTX standards are compatible, with cases designed for one BTX board accommodating any smaller BTX size.

Processor performance can be limited not only by the ability to pull heat out but also by the ability of the motherboard to deliver power into the processor. The power supply of the case converts the AC voltage of a wall socket to standard DC voltages: 3.3, 5, and 12 V. However, the processor itself may require a different voltage. The motherboard **Voltage Regulator (VR)** converts the standard DC voltages into the needed processor voltage.

Early motherboards required switches to be set to determine the voltage delivered by the VR, but this created the risk of destroying your processor by accidentally running it at very high voltage. Modern processors use *voltage identification* (VID) to control the voltage produced by the VR. When the system is first turned on, the motherboard powers a small portion of the microprocessor with a fixed voltage. This allows the processor to read built-in fuses specifying the proper voltage as determined by the manufacturer. This is signaled to the VR, which then powers up the rest of the processor at the right voltage.

Microprocessor power can be over 115 W at voltages as low as 1.4 V, requiring the VR to supply 80 A of current or more. The VR is actually not a single component but a collection of power transistors, capacitors, and inductors. The VR constantly monitors the voltage it is providing to the processor and turns power transistors on and off to keep within a specified tolerance of the desired voltage. The capacitors and inductors help reduce noise on the voltage supplied by the VR.

If the VR cannot react quickly enough to dips or spikes in the processor's current draw, the processor may fail or be permanently damaged. The large currents and fast switching of the VR transistors cause them to become yet another source of heat in the system. Limiting the maximum current they can supply will reduce VR heat and cost, but this may limit the performance of the processor.

To reduce average processor and VR power and extend battery life in portable products, some processors use VID to dynamically vary their voltage. Because the processor controls its own voltage through the VID signals to the VR, it can reduce its voltage to save power. A lower voltage requires running at a lower frequency, so this would typically only be done when the system determines that maximum performance is not currently required. If the processor workload increases, the voltage and frequency are increased back to their maximum levels. This is the mechanism behind Transmeta's LongRun®, AMD's PowerNow!®, and Intel's Enhanced SpeedStep® technologies.

A small battery on the motherboard supplies power to a *Real Time Clock* (RTC) counter that keeps track of the passage of time when the system is powered down. The battery also supplies power to a small memory called the CMOS RAM that stores system configuration information. The name CMOS RAM is left over from systems where the processor and main memory were made using only NMOS transistors, and the CMOS RAM was specially made to use NMOS and PMOS, which allowed it to have extremely low standby power. These days all the chips on the motherboard are CMOS, but the name CMOS RAM persists. Modern chipsets will often incorporate both the real time clock counter and CMOS RAM into the Southbridge chip.

To create clock signals to synchronize all the motherboard components, a quartz crystal oscillator is used. A small sliver of quartz has a voltage applied to it that causes it to vibrate and vary the voltage signal at a specific frequency. The original IBM PC used a crystal with a frequency of 14.318 MHz, and all PC motherboards to this day use a crystal with the same frequency. Multiplying or dividing the frequency of this one crystal creates almost all the clock signals on all the chips in the computer system. One exception is a separate crystal with a frequency of 32.768 kHz, which is used to drive the RTC. This allows the RTC to count time independent of the speed of the buses and prevents an overclocked system from measuring time inaccurately.

The complexity of motherboards and the wide variety of components they use make it difficult to write software to interact directly with more than one type of motherboard. To provide a standard software interface every motherboard provides basic functions through its own Basic Input Output System (BIOS).

Basic Input Output System

Today Microsoft Windows comes with dozens of built-in applications from Internet Explorer to Minesweeper, but at its core the primary function of the operating system is still to load and run programs. However, the operating system itself is a program, which leads to a "chicken-and-egg" problem. If the operating system is used to load programs, what loads the operating system? After the system is powered on the processor's memory state and main memory are both blank. The processor has no way of knowing what type of motherboard it is in or how to load an operating system. The Basic Input Output System (**BIOS**) solves this problem.

After resetting itself, the very first program the processor runs is the BIOS. This is stored in a flash memory chip on the motherboard called the *BIOS ROM*. Using flash memory allows the BIOS to be retained even when the power is off. The first thing the BIOS does is run a Power-On Self-Test (**POST**) check. This makes sure the most basic functions of the motherboard are working. The BIOS program then reads the CMOS RAM configuration information and allows it to be modified if prompted. Finally, the BIOS runs a bootstrap loader program that searches for an operating system to load.

In order to display information on the screen during POST and be able to access storage devices that might hold the operating system, the BIOS includes device drivers. These are programs that provide a standard software interface to different types of hardware. The drivers are stored in the motherboard BIOS as well as in ROM chips built into hardware that may be used during the boot process, such as video adapters and disk drives.

As the operating system boots, one of the first things it will do is load device drivers from the hard drive into main memory for all the hardware that did not have device drivers either in the motherboard BIOS or built-in chips. Most operating systems will also load device drivers to replace all the drivers provided by the BIOS with more sophisticated higher-performance drivers. As a result, the BIOS device drivers are typically only used during the system start-up but still play a crucial role. The drivers stored on a hard drive couldn't be loaded without at least a simple BIOS driver that allows the hard drive to be read in the first place.

In addition to the first few seconds of start-up, the only time Windows XP users will actually be using the BIOS device drivers is when booting Windows in "safe" mode. If a malfunctioning driver is loaded by the operating system, it may prevent the user from being able to load the proper driver. Booting in safe mode causes the operating system to not load it own drivers and to rely upon the BIOS drivers instead. This allows problems with the full boot sequence to be corrected before returning to normal operation.

By providing system initialization and the first level of hardware abstraction, the BIOS forms a key link between the hardware and software.

Memory Hierarchy

Microprocessors perform calculations at tremendous speeds, but this is only useful if the needed data for those calculations is available at similar speeds. If the processor is the engine of your computer, then data would be its fuel, and the faster the processor runs, the more quickly it must be supplied with new data to keep performing useful work. As processor performance has improved, the total capacity of data they are asked to handle has increased. Modern computers can store the text of thousands of books, but it is also critical to provide the processor with the right piece of data at the right time. Without low latency to access the data the processor is like a speed-reader in a vast library, wandering for hours trying to find the right page of a particular book.

Ideally, the data store of a processor should have extremely large capacity and extremely small latency, so that any piece of a vast amount of data could be very quickly accessed for calculation. In reality, this isn't practical because the low latency means of storage are also the most expensive. To provide the illusion of a large-capacity, low-latency memory store, modern computers use a memory hierarchy (Fig. 2-6). This uses progressively larger but longer latency memory stores to hold all the data, which may eventually be needed while providing quick access to the portion of the data currently being used.

The top of the memory hierarchy, the register file, typically contains between 64 and 256 values that are the only numbers on which the processor performs calculations. Before any two numbers are added, multiplied, compared, or used in any calculation, they will first be loaded

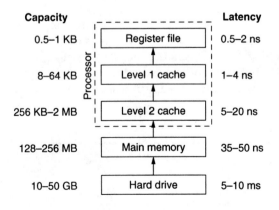

Figure 2-6 Memory hierarchy.

into registers. The register file is implemented as a section of transistors at the heart of the microprocessor die. Its small size and physical location directly next to the portion of the die performing calculations are what make its very low latencies possible. The effective cost of this die area is extremely high because increasing the capacity of the register file will push the other parts of the die farther apart, possibly limiting the maximum processor frequency. Also the latency of the register file will increase if its capacity is increased.

Making any memory store larger will always increase its access time. So the register file is typically kept small to allow it to provide latencies of only a few processor cycles; but operating at billions of calculations per second, it won't be long before the processor will need a piece of data not in the register file. The first place the processor looks next for data is called *cache memory*.

Cache memory is high-speed memory built into the processor die. It has higher capacity than the register file but a longer latency. Cache memories reduce the effective memory latency by storing data that has recently been used. If the processor accesses a particular memory location while running a program, it is likely to access it more than once. Nearby memory locations are also likely to be needed.

By loading and storing memory values and their neighboring locations as they are accessed, cache memory will often contain the data the processor needs. If the needed data is not found in the cache, it will have to be retrieved from the next level of the memory hierarchy, the computer's main memory. The percentage of time the needed data is found when the cache is accessed is called the **hit rate**. A larger cache will provide a higher hit rate but will also take up more die area, increasing the processor cost. In addition, the larger the cache capacity, the longer its latency will be. Table 2-10 shows some of the trade-offs in designing cache memory.

All the examples in Table 2-10 assume an average access time to main memory of 50 processor cycles. The first column shows that a processor with no cache will always have to go to main memory and therefore has an average access time of 50 cycles. The next column shows a 4-kB cache giving a hit rate of 65 percent and a latency of 4 cycles. For each memory access, there is a 65 percent chance the data will be found in the cache (a cache hit) and made available after 4 cycles.

TABLE 2-10 Effective Memory Latency Example

Cache size (kB)	0	4	32	128	4/128
Hit rate	0%	65%	86%	90%	65%/90%
Latency (cycles)	None	4	10	14	4/14
Avg access (cycles)	50	21.5	17.0	19.0	10.7

If the data is not found (a cache miss), it will be retrieved from main memory after 50 cycles. This gives an average access time of 21.5 cycles. Increasing the size of the cache increases the hit rate and the latency of the cache. For this example, the average access time is improved by using a 32-kB cache but begins to increase as the cache size is increased to 128 kB. At the larger cache sizes the improvement in hit rate is not enough to offset the increased latency.

The last column of the table shows the most common solution to this trade-off, a multilevel cache. Imagine a processor with a 4-kB level 1 cache and a 128-kB level 2 cache. The level 1 cache is always accessed first. It provides fast access even though its hit rate is not especially good. Only after a miss in the level 1 cache is the level 2 cache accessed. It provides better hit rate and its higher latency is acceptable because it is accessed much less often than the level 1 cache. Only after misses in both levels of cache is main memory accessed. For this example, the two-level cache gives the lowest overall average access time, and all modern high performance processors incorporate at least two levels of cache memory including the Intel Pentium II/III/4 and AMD Athlon/Duron/Opteron.

If a needed piece of data is not found in any of the levels of cache or in main memory, then it must be retrieved from the hard drive. The hard drive is critical of course because it provides permanent storage that is retained even when the computer is powered down, but when the computer is running the hard drive acts as an extension of the memory hierarchy. Main memory and the hard drive are treated as being made up of fixed-size "pages" of data by the operating system and microprocessor. At any given moment a page of data might be in main memory or might be on the hard drive. This mechanism is called *virtual memory* since it creates the illusion of the hard drive acting as memory.

For each memory access, the processor checks an array of values stored on the die showing where that particular piece of data is being stored. If it is currently on the hard drive, the processor signals a page fault. This interrupts the program currently being run and causes a portion of the operating system program to run in its place. This handler program writes one page of data in main memory back to the hard drive and then copies the needed page from the hard drive into main memory. The program that caused the page fault then continues from the point it left off.

Through this slight of hand the processor and operating system together make it appear that the needed information was in memory all the time. This is the same kind of swapping that goes on between main memory and the processor cache. The only difference is that the operating system and processor together control swapping from the hard drive to memory, whereas the processor alone controls swapping between memory and the cache. All of these levels of storage working together provide the illusion of a memory with the capacity of your hard drive but an effective latency that is dramatically faster.

We can picture a processor using the memory hierarchy the way a man working in an office might use filing system. The registers are like a single line on a sheet of paper in the middle of his desk. At any given moment he is only reading or writing just one line on this one piece of paper. The whole sheet of paper acts like the level 1 cache, containing other lines that he has just read or is about to read. The rest of his desk acts like the level 2 cache holding other sheets of paper that he has worked on recently, and a large table next to his desk might represent main memory. They each hold progressively more information but take longer to access. His filing cabinet acts like a hard drive storing vast amounts of information but taking more time to find anything in it.

Our imaginary worker is able to work efficiently because most of time after he reads one line on a page, he also reads the next line. When finished with one page, most of the time the next page he needs is already out on his desk or table. Only occasionally does he need to pull new pages from the filing cabinet and file away pages he has changed. Of course, in this imaginary office, after hours when the business is "powered down," janitors come and throw away any papers left on his desk or table. Only results that he has filed in his cabinet, like saving to the hard drive, will be kept. In fact, these janitors are somewhat unreliable and will occasionally come around unannounced in the middle of the day to throw away any lose papers they find. Our worker would be wise to file results a few times during the day just in case.

The effective latency of the memory hierarchy is ultimately determined not only by the capacity and latency of each level of the hierarchy, but also by the way each program accesses data. Programs that operate on small data sets have better hit rates and lower average access times than programs that operate on very large data sets. Microprocessors designed for computer servers often add more or larger levels of cache because servers often operate on much more data than typical users require. Computer performance is also hurt by excessive page faults caused by having insufficient main memory. A balanced memory hierarchy from top to bottom is a critical part of any computer.

The need for memory hierarchy has arisen because memory performance has not increased as quickly as processor performance. In DRAMs, transistor scaling has been used instead to provide more memory capacity. This allows for larger more complex programs but limits the improvements in memory frequency. There is no real advantage to running the bus that transfers data from memory to the processor at a higher frequency than the memory supports.

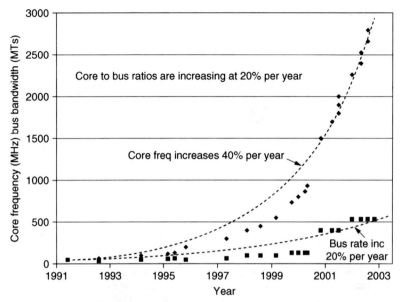

Figure 2-7 The memory gap. (*Source*: Sandpile.org.)

Figure 2-7 shows how processor frequency has scaled over time compared to the processor bus transfer rate. In the 1980s, processor frequency and the bus transfer rate were the same. The processor could receive new data every cycle. In the early 2000s, it was common to have transfer rates of only one-fifth the processor clock rate. To compensate for the still increasing gap between processor and memory performance, processors have added steadily more cache memory and more levels of memory hierarchy.

The first cache memories used in PCs were high-speed SRAM chips added to motherboards in the mid-1980s (Fig. 2-8). Latency for these chips was lower than main memory because they used SRAM cells instead of DRAM and because the processor could access them directly without going through the chipset. For the same capacity, these SRAM chips could be as much as 30 times more expensive, so there was no hope of replacing the DRAM chips used for main memory, but a small SRAM cache built into the motherboard did improve performance.

As transistor scaling continued, it became possible to add a level 1 cache to the processor die itself without making the die size unreasonably large. Eventually this level 1 cache was split into two caches, one for holding instructions and one for holding data. This improved

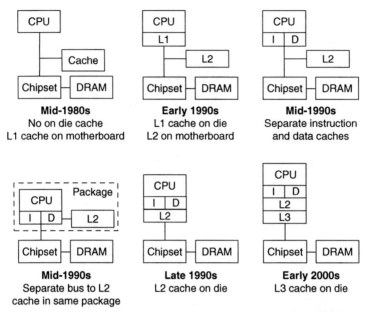

Figure 2-8 Memory hierarchy evolution. (Brey, *INTEL Microprocessors*.)

performance mainly by allowing the processor to access new instructions and data simultaneously.

In the mid-1990s, the memory hierarchy reached an awkward point. Transistor scaling had increased processor frequencies enough that level 2 cache on the motherboard was significantly slower than caches built into the die. However, transistors were still large enough that an on-die level 2 cache would make the chips too large to be economically produced. A compromise was reached in "slot" packaging. These large plastic cartridges contained a small printed circuit board made with the same process as motherboards. On this circuit board were placed the processor and SRAM chips forming the level 2 cache. By being placed in the same package the SRAM chips could be accessed at or near the processor frequency. Manufacturing the dies separately allowed production costs to be controlled.

By the late 1990s, continued shrinking of transistors allowed the in-package level 2 cache to be moved on die, and slot packaging was phased out. In the early 2000s, some processors now include three levels of on-die cache. It seems likely that the gap between memory and processor frequency will continue to grow, requiring still more levels of cache memory, and the die area of future processors may be dominated by the cache memory and not the processor logic.

Conclusion

When looking at a computer, the most noticeable features are things like the monitor, keyboard, mouse, and disk drives, but these are all simply input and output devices, ways of getting information into or out of the computer. For computer performance or compatibility, the components that are most important are those that are the least visible, the microprocessor, chipset, and motherboard. These components and how well they communicate with the rest of the system will determine the performance of the product, and it is the overall performance of the product and not the processor that matters. To create a product with the desired performance, we must design the processor to work well with the other components.

The way a processor will communicate must be considered before starting any design. As processor performance has increased, the components that move data into and out of the processor have become increasingly important. An increasing variety of available components and bus standards have made the flexibility of separate chipsets more attractive, but at the same time the need for lower latencies encourages building more communication logic directly into the processor. The right trade-off will vary greatly, especially since today processors may go into many products very different from a traditional computer.

Handheld devices, entertainment electronics, or other products with embedded processors may have very different performance requirements and components than typical PCs, but they still must support buses for communication and deal with rapidly changing standards. The basic need to support data into and out of a processor, nonvolatile storage, and peripherals is the same for a MP3 player or a supercomputer. Keeping in mind these other components that will shape the final product, we are ready to begin planning the design of the microprocessor.

Key Concepts and Terms

BIOS, POST	MCH, GMCH, ICH
Bus	Motherboard
Chipset	Northbridge, Southbridge
DRAM, SRAM, SDRAM	RAMDAC
FireWire, USB	Super I/O
Hit rate	Voltage regulator (VR)

Review Questions

1. How do bus frequencies limit processor frequencies?
2. How is information read from and written to a one-transistor DRAM cell?

3. What are the advantages and disadvantages of designing a processor with a built-in MCH?
4. What buses does a super I/O chip typically support?
5. Describe the differences between DRAM, SRAM, and SDRAM?
6. What are the important functions of the BIOS?
7. What is the difference between ATX and BTX motherboards?
8. Why is it convenient to a have separate oscillator for a real time clock?
9. What is the purpose of the VID signals?
10. How does memory hierarchy improve performance?
11. How has the design of memory hierarchies changed over time?
12. [Lab] Open the case of a personal computer and identify the following components: motherboard, power supply, hard drive, expansion cards, video adapter, main memory, chipset, and processor.
13. [Discussion] Looking at newspaper or magazine advertisement for computer equipment, identify the different bus standards mentioned in describing the components.

Bibliography

Brey, Barry. *INTEL Microprocessors 8086/8088, 80186/80188, 80286, 80386, 80486, Pentium, Prentium ProProcessor, Pentium II, III*, 4. 7th ed., Englewood Cliffs, NJ: Prentice-Hall, 2005.

Davis, Jim. "Apple Licensing FireWire for a Fee." *CNET News.Com*, January 1999. http://www.news.com.

Hennessy, John, David Patterson, and David Goldberg. *Computer Architecture: A Quantitative Approach*. 3d ed., San Francisco, CA: Morgan Kaufmann, 2002. [One of the only books to show the result of memory hierarchy changes on real benchmarks.]

Mueller, Scott. *Upgrading and Repairing PCs*. 16th ed., Indianapolis, IN: Que Publishing, 2004. [The definitive work on PC hardware. This massive 1600-page book should be sitting on the worktable of anyone who regularly opens up PCs.]

Norton, Peter, and Richard Wilton. *The New Peter Norton Programmer's Guide to the IBM PC & PS/2*. Redmond, WA: Microsoft Press, 1988. [The PS/2 is a distant memory, but I still haven't found a better book for going into the details of how PC BIOS works. As the title says, this book is aimed at programmers rather than hardware engineers.]

"Sandpile.org: The World's Leading Source for Pure Technical x86 Processor Information." http://www.sandpile.org.

Thompson, Robert and Barbara Thompson. *PC Hardware in a Nutshell*. Sebastopol, CA: O'Reilly Publishing, 2003. [Another great PC hardware book. This book is a little less detailed than Mueller's work but perhaps a little more accessible to the beginner.]

White, Ron. *How Computers Work*. 6th ed., Indianapolis, IN: Que Publishing, 2002. [A fine introduction to PC hardware for the layman. Great illustrations!]

Chapter

3

Design Planning

Overview

This chapter presents an overview of the entire microprocessor design flow and discusses design targets including processor roadmaps, design time, and product cost.

Objectives

Upon completion of this chapter, the reader will be able to:

1. Explain the overall microprocessor design flow.
2. Understand the different processor market segments and their requirements.
3. Describe the difference between lead designs, proliferations, and compactions.
4. Describe how a single processor design can grow into a family of products.
5. Understand the common job positions on a processor design team.
6. Calculate die cost, packaging cost, and overall processor cost.
7. Describe how die size and defect density impacts processor cost.

Introduction

Transistor scaling and growing transistor budgets have allowed microprocessor performance to increase at a dramatic rate, but they have also increased the effort of microprocessor design. As more functionality is added to the processor, there is more potential for logic errors. As clock rates increase, **circuit design** requires more detailed simulations. The production of new fabrication generations is inevitably more complex than previous generations. Because of the short lifetime of most microprocessors in the marketplace, all of this must happen under the pressure of an unforgiving schedule. The general steps in processor design are shown in Fig. 3-1.

A microprocessor, like any product, must begin with a plan, and the plan must include not only a concept of what the product will be, but also how it will be created. The concept would need to include the type of applications to be run as well as goals for performance, power, and cost. The planning will include estimates of design time, the size of the design team, and the selection of a general design methodology.

Defining the **architecture** involves choosing what instructions the processor will be able to execute and how these instructions will be encoded. This will determine whether already existing software can be used or whether software will need to be modified or completely rewritten. Because it determines the available software base, the choice of architecture has a huge influence on what applications ultimately run on the processor. In addition, the performance and capabilities of the processor are in part determined by the instruction set. Design planning and defining an architecture is the design specification stage of the project, since completing these steps allows the design implementation to begin.

Although the architecture of a processor determines the instructions that can be executed, the microarchitecture determines the way in which

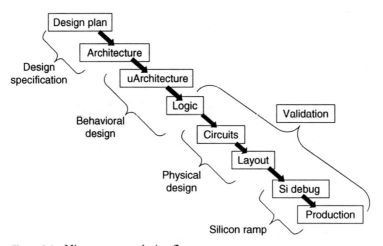

Figure 3-1 Microprocessor design flow.

they are executed. This means that architectural changes are visible to the programmer as new instructions, but microarchitectural changes are transparent to the programmer. The **microarchitecture** defines the different functional units on the processor as well as the interactions and division of work between them. This will determine the performance per clock cycle and will have a strong effect on what clock rate is ultimately achievable.

Logic design breaks the microarchitecture down into steps small enough to prove that the processor will have the correct logical behavior. To do this a computer simulation of the processor's behavior is written in a *register transfer language* (RTL). RTL languages, such as Verilog and VHDL, are high-level programming languages created specifically to simulate computer hardware. It is ironic that we could not hope to design modern microprocessors without high-speed microprocessors to simulate the design. The microarchitecture and logic design together make up the behavioral design of the project.

Circuit design creates a transistor implementation of the logic specified by the RTL. The primary concerns at this step are simulating the clock frequency and power of the design. This is the first step where the real world behavior of transistors must be considered as well as how that behavior changes with each fabrication generation.

Layout determines the positioning of the different layers of material that make up the transistors and wires of the circuit design. The primary focus is on drawing the needed circuit in the smallest area that still can be manufactured. Layout also has a large impact on the frequency and reliability of the circuit. Together circuit design and layout specify the physical design of the processor.

The completion of the physical design is called **tapeout**. In the past upon completion of the layout, all the needed layers were copied onto a magnetic tape to be sent to the fab, so manufacturing could begin. The day the tape went to the fab was tapeout. Today the data is simply copied over a computer network, but the term tapeout is still used to describe the completion of the physical design.

After tapeout the first actual prototype chips are manufactured. Another major milestone in the design of any processor is **first silicon**, the day the first chips arrive from the fab. Until this day the entire design exists as only computer simulations. Inevitably reality is not exactly the same as the simulations predicted. **Silicon debug** is the process of identifying bugs in prototype chips. Design changes are made to correct any problems as well as improving performance, and new prototypes are created. This continues until the design is fit to be sold, and the product is released into the market.

After product release the production of the design begins in earnest. However, it is common for the design to continue to be modified even after sales begin. Changes are made to improve performance or reduce the number of defects. The debugging of initial prototypes and movement into volume production is called the *silicon ramp*.

Throughout the design flow, **validation** works to make sure each step is performed correctly and is compatible with the steps before and after. For a large from scratch processor design, the entire design flow might take between 3 and 5 years using anywhere from 200 to 1000 people. Eventually production will reach a peak and then be gradually phased out as the processor is replaced by newer designs.

Processor Roadmaps

The design of any microprocessor has to start with an idea of what type of product will use the processor. In the past, designs for desktop computers went through minor modifications to try and make them suitable for use in other products, but today many processors are never intended for a desktop PC. The major markets for processors are divided into those for computer servers, desktops, mobile products, and embedded applications.

Servers and workstations are the most expensive products and therefore can afford to use the most expensive microprocessors. Performance and reliability are the primary drivers with cost being less important. Most server processors come with built-in multiprocessor support to easily allow the construction of computers using more than one processor. To be able to operate on very large data sets, processors designed for this market tend to use very large caches. The caches may include parity bits or *Error Correcting Codes* (ECC) to improve reliability. Scientific applications also make floating-point performance much more critical than mainstream usage.

The high end of the server market tends to tolerate high power levels, but the demand for "server farms," which provide very large amounts of computing power in a very small physical space, has led to the creation of low power servers. These "blade" servers are designed to be loaded into racks one next to the other. Standard sizes are 2U (3.5-in thick) and 1U (1.75-in thick). In such narrow dimensions, there isn't room for a large cooling system, and processors must be designed to control the amount of heat they generate. The high profit margins of server processors give these products a much larger influence on the processor industry than their volumes would suggest.

Desktop computers typically have a single user and must limit their price to make this financially practical. The desktop market has further differentiated to include high performance, mainstream, and value processors. The high-end desktop computers may use processors with performance approaching that of server processors, and prices approaching them as well. These designs will push die size and power levels to the limits of what the desktop market will bear. The mainstream desktop market tries to balance cost and performance, and these processor designs must weigh each performance enhancement against the increase in cost or power. Value processors are targeted at low-cost desktop systems, providing less performance but at dramatically lower prices. These designs typically start with a hard cost target and try to provide the most performance possible while keeping cost the priority.

Until recently mobile processors were simply desktop processors repackaged and run at lower frequencies and voltages to reduce power, but the extremely rapid growth of the mobile computer market has led to many designs created specifically for mobile applications. Some of these are designed for "desktop replacement" notebook computers. These notebooks are expected to provide the same level of performance as a desktop computer, but sacrifice on battery life. They provide portability but need to be plugged in most of the time. These processors must have low enough power to be successfully cooled in a notebook case but try to provide the same performance as desktop processors. Other power-optimized processors are intended for mobile computers that will typically be run off batteries. These designs will start with a hard power target and try to provide the most performance within their power budget.

Embedded processors are used inside products other than computers. Mobile handheld electronics such as *Personal Digital Assistants* (PDAs), MP3 players, and cell phones require ultralow power processors, which need no special cooling. The lowest cost embedded processors are used in a huge variety of products from microwaves to washing machines. Many of these products need very little performance and choose a processor based mainly on cost. Microprocessor markets are summarized in Table 3-1.

TABLE 3-1   Microprocessor Markets

Market	Product	Priorities
Server	High-end server	Performance, reliability, and multiprocessing
	2U & 1U server	Performance, reliability, and multiprocessing within power limit
Desktop	High-end desktop	Performance
	Mainstream desktop	Balanced performance and cost
	Value desktop	Lowest cost at required performance
Mobile	Mobile desktop replacement	Performance within power limit
	Mobile battery optimized	Power and performance
Embedded	Mobile handheld	Ultralow power
	Consumer electronics and appliances	Lowest cost at required performance

In addition to targets for performance, cost, and power, software and hardware support are also critical. Ultimately all a processor can do is run software, so a new design must be able to run an existing software base or plan for the impact of creating new software. The type of software applications being used changes the performance and capabilities needed to be successful in a particular product market.

The hardware support is determined by the processor bus standard and chipset support. This will determine the type of memory, graphics cards, and other peripherals that can be used. More than one processor project has failed, not because of poor performance or cost, but because it did not have a chipset that supported the memory type or peripherals in demand for its product type.

For a large company that produces many different processors, how these different projects will compete with each other must also be considered. Some type of product roadmap that targets different potential markets with different projects must be created.

Figure 3-2 shows the Intel roadmap for desktop processors from 1999 to 2003. Each processor has a project name used before completion of the design as well as a marketing name under which it is sold. To maintain name recognition, it is common for different generations of processor design to be sold under the same marketing name. The process generation will determine the transistor budget within a given die size as well as the maximum possible frequency. The frequency range and cache size of the processors give an indication of performance, and the die size gives a sense of relative cost. The *Front-Side Bus* (FSB) transfer rate determines how quickly information moves into or out of the processor. This will influence performance and affect the choice of motherboard and memory.

Figure 3-2 begins with the Katmai project being sold as high-end desktop in 1999. This processor was sold in a slot package that included 512 kB of level 2 cache in the package but not on the processor die. In the same time frame, the Mendocino processor was being sold as a value processor with 128 kB of cache. However, the Mendocino die was actually larger because this was the very first Intel project to integrate the level 2 cache into the processor die. This is an important example of how a larger die does not always mean a higher product cost. By including the cache on the processor die, separate SRAM chips and a multichip package were no longer needed. Overall product cost can be reduced even when die costs increase.

As the next generation Coppermine design appeared, Katmai was pushed from the high end. Later, Coppermine was replaced by the Willamette design that was sold as the first Pentium 4. This design enabled much higher frequencies but also used a much larger die. It became much more profitable when converted to the 130-nm process generation by the Northwood design. By the end of 2002, the Northwood

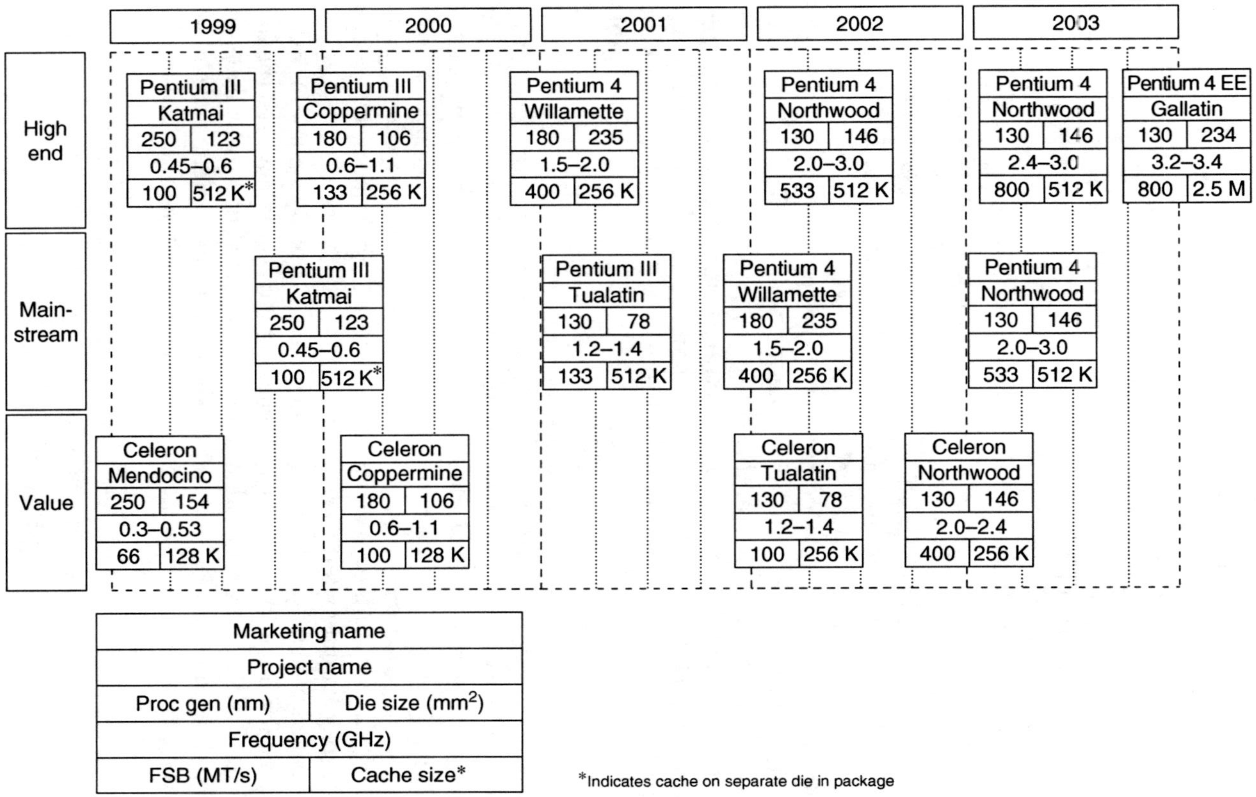

Figure 3-2 Intel's desktop roadmap (1999–2003).

*Indicates cache on separate die in package

Legend:
- Marketing name
- Project name
- Proc gen (nm) | Die size (mm²)
- Frequency (GHz)
- FSB (MT/s) | Cache size*

Tiers: High end, Mainstream, Value

1999
- High end: Pentium III / Katmai / 250 / 123 / 0.45–0.6 / 100 / 512 K*
- Value: Celeron / Mendocino / 250 / 154 / 0.3–0.53 / 66 / 128 K

2000
- High end: Pentium III / Coppermine / 180 / 106 / 0.6–1.1 / 133 / 256 K
- Mainstream: Pentium III / Katmai / 250 / 123 / 0.45–0.6 / 100 / 512 K*
- Value: Celeron / Coppermine / 180 / 106 / 0.6–1.1 / 100 / 128 K

2001
- High end: Pentium 4 / Willamette / 180 / 235 / 1.5–2.0 / 400 / 256 K
- Mainstream: Pentium III / Tualatin / 130 / 78 / 1.2–1.4 / 133 / 512 K

2002
- High end: Pentium 4 / Northwood / 130 / 146 / 2.0–3.0 / 533 / 512 K
- Mainstream: Pentium 4 / Willamette / 180 / 235 / 1.5–2.0 / 400 / 256 K
- Value: Celeron / Tualatin / 130 / 78 / 1.2–1.4 / 100 / 256 K

2003
- High end: Pentium 4 EE / Gallatin / 130 / 234 / 3.2–3.4 / 800 / 2.5 M
- High end: Pentium 4 / Northwood / 130 / 146 / 2.4–3.0 / 800 / 512 K
- Mainstream: Pentium 4 / Northwood / 130 / 146 / 2.0–3.0 / 533 / 512 K
- Value: Celeron / Northwood / 130 / 146 / 2.0–2.4 / 400 / 256 K

design was being sold in all the desktop markets. At the end of 2003, the Gallatin project added 2 MB of level 3 cache to the Northwood design and was sold as the Pentium 4 Extreme Edition.

It is common for identical processor die to be sold into different market segments. Fuses are set by the manufacturer to fix the processor frequency and bus speed. Parts of the cache memory and special instruction extensions may be enabled or disabled. The same die may also be sold in different types of packages. In these ways, the manufacturer creates varying levels of performance to be sold at different prices.

Figure 3-2 shows in 2003 the same Northwood design being sold as a Pentium 4 in the high-end and mainstream desktop markets as well as a Celeron in the value market. The die in the Celeron product is identical to die used in the Pentium 4 but set to run at lower frequency, a lower bus speed, and with half of the cache disabled. It would be possible to have a separate design with only half the cache that would have a smaller die size and cost less to produce. However, this would require careful planning for future demand to make sure enough of each type of design was available. It is far simpler to produce a single design and then set fuses to enable or disable features as needed.

It can seem unfair that the manufacturer is intentionally "crippling" their value products. The die has a full-sized cache, but the customer isn't allowed to use it. The manufacturing cost of the product would be no different if half the cache weren't disabled. The best parallel to this situation might be the cable TV business. Cable companies typical charge more for access to premium channels even though their costs do not change at all based on what the customer is watching. Doing this allows different customers to pay varying amounts depending on what features they are using. The alternative would be to charge everyone the same, which would let those who would pay for premium features have a discount but force everyone else to pay for features they don't really need. By charging different rates, the customer is given more choices and able to pay for only what they want.

Repackaging and partially disabling processor designs allow for more consumer choice in the same way. Some customers may not need the full bus speed or full cache size. By creating products with these features disabled a wider range of prices are offered and the customer has more options. The goal is not to deny good products to customers but to charge them for only what they need.

Smaller companies with fewer products may target only some markets and may not be as concerned about positioning their own products relative to each other, but they must still create a roadmap to plan the positioning of their products relative to competitors. Once a target market and features have been identified, design planning addresses how the design is to be made.

Design Types and Design Time

How much of a previous design is reused is the biggest factor affecting processor design time. Most processor designs borrow heavily from earlier designs, and we can classify different types of projects based on what parts of the design are new (Table 3-2).

Designs that start from scratch are called **lead designs**. They offer the most potential for improved performance and added features by allowing the design team to create a new design from the ground up. Of course, they also carry the most risk because of the uncertainty of creating an all-new design. It is extremely difficult to predict how long lead designs will take to complete as well as their performance and die size when completed. Because of these risks, lead designs are relatively rare.

Most processor designs are **compactions** or **variations**. Compactions take a completed design and move it to a new manufacturing process while making few or no changes in the logic. The new process allows an old design to be manufactured at less cost and may enable higher frequencies or lower power. Variations add some significant logical features to a design but do not change the manufacturing process. Added features might be more cache, new instructions, or performance enhancements. **Proliferations** change the manufacturing process and make significant logical changes.

The simplest way of creating a new processor product is to **repackage** an existing design. A new package can reduce costs for the value market or enable a processor to be used in mobile applications where it couldn't physically fit before. In these cases, the only design work is revalidating the design in its new package and platform.

Intel's Pentium 4 was a lead design that reused almost nothing from previous generations. Its schedule was described at the 2001 Design Automation Conference as approximately 6 months to create a design specification, 12 months of behavioral design, 18 months of physical

TABLE 3-2   Processor Design Types

Design type	Typical design time	Reuse
Lead	4 years	Little to no reuse
Proliferation	3 years	Significant logic changes and new manufacturing process
Compaction	2 years	Little or no logic changes, but new manufacturing process
Variation	2 years	Some logic changes on same manufacturing process
Repackage	6 months	Identical die in different package

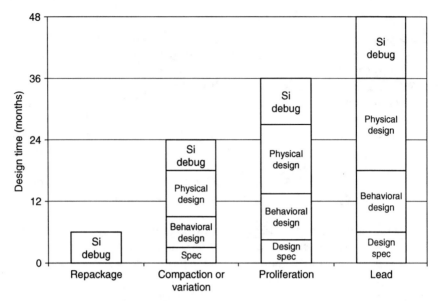

Figure 3-3 Processor design times.

design, and 12 months of silicon debug, for a total of 4 years from design plan to shipping.[1] A compaction or variation design might cut this time in half by reusing significant portions of earlier designs. A proliferation would fall somewhere in between a lead design and a compaction. A repackaging skips all the design steps except for silicon debug, which presumably will go more quickly for a design already validated in a different platform. See Figure 3-3.

Of course, the design times shown in Fig. 3-3 are just approximations. The actual time required for a design will also depend on the overall design complexity, the level of automation being used, and the size of the design team. Productivity is greatly improved if instead of working with individual logic gates, engineers are using larger predesigned blocks in constructing their design. The *International Technology Roadmap for Semiconductors* (ITRS) gives design productivity targets based on the size of the logic blocks being used to build the design.[2] Assuming an average of four transistors per logic gate gives the productivity targets shown in Table 3-3.

Constructing a design out of pieces containing hundreds of thousands or millions of transistors implies that someone has already designed these pieces, but standard libraries of basic logical components are

---

[1] Bentley, "Validating the Intel Pentium 4," 244.
[2] ITRS, "Design," 38.

TABLE 3-3 Designer Productivity

Logic block size (number of transistors)	Design productivity (transistors/engineer year)
Tiny blocks (<10 K)	36 K
Small blocks (10 K–300 K)	160 K
Large blocks (300 K–4 M)	224 K

created for a given manufacturing generation and then assembled into many different designs. Smaller fabless companies license the use of these libraries from manufacturers that sell their own spare manufacturing capacity. The recent move toward dual core processors is driven in part by the increased productivity of duplicating entire processor cores for more performance rather than designing ever-more complicated cores.

The size of the design team needed will be determined both by the type of design and the designer productivity with team sizes anywhere from less than 50 to more than 1000. The typical types of positions are shown in Table 3-4.

The larger the design team, the more additional personnel will be needed to manage and organize the team, growing the team size even more. For design teams of hundreds of people, the human issues of clear communication, responsibility, and organization become just as important as any of the technical issues of design.

The headcount of a processor project typically grows steadily until tapeout when the layout is first sent to be fabricated. The needed headcount drops rapidly after this, but silicon debug and beginning of production may still require large numbers of designers working on refinements for as much as a year after the initial design is completed. One of the most important challenges facing future processor designs is how to enhance productivity to prevent ever-larger design teams even as transistors budgets continue to grow.

The design team and manpower required for lead designs are so high that they are relatively rare. As a result, the vast majority of processor

TABLE 3-4 Processor Design Team Jobs

Position	Responsibilities
Computer architect	Define instruction set and microarchitecture
Logic designer	Convert microarchitecture into RTL
Circuit designer	Convert RTL in transistor level implementation
Mask designer	Convert circuit design into layout
Validation engineer	Verify logical correctness of design at all steps
Design automation engineer	Create and/or support design CAD tools

designs are derived from earlier designs, and a great deal can be learned about a design by looking at its family tree. Because different processor designs are often sold under a common marketing name, tracing the evolution of designs requires deciphering the design project names. For design projects that last years, it is necessary to have a name long before the environment into which the processor will eventually be sold is known for certain. Therefore, the project name is chosen long before the product name and usually chosen with the simple goal of avoiding trademark infringement.

Figure 3-4 shows the derivation of the AMD Athlon® designs. Each box shows the project name and marketing name of a processor design with the left edge showing when it was first sold.

The original Athlon design project was called the *K7* since it was AMD's seventh generation microarchitecture. The K7 used very little of previous AMD designs and was fabricated in a 250-nm fabrication process. This design was compacted to the 180-nm process by the K75 project, which was sold as both a desktop product, using the name Athlon, and a server product with multiprocessing enabled, using the name Athlon MP. Both the K7 and K75 used slot packaging with separate SRAM chips in the same package acting as a level 2 cache.

The Thunderbird project added the level 2 cache to the processor die eventually allowing the slot packaging to be abandoned. A low cost version with a smaller level 2 cache, called *Spitfire*, was also created. To make its marketing as a value product clear, the Spitfire design was given a new marketing name, Duron®.

The Palomino design added a number of enhancements. A hardware prefetch mechanism was added to try and anticipate what data would be used next and pull it into the cache before it was needed. A number of new processor instructions were added to support multimedia operations. Together these instructions were called 3D Now!® Professional. Finally a mechanism was included to allow the processor to dynamically scale its power depending on the amount of performance required by the current application. This feature was marketed as Power Now!®.

The Palomino was first sold as a mobile product but was quickly repackaged for the desktop and sold as the first Athlon XP. It was also marketed as the Athlon MP as a server processor. The Morgan project removed three-fourths of the level 2 cache from the Palomino design to create a value product sold as a Duron and Mobile Duron.

The Thoroughbred and Applebred projects were both compactions that converted the Palomino and Morgan designs from the 180-nm generation to 130 nm. Finally, the Barton project doubled the size of the Thoroughbred cache. The Athlon 64 chips that followed were based on a new lead design, so Barton marked the end of the family of designs based upon the original Athlon. See Table 3-5.

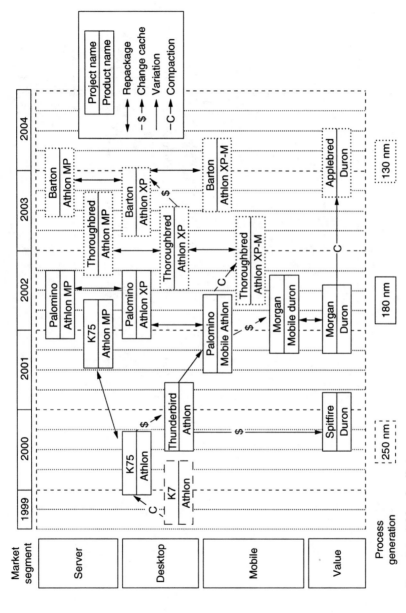

Figure 3-4 Athlon family tree.

TABLE 3-5 Athlon Family Processor Projects[3]

Project name	Marketing name	Process (nm)	Product type	Design type	Base design	Notes
K7	Athlon	250	Desktop	Lead	None	First Athlon design
K75	Athlon Athlon MP	180	Desktop	Compaction	K7	Compact from 250 nm to 180 nm
Thunderbird	Athlon	180	Desktop	Variation	K75	Add 256 kB on-die L2
Spitfire	Duron	180	Value	Variation	K75	Add 64 kB on-die L2
Palomino	Athlon XP Athlon MP Mobile Athlon 4	180	Desktop Server Mobile	Variation	Thunderbird	Add HW prefetch, 3D Now! Pro, Power Now!
Morgan	Duron Mobile Duron	180	Value Mobile	Variation	Palomino	Reduce L2 cache to 64 kB
Thoroughbred	Athlon XP Athlon MP Athlon XP-M	130	Desktop Server Mobile	Compaction	Palomino	Compact from 180 to 130 nm
Applebred	Duron	130	Value	Compaction	Morgan	Compact from 180 to 130 nm
Barton	Athlon XP Athlon MP Athlon XP-M	130	Desktop Server Mobile	Variation	Thoroughbred	Double L2 cache to 512 kB

[3] Sandpile.org.

Because from scratch designs are only rarely attempted, for most processor designs the most important design decision is choosing the previous design on which the new product will be based.

Product Cost

A critical factor in the commercial success or failure of any product is how much it costs to manufacture. For all processors, the manufacturing process begins with blank silicon wafers. The wafers are cut from cylindrical ingots and must be extremely pure and perfectly flat. Over time the industry has moved to steadily larger wafers to allow more chips to be made from each one. In 2004, the most common size used was 200-mm diameter wafers with the use of 300-mm wafers just beginning (Fig. 3-5). Typical prices might be $20 for a 200-mm wafer and $200 for a 300-mm wafer.[4] However, the cost of the raw silicon is typically only a few percent of the final cost of a processor.

Much of the cost of making a processor goes into the fabrication facilities that produce them. The consumable materials and labor costs of operating the fab are significant, but they are often outweighed by the

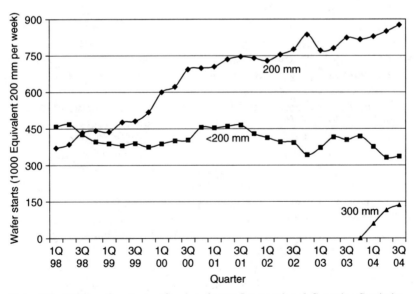

Figure 3-5 MOS wafer starts. (Semiconductor International Capacity Statistics, Q3'2004.)

---

[4] Jones, "Cost of Wafers."

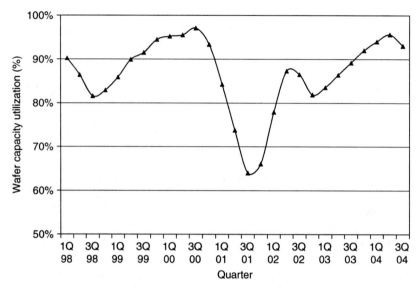

Figure 3-6 MOS fab utilization. (ibid.)

cost of depreciation. These factories cost billions to build and become obsolete in a few years. This means the depreciation in value of the fab can be more than a million dollars every day. This cost must be covered by the output of the fab but does not depend upon the number of wafers processed. As a result, the biggest factor in determining the cost of processing a wafer is typically the utilization of the factory. The more wafers the fab produces, the lower the effective cost per wafer.

Balancing **fab utilization** is a tightrope all semiconductor companies must walk. Without sufficient capacity to meet demand, companies will lose market share to their competitors, but excess capacity increases the cost per wafer and hurts profits. Because it takes years for a new fab to be built and begin producing, construction plans must be based on projections of future demand that are uncertain. From 1999 to 2000, demand grew steadily, leading to construction of many new facilities (Fig. 3-6). Then unexpectedly low demand in 2001 left the entire semiconductor industry with excess capacity. Matching capacity to demand is an important part of design planning for any semiconductor product.

The characteristics of the fab including utilization, material costs, and labor will determine the cost of processing a wafer. In 2003, a typical cost for processing a 200-mm wafer was $3000.[5] The size of the die will

---

[5]"AMD's Athlon 64 Has Arrived."

determine the cost of an individual chip. The cost of processing a wafer does not vary much with the number of die, so the smaller the die, the lesser the cost per chip. The total number of die per wafer are estimated as:[6]

$$\text{Die per wafer} = \frac{\pi(\text{wafer diameter}/2)^2}{\text{die area}} - \frac{\pi \times \text{wafer diameter}}{\sqrt{2 \times \text{die area}}}$$

The first term just divides the area of the wafer by the area of a single die. The second term approximates the loss of rectangular die that do not entirely fit on the edge of the round wafer. The 2003 International Technology Roadmap for Semiconductors (ITRS) suggests a target die size of 140 mm^2 for a mainstream microprocessor and 310 mm^2 for a server product. On 200-mm wafers, the equation above predicts the mainstream die would give 186 die per wafer whereas the server die size would allow for only 76 die per wafer. The 310-mm^2 die on 200-mm wafer is shown in Fig. 3-7.

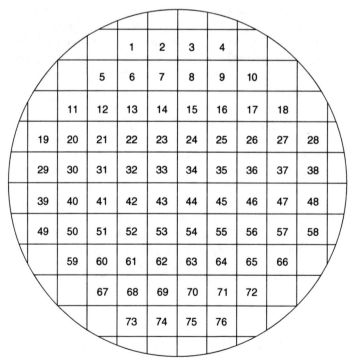

Figure 3-7   310-mm^2 die on 200-mm wafer.

---

[6]Hennessy et al., *Computer Architecture*, 19.

Unfortunately not all the die produced will function properly. In fact, although it is something each factory strives for, in the long run 100 percent yield will not give the highest profits. Reducing the on-die dimensions allows more die per wafer and higher frequencies that can be sold at higher prices. As a result, the best profits are achieved when the process is always pushed to the point where at least some of the die fail. The density of defects and complexity of the manufacturing process determine the **die yield**, the percentage of functional die. Assuming defects are uniformly distributed across the wafer, the die yield is estimated as

$$\text{Die yield} = \text{wafer yield} \times \left(1 + \frac{\text{defects per area} \times \text{die area}}{\alpha}\right)^{-\alpha}$$

The **wafer yield** is the percentage of successfully processed wafers. Inevitably the process flow fails altogether on some wafers preventing any of the die from functioning, but wafer yields are often close to 100 percent. On good wafers the failure rate becomes a function of the frequency of defects and the size of the die. In 2001, typical values for defects per area were between 0.4 and 0.8 defects per square centimeter.[7] The value α is a measure of the complexity of the fabrication process with more processing steps leading to a higher value. A reasonable estimate for modern CMOS processes is $\alpha = 4$.[8] Assuming this value for α and a 200-mm wafer, the calculation of the relative die cost for different defect densities and die sizes.

Figure 3-8 shows how at very low defect densities, it is possible to produce very large die with only a linear increase in cost, but these die quickly become extremely costly if defect densities are not well controlled. At 0.5 defects per square centimeter and $\alpha = 4$, the target mainstream die size gives a yield of 50 percent while the server die yields only 25 percent.

Die are tested while still on the wafer to help identify failures as early as possible. Only the die that pass this sort of test will be packaged. The assembly of die into package and the materials of the package itself add significantly to the cost of the product. Assembly and **package costs** can be modeled as some base cost plus some incremental cost added per package pin.

Package cost = base package cost + cost per pin × number of pins

The base package cost is determined primarily by the maximum power density the package can dissipate. Low cost plastic packages might have

---

[7]Ibid.
[8]Ibid.

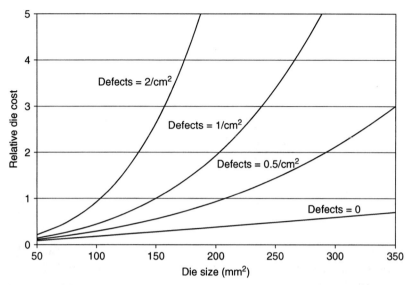

Figure 3-8 Die cost.

a base cost of only a few dollars and add only 0.5 cent per pin, but limit the total power to less than 3 W. High-cost, high-performance packages might allow power densities up to 100 W/cm^2, but have base costs of $10 to $20 plus 1 to 2 cents per pin.[9] If high performance processor power densities continue to rise, packaging could grow to be an even larger percentage of total product costs.

After packaging the die must again be tested. Tests before packaging cannot screen out all possible defects, and new failures may have been created during the assembly step. Packaged part testing identifies parts to be discarded and the maximum functional frequency of good parts. Testing typically takes less than 1 min, but every die must be tested and the testing machines can cost hundreds of dollars per hour to operate. All modern microprocessors add circuitry specifically to reduce test time and keep **test costs** under control.

$$\text{Test cost} = \text{test time} \times \text{test cost per hour}$$

The final cost of the processor is the sum of the die, packaging, and testing costs, divided by the yield of the packaged part testing.

$$\text{Processor cost} = \frac{\text{die cost} + \text{package cost} + \text{test cost}}{\text{test yield}}$$

---

[9]ITRS, "Assembly and Packaging," 4.

TABLE 3-6 Examples of Mainstream Product Costs

Assumptions		Calculations	
Die area	140 mm^2		
Wafer diameter	200 mm	Die per wafer	186
Defect density	0.5 cm^{-2}		
Process complexity (α)	4		
Wafer yield	95%	Die yield	50%
Processed wafer cost	$3000	Die cost	$33
Base package cost	$10		
Cost per pin	$0.01		
Number pins	500	Package cost	$15
Test time	30 s		
Test cost per hour	$400/h	Test cost	$3
Test yield	95%	Processor cost	$54

Assuming typical values we can calculate the product cost of the ITRS mainstream and server die sizes, as shown in Tables 3-6 and 3-7.

Calculating the percentage of different costs from these two examples gives a sense of the typical contributions to overall processor cost.

Table 3-8 shows that the relative contributions to cost can be very different from one processor to another. Server products will tend to be dominated by the cost of the die itself, but for mainstream processors and especially value products, the cost of packaging, assembly, and test cannot be overlooked. These added costs mean that design changes that grow the die size do not always increase the total processor cost. Die growth that allows for simpler packaging or testing can ultimately reduce costs.

Whether a particular processor cost is reasonable depends of course on the price of the final product the processor will be used in. In 2001,

TABLE 3-7 Examples of Server Product Costs

Assumptions		Calculations	
Die area	310 mm^2		
Wafer diameter	200 mm	Die per wafer	76
Defect density	0.5 cm^{-2}		
Process complexity (α)	4		
Wafer yield	95%	Die yield	25%
Processed wafer cost	$3000	Die cost	$158
Base package cost	$15		
Cost per pin	$0.01		
Number pins	1000	Package cost	$25
Test time	45 s		
Test cost per hour	$400/h	Test cost	$5
Test yield	95%	Process cost	$198

TABLE 3-8 Examples of Cost Breakdown

	Mainstream processor	Server processor
Die cost	64%	84%
Package & assembly	29%	13%
Test	7%	3%

the processor contributed approximately 20 percent to the cost of a typical $1000 PC.[10] If sold at $200 our desktop processor example costing only $54 would show a large profit, but our server processor example at $198 would give almost no profit. Producing a successful processor requires understanding the products it will support.

Conclusion

Every processor begins as an idea. Design planning is the first step in processor design and it can be the most important. Design planning must consider the entire design flow from start to finish and answer several important questions.

- What type of product will use the processor?
- What is the targeted performance and power of the design?
- What will be the performance and power of competing processors?
- What previous design (if any) will be used as a starting point and how much will be reused?
- How long will the design take and how many designers are needed?
- What will the final processor cost be?

Errors or poor trade-offs in any of the later design steps can prevent a processor from meeting its planned goals, but just as deadly to a project is perfectly executing a poor plan or failing to plan at all.

The remaining chapters of this book follow the implementation of a processor design plan through all the needed steps to reach manufacturing and ultimately ship to customers. Although in general these steps do flow from one to the next, there are also activities going on in parallel and setbacks that force earlier design steps to be redone. Even planning itself will require some work from all the later design steps to estimate what performance, power, and die area are possible. No single design step is performed entirely in isolation. The easiest solution at one

---

[10]Hennessy et al., *Computer Architecture*, 21.

step may create insurmountable problems for later steps in the design. The real challenge of design is to understand enough of the steps before and after your own specialty to make the right choices for the whole design flow.

Key Concepts and Terms

Architecture	Package and test cost
Circuit design	Proliferation and compaction
Fab utilization	Silicon debug
Layout	Tapeout and first silicon
Lead design	Validation
Logic design	Variation and repackage
Microarchitecture	Wafer and die yield

Review Questions

1. Describe the overall processor design flow.
2. What are the significance of tapeout and first silicon?
3. What are some common positions on a design team and their responsibilities?
4. What are some common differences between high-end, mainstream, and value desktop processors?
5. How are proliferations, compactions, and variations different from lead designs?
6. Why is fab utilization important?
7. Using the assumptions of Table 3-6, what is the maximum die size that gives a total processor cost below $100? What die size would give the same processor cost using 300-mm wafers with processed wafer cost of $4000?
8. Why are server processor die typically larger than desktop processor die?
9. [Discussion] Plan a processor design project. Describe the project with the following: target market, previous design used as starting point (if any), fabrication technology, total number of transistors, die size, cost, design time.
10. [Discussion] Identify processor products currently being sold in each of the following markets: high-end server, 1U server, high-end desktop, value desktop, mobile desktop replacement, and mobile handheld.

Bibliography

"AMD's Athlon 64 Has Arrived: the Athlon 64 FX and Athlon 64 (and Intel's P4 Extreme)." *Tom's Hardware Guide*, http://www.tomshardware.com/.

Bentley, Bob. "Validating the Intel® Pentium® 4 Microprocessor." *Design Automation Conference*, 2001, pp. 244–248. [Describes the timeline of the Pentium 4 design and the presilicon validation effort.]

Hennessy, John, David Patterson, and David Goldberg. *Computer Architecture: A Quantitative Approach*. 3d ed., San Francisco, CA: Morgan Kaufmann, 2002.

Jones, Scotten. "A Simulation Study of the Cost and Economics of 450mm Wafers." *Semiconductor International*, August 2005, http://www.reed-electronics.com/semiconductor/.

International Technology Roadmap for Semiconductors (ITRS), 2003. http://public.itrs.net. [The ITRS reports are incredibly detailed assessments of all the technical challenges facing the entire semiconductor industry out as far as 2018.]

Miraglia, Stephanie et al. "Beyond Cost-of-Ownership: A Causal Methodology for Costing Wafer Processing." *IEEE Advanced Semiconductor Manufacturing Conference*, Boston, MA: 1998, pp. 289–293. [Describes all the different factors contributing to processed wafer cost and how to properly track those costs.]

Mueller, Scott. *Upgrading and Repairing PCs*. 16th ed., Indianapolis, IN: Que Publishing, 2004. [The definitive work PC hardware. This massive 1600-page book should be sitting on the worktable of anyone who regularly opens up PCs.]

"Sandpile.org: The World's Leading Source for Pure Technical x86 Processor Information." http://www.sandpile.org.

Semiconductor International Capacity Statistics (SICAS). http://www.sicas.info.

Skinner, Richard. "What GaAs Chips Should Cost." *IEEE GaAs IC Symposium*, Monterey, CA: 1991, pp. 273–276. [Compares the contributions to cost for GaAs and silicon chips. Discusses the impact of fab utilization on cost.]

Chapter

4

Computer Architecture

Overview

This chapter discusses the trade-offs in defining an instruction set architecture, including operations, operands types, and instruction encoding.

Objectives

Upon completion of this chapter, the reader will be able to:

1. Understand the design choices that define computer architecture.
2. Describe the different types of operations typically supported.
3. Describe common operand types and addressing modes.
4. Understand different methods for encoding data and instructions.
5. Explain control flow instructions and their types.
6. Be aware of the operation of virtual memory and its advantages.
7. Understand the difference between CISC, RISC, and VLIW architectures.
8. Understand the need for architectural extensions.

Introduction

In 1964, IBM produced a series of computers beginning with the IBM 360. These computers were noteworthy because they all supported the same instructions encoded in the same way; they shared a common computer architecture. The IBM 360 and its successors were a critical development because they allowed new computers to take advantage of the already existing software base written for older computers. With the advance of the microprocessor, the processor now determines the architecture of a computer.

Every microprocessor is designed to support a finite number of specific instructions. These instructions must be encoded as binary numbers to be read by the processor. This list of instructions, their behavior, and their encoding define the processors' architecture. All any processor can do is run programs, but any program it runs must first be converted to the instructions and encoding specific to that processor architecture. If two processors share the same architecture, any program written for one will run on the other and vice versa. Some example architectures and the processors that support them are shown in Table 4-1.

The VAX architecture was introduced by Digital Equipment Corporation (DEC) in 1977 and was so popular that new machines were still being sold through 1999. Although no longer being supported, the VAX architecture remains perhaps the most thoroughly studied computer architecture ever created.

The most common desktop PC architecture is often called simply *x86* after the numbering of the early Intel processors, which first defined this architecture. This is the oldest computer architecture for which new processors are still being designed. Intel, AMD, and others carefully design new processors to be compatible with all the software written for this architecture. Companies also often add new instructions while still supporting all the old instructions. These architectural extensions mean that the new processors are not identical in architecture but are backward compatible. Programs written for older processors will run on the newer implementations, but the reverse may not be true. Intel's Multi-Media Extension (MMX^{TM}) and AMD's 3DNow!^{TM} are examples of "x86" architectural extensions. Older programs still run on processors supporting these extensions, but new software is required to take advantage of the new instructions.

In the early 1980s, research began into improving the performance of microprocessors by simplifying their architectures. Early implementation efforts were led at IBM by John Cocke, at Stanford by John Hennessy, and at Berkeley by Dave Patterson. These three teams produced the IBM 801, MIPS, and RISC-I processors. None of these were

TABLE 4-1 Computer Architectures

Architecture	Processors	Manufacturer
VAX	MicroVax 78032	DEC
x86	Pentium 4, Athlon XP	Intel, AMD
SPARC	UltraSPARC IV	Sun
PA-RISC	PA 8800	Hewlett Packard
PowerPC	PPC 970 (G5)	IBM
JVM	PicoJava	Sun
EPIC	Itanium 2	Intel

ever sold commercially, but they inspired a new wave of architectures referred to by the name of the Berkeley project as Reduce Instruction Set Computers (**RISC**).

Sun (with direct help from Patterson) created Scalable Processor Architecture (SPARC®), and Hewlett Packard created the Precision Architecture RISC (PA-RISC). IBM created the POWER™ architecture, which was later slightly modified to become the PowerPC architecture now used in Macintosh computers. The fundamental difference between Macintosh and PC software is that programs written for the Macintosh are written in the PowerPC architecture and PC programs are written in the x86 architecture. SPARC, PA-RISC, and PowerPC are all considered RISC architectures. Computer architects still debate their merits compared to earlier architectures like VAX and x86, which are called Complex Instruction Set Computers (**CISC**) in comparison.

Java is a high-level programming language created by Sun in 1995. To make it easier to run programs written in Java on any computer, Sun defined the Java Virtual Machine (JVM) architecture. This was a virtual architecture because there was not any processor that actually could run JVM code directly. However, translating Java code that had already been compiled for a "virtual" processor was far simpler and faster than translating directly from a high-level programming language like Java. This allows JVM code to be used by Web sites accessed by machines with many different architectures, as long as each machine has its own translation program. Sun created the first physical implementation of a JVM processor in 1997.

In 2001, Intel began shipping the Itanium processor, which supported a new architecture called Explicitly Parallel Instruction Computing (**EPIC**). This architecture was designed to allow software to make more performance optimizations and to use **64-bit addresses** to allow access to more memory. Since then, both AMD and Intel have added architectural extensions to their x86 processors to support 64-bit memory addressing.

It is not really possible to compare the performance of different architectures independent of their implementations. The Pentium® and Pentium 4 processors support the same architecture, but have dramatically different performance. Ultimately processor microarchitecture and fabrication technologies will have the largest impact on performance, but the architecture can make it easier or harder to achieve high performance for different applications. In creating a new architecture or adding an extension to an existing architecture, designers must balance the impact to software and hardware. As a bridge from software to hardware, a good architecture will allow efficient bug-free creation of software while also being easily implemented in high-performance hardware. In the end, because software applications and hardware implementations are always changing, there is no "perfect" architecture.

Instructions

Today almost all software is written in "high-level" programming languages. Computer languages such as C, Perl, and HTML were specifically created to make software more readable and to make it independent of a particular computer architecture. High-level languages allow the program to concisely specify relatively complicated operations. A typical instruction might look like:

    If (A + B = 5) then...   # Jump if sum of A and B is 5

To perform the same operation in instructions specific to a particular processor might take several instructions.

    ADD AX, BX   # Add BX to AX and store in AX
    CMP AX, 5   # Compare value in AX to 5
    BRE 16   # Branch to instruction 16 bytes ahead if equal

These are assembly language instructions, which are specific to a particular computer architecture. Of course, even assembly language instructions are just human readable mnemonics for the binary encoding of instructions actually understood by the processor. The encoded binary instructions are called *machine language* and are the only instructions a processor can execute. Before any program is run on a real processor, it must be translated into machine language. The programs that perform this translation for high-level languages are called *compilers*. Translation programs for assembly language are called *assemblers*. The only difference is that most assembly language instructions will be converted to a single machine language instruction while most high-level instructions will require multiple machine language instructions.

Software for the very first computers was written all in assembly and was unique to each computer architecture. Today almost all programming is done in high-level languages, but for the sake of performance small parts of some programs are still written in assembly. Ideally, any program written in a high-level language could be compiled to run on any processor, but the use of even small bits of architecture specific code make conversion from one architecture to another a much more difficult task.

Although architectures may define hundreds of different instructions, most processors spend the vast majority of their time executing only a handful of basic instructions. Table 4-2 shows the most common types of operations for the x86 architecture for the five SPECint92 benchmarks.[1]

---

[1]Hennessy and Patterson, *Computer Architecture*, 109.

TABLE 4-2 Common x86 Instructions

Instruction	Instruction type	Percent of instructions executed	Instruction type	Overall percentage
Load	Data transfer	22%	Data transfer	38%
Branch	Control flow	20%	Computation	35%
Compare	Computation	16%	Control flow	22%
Store	Data transfer	12%		
Add	Computation	8%		
And	Computation	6%		
Sub	Computation	5%		
Move	Data transfer	4%		
Call	Control flow	1%		
Return	Control flow	1%		
	Total	95%		

Table 4-2 shows that for programs that are considered important measures of performance, the 10 most common instructions make up 95 percent of the total instructions executed. The performance of any implementation is determined largely by how these instructions are executed.

Computation instructions

Computational instructions create new results from operations on data values. Any practical architecture is likely to provide the basic arithmetic and logical operations shown in Table 4-3.

A compare instruction tests whether a particular value or pair of values meets any of the defined conditions. Logical operations typically treat each bit of each operand as a separate boolean value. Instructions to shift all the bits of an operand or reverse the order of bytes make it easier to encode multiple booleans into a single operand.

The actual operations defined by different architectures do not vary that much. What makes different architectures most distinct from one

TABLE 4-3 Computation Instructions[2]

Arithmetic operations		Logical operations	
ADD	Add	AND	True if A and B true
SUB	Subtract	OR	True if A or B true
MUL	Multiply	NOT	True if A is false
DIV	Divide	XOR	True if only one of
INC	Increment		A and B is true
DEC	Decrement	SHL	Shift bits left
CMP	Compare	SHR	Shift bits right
		BSWAP	Reverse byte order

[2]"IA-32 Software Developer's Manual."

another is not the operations they allow, but the way in which instructions specify the inputs and outputs of their instructions. Input and output operands are implicit or explicit. An implicit destination means that a particular type of operation will always write its result to the same place. Implicit operands are usually the top of the stack or a special accumulator register. An explicit destination includes the intended destination as part of the instruction. Explicit operands are general-purpose registers or memory locations. Based on the type of destination operand supported, architectures can be classified into four basic types: **stack, accumulator, register**, or **memory**. Table 4-4 shows how these different architectures would implement the adding of two values stored in memory and writing the result back to memory.

Instead of registers, the architecture can define a "stack" of stored values. The stack is a first-in last-out queue where values are added to the top of the stack with a push instruction and removed from the top with a pop instruction. The concept of a stack is useful when passing many pieces of data from one part of a program to another. Instead of having to specify multiple different registers holding all the values, the data is all passed on the stack. The calling subroutine pushes as many values as needed onto the stack, and the procedure being called pops, the appropriate number of times to retrieve all the data. Although it would be possible to create an architecture with only load and store instructions or with only push and pop instructions, most architectures allow for both.

A stack architecture uses the stack as an implicit source and destination. First the values A and B, which are stored in memory, are pushed on the stack. Then the Add instruction removes the top two values on the stack, adds them together, and pushes the result back on the stack. The pop instruction then places this value into memory. The stack architecture Add instruction does not need to specify any operands at all since all sources come from the stack and all results go to the stack. The Java Virtual Machine (JVM) is a stack architecture.

An accumulator architecture uses a special register as an implicit destination operand. In this example, it starts by loading value A into the accumulator. Then the Add instruction reads value B from memory and adds it to the accumulator, storing the result back in the accumulator. A store instruction then writes the result out to memory.

TABLE 4-4   Operands Types

Stack	Accumulator	Register	Memory
Push A	Ld A	Ld R1, A	Add C, B, A
Push B	Add B	Ld R2, B	
Add	St C	Add R3, R2, R1	
Pop C		St C, R3	

Register architectures allow the destination operand to be explicitly specified as one of number of general-purpose registers. To perform the example operation, first two load instructions place the values A and B in two general-purpose registers. The Add instruction reads both these registers and writes the results to a third. The store instruction then writes the result to memory. RISC architectures allow register destinations only for computations.

Memory architectures allow memory addresses to be given as destination operands. In this type of architecture, a single instruction might specify the addresses of both the input operands and the address where the result is to be stored. What might take several separate instructions in the other architectures is accomplished in one. The x86 architecture supports memory destinations for computations.

Many early computers were based upon stack or accumulator architectures. By using implicit operands they allow instructions to be coded in very few bits. This was important for early computers with extremely limited memory capacity. These early computers also executed only one instruction at a time. However, as increased transistor budgets allowed multiple instructions to be executed in parallel, stack and accumulator architectures were at a disadvantage. More recent architectures have all used register or memory destinations. The JVM architecture is an exception to this rule, but because it was not originally intended to be implemented in silicon, small code size and ease of translation were deemed far more important than the possible impact on performance.

The results of one computation are commonly used as a source for another computation, so typically the first source operand of a computation will be the same as the destination type. It wouldn't make sense to only support computations that write to registers if a register could not be an input to a computation. For two source computations, the other source could be of the same or a different type than the destination. One source could also be an **immediate value**, a constant encoded as part of the instruction. For register and memory architectures, this leads to six types of instructions. Table 4-5 shows which architectures discussed so far provide support for which types.

TABLE 4-5 Computation Instruction Types

Destination and first source type	Second source type		
	Immediate	Register	Memory
Register	VAX, x86, SPARC, PA-RISC, PowerPC, EPIC	VAX, x86, SPARC, PA-RISC, PowerPC, EPIC	VAX, x86
Memory	VAX, x86	VAX, x86	VAX

TABLE 4-6 Architectural Types

Type	Most complex operands	Examples
Pure register	Register	SPARC, PA-RISC, PowerPC, EPIC
Register/memory	Mix of register and memory	x86
Pure memory	All memory	VAX

The VAX architecture is the most complex, supporting all these possible combinations of source and destination types. The RISC architectures are the simplest, allowing only register destinations for computations and only immediate or register sources. The x86 architecture allows one of the sources to be of any type but does not allow both sources to be memory locations. Like most modern architectures, the examples in Table 4-5 fall into three basic types shown in Table 4-6.

RISC architectures are pure register architectures, which allow register and immediate arguments only for computations. They are also called *load/store architectures* because all the movement of data to and from memory must be accomplished with separate load and store instructions. Register/memory architectures allow some memory operands but do not allow all the operands to be memory locations. Pure memory architectures support all operands being memory locations as well as registers or immediates.

The time it takes to execute any program is the number of instructions executed times the average time per instruction. Pure register architectures try to reduce execution time by reducing the time per instruction. Their very simple instructions are executed quickly and efficiently, but more of them are necessary to execute a program. Pure memory architectures try to use the minimum number of instructions, at the cost of increased time per instruction.

Comparing the dynamic instruction count of different architectures to an imaginary ideal high-level language execution, Jerome Huck found pure register architectures executing almost twice as many instructions as a pure memory architecture implementation of the same program (Table 4-7).[3] Register/memory architectures fell between these two extremes. The highest performance of architectures will ultimately depend upon the implementation, but pure register architectures must execute their instructions on average twice as fast to reach the same performance.

In addition to the operand types supported, the maximum number of operands is chosen to be two or three. Two-operand architectures use one source operand and a second operand which acts as both a source and the destination. Three-operand architectures allow the destination to be distinct from both sources. The x86 architecture is a two-operand

---

[3]Huck and Flynn, *Analyzing Computer Architectures.*

TABLE 4-7 Relative Instruction Count

Architectural type	Instruction count
Pure register	198%
Register/memory	178%
Pure memory	114%

architecture, which can provide more compact code. The RISC architectures are three-operand architectures. The VAX architecture, seeking the greatest possible flexibility in instruction type, provides for both two- and three-operand formats.

The number and type of operands supported by different instructions will have a great effect on how these instructions can be encoded. Allowing for different operand encoding can greatly increase the functionality and complexity of a computer architecture. The resulting size of code and complexity in decoding will have an impact on performance.

Data transfer instructions

In addition to computational instructions, any computer architecture will have to include data transfer instructions for moving data from one location to another. Values may be copied from main memory to the processor or results written out to memory. Most architectures define registers to hold temporary values rather than requiring all data to be accessed by a memory address. Some common data transfer instructions and their mnemonics are listed in Table 4-8.

Loads and stores move data to and from registers and main memory. Moves transfer data from one register to another. The conditional move only transfers data if some specific condition is met. This condition might be that the result of a computation was 0 or not 0, positive or not positive, or many others. It is up to the computer architect to define all the possible conditions that can be tested. Most architectures define a special flag register that stores these conditions. Conditional moves can

TABLE 4-8 Data Transfer Instructions

LD	Load value from memory to a register
ST	Store value from a register to memory
MOV	Move value from register to register
CMOV	Conditionally move value from register to register if a condition is met
PUSH	Push value onto top of stack
POP	Pop value from top of stack

TABLE 4-9 Integer Formats (16-Bit Width)

Format	Binary range	Decimal range
Unsigned integer	0 to $2^{16} - 1$	0 to 65,535
Signed integer	-2^{15} to $2^{15} - 1$	−32,768 to 32,767
Binary coded decimal (BCD)	0 to $\approx 2^{13}$	0 to 9,999

improve performance by taking the place of instructions controlling the program flow, which are more difficult to execute in parallel with other instructions.

Any data being transferred will be stored as binary digits in a register or memory location, but there are many different formats that are used to encode a particular value in binary. The simplest formats only support integer values. The ranges in Table 4-9 are all calculated for 16-bit integers, but most modern architectures also support 32- and 64-bit formats.

Unsigned format assumes every value stored is positive, and this gives the largest positive range. Signed integers are dealt with most simply by allowing the most significant bit to act as a sign bit, determining whether the value is positive or negative. However, this leads to the unfortunate problem of having representations for both a "positive" 0 and a "negative" 0. As a result, signed integers are instead often stored in **two's complement** format where to reverse the sign, all the bits are negated and 1 is added to the result. If a 0 value (represented by all 0 bits) is negated and then has 1 added, it returns to the original zero format.

To make it easier to switch between binary and decimal representations some architectures support binary coded decimal (**BCD**) formats. These treat each group of 4 bits as a single decimal digit. This is inefficient since 4 binary digits can represent 16 values rather than only 10, but it makes conversion from binary to decimal numbers far simpler.

Storing numbers in **floating-point** format increases the range of values that can be represented. Values are stored as if in scientific notation with a fraction and an exponent. IEEE standard 754 defines the formats listed in Table 4-10.[4]

The total number of discrete values that can be represented by integer or floating-point formats is the same, but treating some of the bits as an exponent increases the range of values. For exponents below 1, the possible values are closer together than an integer representation; for exponents greater than 1, the values are farther apart. The IEEE standard reserves an exponent of all ones to represent special values like infinity and "Not-A-Number."

---

[4]"IEEE 754."

TABLE 4-10 Floating-Point Formats

Precision	Size (bits)	Sign (bits)	Exponent (bits)	Fraction (bits)	Binary range	Decimal range
Single	32	1	8	23	2^{-126} to 2^{127}	1.18×10^{-38} to 3.40×10^{38}
Double	64	1	11	52	2^{-1022} to 2^{1023}	2.23×10^{-308} to 1.79×10^{308}
Extended	80	1	16	63	2^{-16382} to 2^{16383}	3.37×10^{-4932} to 1.18×10^{4932}

Value = 1.<Fraction> $\times 2^{\text{Exponent}}$

Working with floating-point numbers requires more complicated hardware than integers; as a result the latency of floating-point operations is longer than integer operations. However, the increased range of possible values is required for many graphics and scientific applications. As a result, when quoting performance, most processors provide separate integer and floating-point performance measurements. To improve both integer and floating-point performance many architectures have added single instruction multiple data (**SIMD**) operations.

SIMD instructions simultaneously perform the same computation on multiple pieces of data (Fig. 4-1). In order to use the already defined instruction formats, the SIMD instructions still have only two- or three-operand instructions. However, they treat each of their operands as a vector containing multiple pieces of data.

For example, a 64-bit register could be treated as two 32-bit integers, four 16-bit integers, or eight 8-bit integers. Instead, the same 64-bit register could be interpreted as two single precision floating-point numbers. SIMD instructions are very useful in multimedia or scientific applications where very large amounts of data must all be processed in the same way. The Intel MXX and AMD 3DNow! extensions both allow operations on 64-bit vectors. Later, the Intel Streaming SIMD Extension

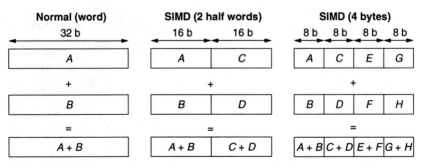

Figure 4-1 SIMD instructions.

(SSE) and AMD 3DNow! Professional extensions provide instructions for operating on 128-bit vectors. RISC architectures have similar extensions including the SPARC VIS, PA-RISC MAX2, and PowerPC AltiVec.

Integer, floating-point, and vector operands show how much computer architecture is affected not just by the operations allowed but by operands allowed as well.

Memory addresses. In *Gulliver's Travels* by Jonathan Swift, Gulliver finds himself in the land of Lilliput where the 6-in tall inhabitants have been at war for years over the trivial question of how to eat a hard-boiled egg. Should one begin by breaking open the little end or the big end? It is unfortunate that Gulliver would find something very familiar about one point of contention in computer architecture.

Computers universally divide their memory into groups of 8 bits called *bytes*. A byte is a convenient unit because it provides just enough bits to encode a single keyboard character. Allowing smaller units of memory to be addressed would increase the size of memory addresses with address bits that would be rarely used. Making the minimum addressable unit larger could cause inefficient use of memory by forcing larger blocks of memory to be used when a single byte would be sufficient. Because processors address memory by bytes but support computation on values of more than 1 byte, a question arises: For a number of more than 1 byte, is the byte stored at the lowest memory address the least significant byte (the little end) or the most significant byte (the big end)? The two sides of this debate take their names from the two factions of Lilliput: Little Endian and Big Endian. Figure 4-2 shows how this choice leads to different results.

There are a surprising number of arguments as to why little endian or big endian is the correct way to store data, but for most people none of these arguments are especially convincing. As a result, each architecture has made a choice more or less at random, so that today different computers answer this question differently. Table 4-11 shows architectures that support little endian or big endian formats.

To help the sides of this debate reach mutual understanding, many architectures support a byte swap instruction, which reverses the byte

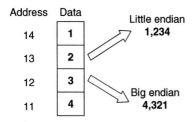

Figure 4-2 Little endian vs. big endian.

TABLE 4-11 Little Endian versus Big Endian

Little endian	Big endian
VAX, x86, EPIC	SPARC, PA-RISC, PowerPC

order of a number to convert between the little endian and big endian formats. In addition, the EPIC, PA-RISC, and PowerPC architectures all support special modes, which cause them to read data in the opposite format from their default assumption. Any new architecture will have to pick a side or build in support for both.

Architectures must also decide whether to support unaligned memory accesses. This would mean allowing a value of more than 1 byte to begin at any byte in memory. Modern memory bus standards are all more than 1-byte wide and for simplicity allow only accesses aligned on the bus width. In other words, a 64-bit data bus will always access memory at addresses that are multiples of 64 bits. If the architecture forces 64 bit and smaller values to be stored only at addresses that are multiples of their width, then any value can be retrieved with a single memory access. If the architecture allows values to start at any byte, it may require two memory accesses to retrieve the entire value. Later accesses of misaligned data from the cache may require multiple cache accesses. Forcing aligned addresses improves performance, but by restricting where values can be stored, the use of memory is made less efficient.

Given an address, the choice of little endian or big endian will determine how the data in memory is loaded. This still leaves the question of how the address itself is generated. For any instruction that allows a memory operand, it must be decided how the address for that memory location will be specified. Table 4-12 shows examples of different addressing modes.

The simplest possible addressing is absolute mode where the memory address is encoded as a constant in the instruction. Register indirect addressing provides the number of a register that contains the address. This allows the address to be computed at run time, as would be the case

TABLE 4-12 Data Addressing Modes

Address mode	Address	Typical use
Absolute	Const	Global variable
Register Indirect	Reg	Dynamically allocated variable
Displacement	Const + Reg(× Size)	Array
Indexed	Reg + Reg(× Size)	Dynamically allocated array
Scaled	Const + Reg + Reg(× Size)	Two dimensional array
Memory Indirect	Mem[Reg]	Memory pointers

for dynamically allocated variables. Displacement mode calculates the address as the sum of a constant and a register value. Some architectures allow the register value to be multiplied by a size factor. This mode is useful for accessing arrays. The constant value can contain the base address of the array while the registers hold the index. The size factor allows the array index to be multiplied by the data size of the array elements. An array of 32-bit integers will need to multiply the index by 4 to reach the proper address because each array element contains 4 bytes.

The indexed mode is the same as the displacement mode except the base address is held in a register rather than being a constant. The scaled address mode sums a constant and two registers to form an address. This could be used to access a two-dimensional array. Some architectures also support auto increment or decrement modes where the register being used as an index is automatically updated after the memory access. This supports serially accessing each element of an array. Finally, the memory indirect mode specifies a register that contains the address of a memory location that contains the desired address. This could be used to implement a memory pointer variable where the variable itself contains a memory address.

In theory, an architecture could function supporting only register indirect mode. However, this would require computation instructions to form each address in a register before any memory location could be accessed. Supporting additional addressing modes can greatly reduce the total number of instructions required and can limit the number of registers that are used in creating addresses. Allowing a constant or a constant added to a register to be used as an address is ideal for static variables allocated during compilation. Therefore, most architectures support at least the first three address modes listed in Table 4-12. RISC architectures typically support only these three modes.

The more complicated modes further simplify coding but make some memory accesses much more complex than others. Memory indirect mode in particular requires two memory accesses for a single memory operand. The first access retrieves the address, and the second gets the data. VAX is one of the only architectures to support all the addressing modes shown in Table 4-12. The x86 architecture supports all these modes except for memory indirect. In addition to addressing modes, modern architectures also support an additional translation of memory addresses to be controlled by the operating system. This is called *virtual memory*.

Virtual memory. Early architectures allowed each program to calculate its own memory addresses and to access memory directly using those addresses. Each program assumed that its instructions and data would always be located in the exact same addresses every time it ran. This created problems when running the same program on computers with varying amounts of memory. A program compiled assuming a certain amount of memory might try to access more memory than the user's computer had. If instead, the program had been compiled assuming a very small amount of memory, it would be unable to make use of extra memory when running on machines that did have it.

Even more problems occurred when trying to run more than one program simultaneously. Two different programs might both be compiled to use the same memory addresses. When running together they could end up overwriting each other's data or instructions. The data from one program read as instructions by another could cause the processor to do almost anything. If the operating system were one of the programs overwritten, then the entire computer might lock up.

Virtual memory fixes these problems by translating each address before memory is accessed. The address generated by the program using the available addressing modes is called the *virtual address*. Before each memory access the virtual address is translated to a physical address. The translation is controlled by the operating system using a lookup table stored in memory.

The lookup table needed for translations would become unmanageable if any virtual address could be assigned any physical address. Instead, some of the least significant virtual address bits are left untranslated. These bits are the page offset and determine the size of a memory page. The remaining virtual address bits form the virtual page number and are used as an index into the lookup table to find the physical page number. The physical page number is combined with the page offset to make up the physical address.

The translation scheme shown in Fig. 4-3 allows every program to assume that it will always use the exact same memory addresses, it is the only program in memory, and the total memory size is the maximum amount allowed by the virtual address size. The operating system determines where each virtual page will be located in physical memory. Two programs using the same virtual address will have their addresses

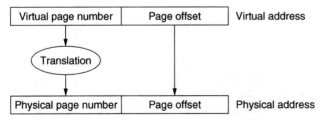

Figure 4-3 Virtual to physical address translation.

translated to different physical addresses, preventing any interference. Virtual memory cannot prevent programs from failing or having bugs, but it can prevent these errors from causing problems in other programs.

Programs can assume more virtual memory than there is physical memory available because not all the virtual pages need be present in physical memory at the same time. If a program attempts to access a virtual page not currently in memory, this is called a *page fault*. The program is interrupted and the operating system moves the needed page into memory and possibly moves another page back to the hard drive. Once this is accomplished the original program continues from where it was interrupted.

This slight of hand prevents the program from needing to know the amount of memory really available. The hard drive latency is huge compared to main memory, so there will be a performance impact on programs that try to use much more memory than the system really has, but these programs will be able to run. Perhaps even more important, programs will immediately be able to make use of new memory installed in the system without needing to be recompiled.

The architecture defines the size of the virtual address, virtual page number, and page offset. This determines the size of a page as well as the maximum number of virtual pages. Any program compiled for this architecture cannot make use of more memory than allowed by the virtual address size. A large virtual address makes very large programs possible, but it also requires the processor and operating system to support these large addresses. This is inefficient if most of the virtual address bits are never used. As a result, each architecture chooses a virtual address size that seems generous but not unreasonable at the time.

As Moore's law allows the cost of memory per bit to steadily drop and the speed of processors to steadily increase, the size of programs continues to grow. Given enough time any architecture begins to feel constrained by its virtual address size. A 32-bit address selects one of 2^{32} bytes for a total of 4 GB of address space. When the first 32-bit processors were designed, 4 GB seemed an almost inconceivably large amount, but today some high-performance servers already have more than 4 GB of memory storage. As a result, the x86 architecture was extended in 2004 to add support for 64-bit addresses. A **64-bit address** selects one of 2^{64} bytes, an address space 4 billion times larger than the 32-bit address space. This will hopefully be sufficient for some years to come.

The processor, chipset, and motherboard implementation determine the maximum physical address size. It can be larger or smaller than the virtual address size. A physical address larger than the virtual address means a computer system could have more physical memory than any one program could access. This could still be useful for running multiple programs simultaneously. The Pentium III supported 32-bit virtual addresses, limiting each program to 4 GB, but it used 36-bit physical addresses, allowing systems to use up to 64 GB of physical memory.

A physical address smaller than the virtual address simply means a program cannot have all of its virtual pages in memory at the same time. The EPIC architecture supports 64-bit virtual addresses, but only 50-bit physical addresses.[5] Luckily the physical address size can be increased from one implementation to the next while maintaining software compatibility. Increasing virtual addresses requires recompiling or rewriting programs if they are to make use of the larger address space. The operating system must support both the virtual and physical address sizes, since it will determine the locations of the pages and the permissions for accessing them.

Virtual memory is one of the most important innovations in computer architecture. Standard desktops today commonly run dozens of programs simultaneously; this would not be possible without virtual memory. However, virtual memory makes very specific requirements upon the processor. Registers as well as functional units used in computing addresses must be able to support the virtual address size. In the worst case, virtual memory would require two memory accesses for each memory operand. The first would be required to read the translation from the virtual memory lookup table and the second to access the correct physical address. To prevent this, all processors supporting virtual memory include a cache of the most recently accessed virtual pages and their physical page translations. This cache is called the **translation lookaside buffer (TLB)** and provides translations without having to access main memory. Only on a TLB miss, when a needed translation is not found, is an extra memory access required. The operating system manages virtual memory, but it is processor support that makes it practical.

Control flow instructions

Control flow instructions affect which instructions will be executed next. They allow the linear flow of the program to be altered. Some common control flow instructions are shown in Table 4-13.

---

[5]McNairy and Soltis, "Itanium 2 Processor Microarchitecture," 51.

TABLE 4-13   Control Flow Instructions

JMP	Unconditional jump to another instruction
BR	Branch to instruction if condition is met
CALL	Call a procedure
RET	Return from procedure
INT	Software interrupt

TABLE 4-14 Control Flow Relative Frequency

Instruction	Integer programs	Floating-point programs
Branch	75%	82%
Jump	6%	10%
Call & return	19%	8%

Unconditional **jumps** always direct execution to a new point in the program. Conditional jumps, also called **branches**, redirect or not based on defined conditions. The same subroutines may be needed by many different parts of a program. To make it easy to transfer control and then later resume execution at the same point, most architectures define **call** and **return** instructions. A call instruction saves temporary values and the **instruction pointer (IP)**, which points to next instruction address, before transferring control. The return instruction uses this information to continue execution at the instruction after the call, with the same architectural state. When requesting services of the operating system, the program needs to transfer control to a subroutine that is part of the operating system. An **interrupt** instruction allows this without requiring the program to be aware of the location of the needed subroutine.

The distribution of control flow instructions measured on the SpecInt 2000 and SpecFP2000 benchmarks for the DEC Alpha architecture is shown in Table 4-14.[6] Branches are by far the most common control flow instruction and therefore the most important for performance.

The performance of a branch is affected by how it determines whether it will be taken or not. Branches must have a way of explicitly or implicitly specifying what value is to be tested in order to decide the outcome of the branch. The most common methods of evaluating branch conditions are shown in Table 4-15.

Many architectures provide an implicit condition code register that contains flags specifying important information about the most recently calculated result. Typical flags would show whether the results were positive or negative, zero, an overflow, or other conditions. By having all computation instructions set the condition codes based on their result, the comparison needed for a branch is often performed automatically. If needed, an explicit compare instruction is used to set the condition codes based on the comparison. The disadvantage of condition codes is they make reordering of instructions for better performance more difficult because every branch now depends upon the value of the condition codes.

Allowing branches to explicitly specify a condition register makes reordering easier since different branches test different registers.

---

[6]Hennessy and Patterson, p. 113.

TABLE 4-15 Branch Condition Testing

Type	Evaluation	Example architecture
Condition code	Test implicit flag bits	x86, PowerPC, SPARC
Condition register	Test explicit register	EPIC
Compare and branch	Compare two registers to each other	PA-RISC, VAX

However, this approach does require more registers. Some architectures provide a combined compare and branch instruction that performs the comparison and switches control flow all in one instruction. This eliminates the need for either condition codes or using condition registers but makes the execution of a single branch instruction more complex.

All control flow instructions must also have a way to specify the address of the target instruction to which control is being transferred. The common methods are listed in Table 4-16.

Absolute mode includes the target address in the control flow instruction as a constant. This works well for destination instructions with a known address during compilation. If the target address is not known during compilation, register indirect mode allows it to be written to a register at run time.

The most common control flow addressing mode is IP relative addressing. The vast majority of control flow instructions have targets that are very close to themselves. It is far more common to jump over a few dozen instructions than millions. As a result, the typical size of the constant needed to specify the target address is dramatically reduced if it represents only the distance from branch to target. In IP relative addressing, the constant is added to the current instruction pointer to generate the target address.

Return instructions commonly make use of stack addressing, assuming that the call instruction has placed the target address on the stack. This way the same procedure can be called from many different locations within a program and always return to the appropriate point.

Finally, software interrupt instructions typically specify a constant that is used as an index into a global table of target addresses stored in

TABLE 4-16 Control Flow Addressing Modes

Address mode	Address	Typical use
Absolute	Const	Target known at compile time
Register Indirect	Reg	Target unknown at compile time
IP Relative	IP + Const	Relative target known at compile time
Stack	Stack	Procedure returns
Global Lookup	Interrupt Table + Const	Software interrupts

memory. These interrupt instructions are used to access procedures within other applications such as the operating system. Requests to access hardware are handled in this way without the calling program needing any details about the type of hardware being used or even the exact location of the handler program that will access the hardware. The operating system maintains a global table of pointers to these various handlers. Different handlers are loaded by changing the target addresses in this global table.

There are three types of control flow changes that typically use global lookup to determine their target address: software interrupts, hardware interrupts, and exceptions. Software interrupts are caused by the program executing an interrupt instruction. A software interrupt differs from a call instruction only in how the target address is specified. Hardware interrupts are caused by events external to the processor. These might be a key on the keyboard being pressed, a USB device being plugged in, a timer reaching a certain value, or many others. An architecture cannot define all the possible hardware causes of interrupts, but it must give some thought as to how they will be handled. By using the same mechanism as software interrupts, these external events are handled by the appropriate procedure before returning control to the program that was running when they occurred.

Exceptions are control flow events triggered by noncontrol flow instructions. When a divide instruction attempts to divide by 0, it is useful to have this trigger a call to a specific procedure to deal with this exceptional event. It makes sense that the target address for this procedure should be stored in a global table, since exceptions allow any instruction to alter the control flow. An add that produced an overflow, a load that caused a memory protection violation, or a push that overflowed the stack could all trigger a change in the program flow. Exceptions are classified by what happens after the exception procedure completes (Table 4-17).

Fault exceptions are caused by recoverable events and return to retry the same instruction that caused the exception. An example would be a push instruction executed when the stack had already used all of its available memory space. An exception handler might allocate more memory space before allowing the push to successfully execute.

TABLE 4-17 Exception Types

Type	Return to	Example
Fault	Instruction causing exception	Stack overflow
Trap	Instruction after cause of exception	Divide by 0
Abort	Shutdown procedure	Invalid instruction

Trap exceptions are caused by events that cannot be easily fixed but do not prevent continued execution. They return to the next instruction after the cause of the exception. A trap handler for a divide by 0 might print a warning message or set a variable to be checked later, but there is no sense in retrying the divide. Abort exceptions occur when the execution can no longer continue. Attempting to execute invalid instructions, for example, would indicate that something had gone very wrong with the program and make the correct next action unclear. An exception handler could gather information about what had gone wrong before shutting down the program.

Instruction Encoding

Once the types of operations and operands are chosen, a scheme for encoding these instructions as binary values must be decided upon. How an architecture encodes its instructions has a tremendous impact on future implementations.

Instruction fields to be encoded are as follows:

- Operation
- Register sources and destinations
- Computation immediates
- Data address and address mode
- Control address and address mode

Every instruction must have an opcode, which determines the particular type of operation. To keep code size to a minimum, the opcode should only be as many bits as are needed for the number of instructions, but running out of opcodes can be deadly for any architecture. Successful architectures inevitably end up needing to add new instructions as implementations and applications change. Any prudent architect will reserve some number of opcodes when the architecture is first defined in order to allow for later extensions.

A common tactic to keep opcodes small is to use a single opcode for a whole class of instructions that do not use all the other fields in the instruction encoding. The unused fields are then be used as extended opcodes to determine the precise operation. For example, if arithmetic instructions never use memory addresses, then a single opcode could be used for all of them, and the address field used to distinguish between the different types of arithmetic instructions.

Each explicit register source or destination must be encoded in the instruction. The number of architecturally defined registers determines the number of bits needed. More registers can improve performance by reducing the number of loads and stores needed and by allowing for more instruction reordering. However, more sophisticated compilers are required to make use of more registers and a large number of registers will increase the instruction size. Doubling the number of registers increases the needed register address size by 1 bit. For an instruction with two source registers and one destination register encoded as 32 bits, this would be about a 10 percent increase in code size.

Many computations are performed on constants that can be encoded in the instruction as immediate values. This prevents the need to keep the immediate value in a separate memory location and perhaps eliminates a load command to move it to a register. The more bits that are available for immediates, the larger the immediate values that can be represented. Luckily most immediates are relatively small with typically 75 percent represented in less than 16 bits.[7]

Any instruction that accesses memory must encode the data address as well as what addressing mode should be used to interpret the address. Larger address constants enable memory accesses to a larger portion of memory without having to use registers to store addresses. More addressing modes allow more flexibility in how addresses are generated. However, both of these will lead to larger instruction sizes, especially in cases where a single instruction may have multiple memory operands.

Control flow instructions must specify the address of the target instruction and possibly an addressing mode. If the number of addressing modes is small, it may be encoded as part of the opcode. Using IP relative addressing, an 8-bit offset is sufficient for about 90 percent of branches.[8] The remaining 10 percent may be converted by the compiler into a shorter branch that executes or skips an unconditional jump to the desired instruction.

Architectures can encode all their instructions to be the same number of bits or use encodings of fewer bits for more common instructions. Code size is improved by allowing variable length instructions. This textbook would become much longer if every word had to be the same number of characters. Real world languages like English inevitably use fewer characters on average for common words, which allows more words overall in the same number of total characters. However, variable length instructions make instruction decode more difficult, especially for processors that attempt to read multiple instructions simultaneously. Fixed length instructions are simpler to work with since the beginning of the next instruction is known even before the current instruction is

---

[7]ibid., 102.

[8]ibid., 163.

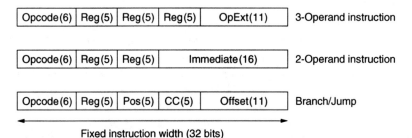

Figure 4-4 Example instruction encodings.

decoded. The disadvantage is the more memory space will be required to encode the same number of instructions. The instruction encodings for a fixed 32-bit instruction length pure register architecture might look like those shown in Fig. 4-4.

The example in Fig. 4-4 shows three basic instruction formats and the number of bits used to encode each field. The three-operand format could be used for computation instructions that specify a destination register and two source registers. The register fields are 5 bits wide, setting a limit of 32 architectural registers. An 11-bit opcode extension would allow a single opcode to be used for all instructions of this type and the specific operation to be encoded in this field.

The two-operand format could be used for computation instructions that include an immediate value as a source instead of two source registers. This same format could be used for loads and stores where the first register field points to the source or destination register, and the second register field specifies a register to be added to the immediate value to form the address.

The branch/jump format allows the register addressed by the first field to be tested in order to determine if the branch should be taken or not. The Pos field specifies the position of the bit to be tested; the condition to be tested is specified by the CC field. The offset field provides an 11-bit offset from the current instruction pointer to the target address. If the opcode indicates a jump, only the first register field would be used. This would specify the register containing the target address, and the remaining fields would be ignored.

This simple example shows how added bits for any one field must come at the expense of simplifying other fields or increasing instruction size. A few long instruction formats may not be particularly significant for architectures that support variable length instructions, but for fixed length architectures, growing any type instructions requires increasing the size of all instructions.

CISC versus RISC

One question in modern computer architecture that made the little endian versus big endian war seem like a minor skirmish was the debate of CISC versus RISC. When RISC architectures were first being promoted in the 1980s, many grand pronouncements were made on both sides; this question of computer design sometimes took on the tones of a religious war. Even at the time there wasn't clear agreement on exactly what constituted a RISC architecture. Some products were labeled RISC based more on marketing strategy than their actual design, which caused Steven Przybylski, one of pioneers of RISC architectures, to jokingly define RISC as "Any computer announced after 1985."[9] Using the concepts from this chapter, we can list some basic differences between CISC and RISC architectures.

The VAX and x86 architectures were created in the late 1970s when program size was a large concern. DRAM memory was extremely expensive and some programs were not practical because systems with enough memory to run them were too expensive. Supporting memory and register operands with many addressing modes allowed programs to use as few total instructions as possible. In addition, variable length instruction formats allowed the most common instructions to be encoded in fewer bits while using only longer formats for uncommon instructions. By keeping the number of instructions and the average number of bytes per instruction to a minimum, these architectures allowed the most functionality using the least amount of instruction memory.

At the time transistor budgets did not allow for large numbers of registers. An architecture defining 32 registers or more would force any processor implementing that architecture to dedicate an unreasonable fraction of its total transistors to registers. In any case, serial execution of instructions did not require a large number of registers.

As Moore's law continued to drive DRAM and processor development forward, RISC architectures made different choices to take advantage of new possible implementations. As DRAM became more affordable, minimizing code size became less important. Supporting fewer operand types and allowing only fixed instruction sizes made RISC programs much larger. However, it also made their implementation much simpler.

In the late 1980s, transistor budgets were sufficient to allow some of the first pipelined processors. These began executing one instruction before the previous was completed. To keep the pipeline full, it is important that most instructions are executed in the same amount of time. Every slow instruction causes faster instructions to pile up behind them in the pipeline. RISC architectures allowed efficient execution not just

---

[9]ibid., Appendix C.

by having simple instructions but also by having instructions of uniform complexity. Transistor budgets also allowed for a larger number of registers. Making these visible to the programmer meant fewer memory accesses were needed, therefore addressing modes were less important.

The continuing progress of Moore's law means that today there are new reasons to favor either the RISC or CISC approach. Although having sufficient DRAM for instructions is rarely an issue today, the widening gap between processor and main memory speed means that memory bandwidth is a performance limiter. By having smaller code size, CISC programs reduce the bandwidth to memory required to fetch instructions and make better use of their instruction caches.

Processors' transistor budgets now allow multiple parallel pipelines. For CISC architectures, with variable length instructions, one instruction must first be decoded before the starting byte of the next instruction can be determined. RISC architectures, with fixed length instructions, allow multiple instructions to be decoded easily in parallel.

Today most architects would agree RISC is the better choice, but there hasn't been a clear winner in the marketplace. All successful new architectures created since the 1980s have used the RISC features described in Table 4-18. Even Digital Equipment Corporation, the creator of the VAX architecture, which is considered the epitome of CISC, eventually converted their computers to the Alpha architecture, which was clearly a RISC design. However, the x86 architecture, which is CISC, remains the most widely used desktop architecture. The performance features of RISC have not yet been enough to overcome the enormous value of binary compatibility.

In addition, the most recent x86 implementations have found ways to use the same performance tricks as RISC processors. The Pentium 4 and Athlon 64 use hardware to translate x86 instructions into simpler microinstructions before execution. These microinstructions bare a striking resemblance to RISC instructions. Converting complex instructions into a stream of simple instructions in hardware, these processors are able to take advantage of RISC features while maintaining compatibility. They pay a price in die area and complexity to do this, but maintaining backward compatibility makes it more than worth it.

TABLE 4-18 CISC versus RISC Architectures

Feature	CISC (VAX, x86)	RISC (SPARC, PA-RISC, PowerPC)
Computation operands	Memory operands allowed	Register & immediate only
Instruction size	Variable	Fixed
Addressing modes	Many	Few
Architectural registers	Few	Many

RISC versus EPIC

Probably the most important new architecture since the development of RISC is the Explicitly Parallel Instruction Computing (EPIC) architecture. The first implementation of this architecture was the Intel Itanium, which began shipping in 2001. EPIC architecture was designed for implementations with even higher transistor counts than the original RISC processors. Superscalar processors of the 1990s had the functional units to execute multiple instructions in parallel. However, they used a great deal of die area on scheduling circuits used to determine which instructions could execute in parallel. One suggested solution to this was Very Long Instruction Word (**VLIW**) architectures.

VLIW architectures bundle multiple instructions that can be executed in parallel into a single long instruction. The compiler performs the scheduling, so that the processor avoids wasting run time and silicon area determining which instructions to execute in parallel. The compiler has the ability to look at hundreds of instructions to find independent operations, so in theory it can do a much better job scheduling than hardware implementations. VLIW approaches had two significant shortcomings.

One problem was a dramatic increase in code size. If enough independent instructions could not be found to fill all the slots in a VLIW instruction, the remaining slots are wasted. If fixed instruction size is to be maintained, then each long instruction must be the same length no matter how many or few of its execution slots are used. This caused programs compiled for VLIW architectures to become much larger than even RISC code.

Another problem was the potential lack of binary compatibility from one processor implementation to the next. Because superscalar machines use hardware to determine instruction scheduling, new implementations can increase the number of instructions that are executed in parallel while maintaining compatibility. If a new VLIW processor had a larger instruction width, all the code for the previous generation would need to be recompiled for proper scheduling. The EPIC architecture tries to take the good points of VLIW architectures while addressing these problems.

The EPIC architecture encodes its instructions into 128-bit-wide bundles. Each bundle contains three instructions encoded in 41 bits each and a 5-bit template field. The template field contains information about the types of instructions in the bundle and which instructions can be executed in parallel. This allows all the slots of an instruction to be filled even if enough independent instructions cannot be found. The template also specifies whether one or more instructions in this bundle can be executed in parallel with at least the first instruction of the next bundle. This means there is no limit on how many instructions the compiler could mark as being able to execute in parallel. Future processors may include additional functional units and immediately make use of them without recompiling code.

The EPIC architecture adds other features specifically intended to allow the compiler to expose more instructions that can be executed in parallel.

EPIC architecture features are as follows:

- 128 64-bit integer registers
- 128 82-bit floating-point registers
- 64 1-bit predicate registers
- Speculative loads

Most RISC architectures defined 32 integer and 32 floating-point registers. Increased transistor budgets have made implementation of more registers possible, so the EPIC architecture defines a full 128 integer and 128 floating-point registers. The implementation of this many registers alone takes several times the total transistor budget of Intel's original 8086 processor, but with modern fabrication processes this number of registers is no longer unreasonable. With few architectural registers, the compiler must create false dependencies between instructions when it reuses registers. More registers allow the compiler to create more parallel instructions and perform more reordering.

Branches present a special problem when trying to find parallelism. The compiler cannot determine which instructions can be executed in parallel because the control flow is determined at run time. Some RISC architectures support conditional move instructions, which only move a value from one register to another if a condition is met. This has the effect of a branch without altering the control flow of the program. The EPIC architecture builds upon this idea with predicated execution.

Almost any EPIC instruction can specify one of 64 single-bit predicate registers. The instruction will be allowed only to write back its result if its predict bit is true. This means that any instruction can be a conditional instruction. Compare instructions specify two predicate registers to which to write. One is written with the result of the comparison and the other with the complement. Executing instructions on both sides of the branch with complementary predicate registers eliminates branches altogether. There is wasted effort because not all the instructions being executed are truly needed, but the entire premise of the EPIC architecture is that its implementations will have many functional units available. For branches that are difficult to predict accurately, executing both sides of the branch may give the best performance.

Loads also pose a problem for compilers attempting reordering. Because loads have a very long latency if they miss in the cache, moving them much earlier in a program than when their results will be used can give a large performance benefit. However, moving a load before a branch may cause exceptions to occur, which should never have happened. A branch may be specifically checking to see if a particular load would cause a memory protection violation or other exception before executing it. If the compiler moves the load before the branch, the program will not execute in the way it should on an in-order machine.

Another problem with moving loads is possible dependencies on stores. Many load and store instructions use registers as part of their address, making it impossible for the compiler to determine whether they might access the same memory location. In theory, any load could depend upon any store. If the compiler cannot move loads before the earlier branch or stores, then very little reordering is going to be possible.

The EPIC architecture addresses this by supporting speculative loads. The speculative load executes like a normal load, but if it would trigger an exception, it merely marks a flag bit in its destination register. Also the address and destination register of the speculative load are stored in a table. Each store checks the table to see if any speculative loads have read from the same address the store is writing to. If so, the destination register is marked as invalid.

When the program flow reaches the point of the original load, a load check instruction is executed instead. If the speculative load caused an exception, then the exception is executed at this point. If the program flow never reaches the load check, then the exception will correctly never be called. The load check also sees if the load result has been marked invalid by a later store. If so, the load is executed again. Most of the time, the load check will pass. The correct value will already have been loaded into a register, ready for use by other instructions.

The EPIC architecture has even more potential performance advantages over older CISC architecture than RISC architectures do, but it has had difficulty achieving widespread acceptance. The same desire for compatibility that has kept RISC architecture from becoming more widely used has limited the use of EPIC architecture processors. Initially one important selling point for EPIC was support of 64-bit registers and virtual addresses, but recent extensions to the x86 architecture have duplicated much of this functionality. The future of EPIC remains uncertain, but it shows how much Moore's law has changed the kinds of architectures that are possible.

Recent x86 extensions

Since 2003, increasing power densities have slowed the rate of frequency improvement in x86 implementations. This had led to both Intel and AMD placing more emphasis on architecture as a way of providing added features and performance. One example is the extension of virtual addresses to 64 bits. Intel calls this Extended Memory 64-bit Technology (EM64T), and AMD refers to it as AMD64 Technology. Larger virtual addresses do not by themselves offer any improvement in performance, but they allow new programs to be written with new capabilities, such as accessing more than 4 GB of data.

Another recent extension allows pages of virtual memory to be marked as "data only," preventing them from being accidentally, or perhaps maliciously, executed as instructions. Intel calls this Execute Disable Bit, and AMD calls it Enhanced Virus Protection. Some computer viruses are spread by overflowing data buffers. A program expecting data allocates a certain amount of memory space for the response. A response that is far larger than expected may end up writing data in unexpected locations. Some programs do this accidentally whereas others do it intentionally to allow them to place new instructions in memory. Marking these areas of memory as "no-execute" prevents this type of attack from succeeding.

One recent x86 architecture extension that can improve performance is Intel's **HyperThreading (HT)**. HT allows more efficient multitasking by allowing one processor to divide its resources to act as two processors. Technically processors without HT are not capable of running more than one application at a time. They give the illusion of running multiple programs by switching between them very quickly. Every 10 to 15 ms the operating system interrupts the current application and checks to see if there are other applications waiting. Switching many times a second between programs gives the illusion that they are running simultaneously, but at any particular instant only one program is actually running.

Dividing the processor's operation into these time slices dedicated to different programs causes some inefficiency. If the time slice is very long, the program currently running may stall and waste the rest of its time slice, which other programs could have used. Making the time slice shorter causes more overhead. Each time the processor switches between tasks, the register values of one program must be saved and the values of another loaded. With very short time slices, the processor could end up spending most of its time just switching between tasks, rather than making any forward progress on the tasks themselves.

HT reduces this wasted time by allowing two programs to run on the processor at the same instant. The hardware resources are divided between the two threads, so that one does not have to be stopped for the other to make progress, but if one thread does come to a halt while waiting for data or some other dependence, the other thread makes use of that idle time which before would have been wasted. Both threads have all their needed register values already on the processor, so there is no overhead to switch between them. Any application written to use HT can see some performance improvement by making more efficient use of the processor's time.

These recent extensions are examples of AMD and Intel, both trying to add value to the x86 architecture in ways other than increasing frequency. The enormous established software base written for the x86 architecture gives processors a huge advantage, but without change any architecture eventually begins to show its age. By continuing to add new capabilities AMD and Intel hope to keep the x86 architecture the leading desktop choice for many years to come.

Conclusion

A processor's architecture is all of the features of a processor that are visible to the programmer. The processor architect is faced with a great many important decisions before the implementation can even begin. Some choices are simply a matter of taste, but most will have a serious impact on the performance and complexity of any processor supporting that architecture. Some of the more important decisions are listed in Table 4-19.

Early architectures made choices to give the most compact code size. More recent architectures have deliberately sacrificed code size in order to make it easier to execute instructions in parallel. In the end, there cannot be a perfect architecture because the technology for processor implementation is always changing. Any new architecture should be made with a sense of the likely future technological advances. However, designing an architecture to be too forward looking can also be fatal. Each architecture must make trade-offs between the best choices for future implementations and what is practical and efficient in the near term.

Most of all it is impossible to overstate the importance of software compatibility. Elegant code and efficient implementations are worthless if a processor can not run the software the customer wants to run. Extensions

TABLE 4-19 Architectural Choices

Design decision	Possible choices
Operand types for computation	Pure register, register/memory, pure memory
Data formats supported	Integer, BCD, floating point (single, double, extended), SIMD
Data Address Modes	Absolute, register indirect, displacement, indexed, scaled
Virtual memory implementation	Virtual address size, allowed page sizes, page properties
Instruction encoding	Fixed or variable, number registers, size of immediates & addresses

to architectures improve performance or add new capabilities, but only if they are supported by new software. Very few architects would ever think to describe the x86 architecture as elegant or efficient, but by maintaining software compatibility it has become the most widely used architecture. To displace it, any new architecture will have to exceed x86 capabilities or performance by such a wide margin as to overcome the inertia of the established software base. Meanwhile, all the currently supported architectures will continue to expand their capabilities to try to avoid fading into computer history like the VAX architecture and many others.

Key Concepts and Terms

64-bit addressing	Registers
Floating point	Return
HyperThreading (HT)	RISC and CISC
Immediate value	SIMD
Instruction pointer (IP)	Stack and accumulator
Interrupt and exception	Translation lookaside buffer (TLB)
Jump, branch, and call	Two's complement and BCD
Memory	VLIW and EPIC

Review Questions

1. What are the differences between high-level programming languages, assembly code, and machine language?
2. What are the functions of computation, data transfer, and control flow instructions?
3. Describe the differences between RISC and CISC architectures.
4. Describe the differences between VLIW and EPIC architectures.
5. Explain the operation of jumps, branches, calls, and interrupts.
6. What are the advantages and disadvantages of fixed instruction encoding length?
7. What are the advantages and disadvantages of floating-point data compared to integer formats?
8. What are the sources of software interrupts, hardware interrupts, and exceptions?
9. Describe how virtual memory works and its advantages.
10. How can SIMD instructions improve performance?
11. [Bonus] Create a pure register instruction set architecture that contains the minimum number of instruction types needed to create a

program that reads an integer value from memory, computes the factorial for that number, and then writes the result back to the same location in memory. Show the assembly code for this program.

12. [Bonus] What would be the minimum number of bits per instruction required to encode the instruction set devised for the previous question. Demonstrate your encoding scheme by converting the program assembly code in machine language.

Bibliography

Flynn, Michael. *Computer Architecture: Pipelined and Parallel Processor Design*. Boston, MA: Jones & Bartlett, 1995. [Filled with tons of quantitative data and mathematical models, Flynn's book make the trade-offs of architecture and processor design real by enabling the reader to reasonably estimate the positive and negative impacts of his design decisions.]

Hennessy, John and David Patterson. *Computer Architecture: A Quantitative Approach*. 3d ed., San Francisco, CA: Morgan Kaufmann, 2003. [This book by two of the three original creators of the concept of RISC architecture is an absolute must for anyone studying computer architecture; the "Fallacies and Pitfalls" sections at the end of each chapter should be required reading for anyone working in the computer industry.]

Huck, Jerome and Michael Flynn. *Analyzing Computer Architectures*. New York: IEEE Computer Society Press, 1989. [An extensive comparison of ideal "high-level language" program implementations and real world architectures.]

"IA-32 Intel® Architecture Software Developer's Manual." Volume 1: Basic Architecture, Volume 2A & 2B: Instruction Set Reference, 2004, http://www.intel.com/design/pentium4/manuals/index_new.htm. [These online manuals describe in extreme detail the Intel's architecture as well as the encoding and operation of every instruction supported by their processors.]

"IEEE 754: Standard for Binary Floating-Point Arithmetic," 1985, http://grouper.ieee.org/groups/754/. [The most widely accepted standard for floating-point number formats. Every processor today describes its floating-point support by how it matches or diverges from this standard.]

McNairy, Cameron and Don Soltis. "Itanium 2 Processor Microarchitecture." *IEEE Micro*, 2003, pp. 44–55.

Patterson, David and John Hennessy. *Computer Organization & Design: The Hardware/Software Interface*. 2d ed., San Mateo, CA: Morgan Kaufmann, 1998. [This book by the same authors as *Computer Architecture: A Quantitative Approach* focuses more on programming processors rather than designing them but manages to give many of the same insights in a simpler format, a great lead into their other book.]

Petzold, Charles. *Code: The Hidden Language of Computer Hardware and Software*. Redmond, WA: Microsoft Press, 1999. [A good introduction to the basic concepts of binary, floating-point numbers, and machine language programming.]

Smotherman, Mark. "Understanding EPIC Architectures and Implementations." Association of Computing Machinery (ACM) Southeast Conference, Raleigh, NC: April 2002. [An excellent description of the EPIC architecture, which points out that many of its basic concepts can be traced back to computers built as early as the 1950s and 1960s.]

Waser, Shlomo and Michael Flynn. *Introduction to Arithmetic for Digital Systems Designers*. Austin, TX:: Holt, Rinehart and Winston, 1982. [One of very few books to specifically address not only the theory but also the implementation of computer arithmetic.]

Chapter 5

Microarchitecture

Overview

This chapter discusses the trade-offs in the high-level implementation of a microprocessor and how they affect performance. The different measures of processor performance are compared along with methods for improving performance. Cache operation is described in more detail, and the life of an instruction in the Pentium 4 pipeline is presented as a detailed example of modern processor microarchitecture.

Objectives

Upon completion of this chapter, the reader will be able to:

1. Explain how a processor pipeline improves performance.
2. Describe the causes of pipeline breaks.
3. Explain how branch prediction, register renaming, out-of-order execution, and HyperThreading improve performance.
4. Be aware of the limitations of different measures of processor performance.
5. Understand the impacts of pipeline depth on processor performance.
6. Describe the causes of cache misses and how different cache parameters affect them.
7. Explain the need for cache coherency and understand its basic operation.
8. Understand the difference between macroinstructions and microinstructions.
9. Be familiar with the pipeline of the Pentium 4 processor.

Introduction

Although a processor's architecture defines the instructions it can execute, its microarchitecture determines the way in which those instructions are executed. Of all the choices made in a processor design flow, microarchitectural decisions will have the greatest impact on the processor's performance and die area. Chapter 4 discussed how changes in architecture affect performance, but it is possible to build multiple processors that all support the same architecture with enormously different levels of performance. The difference between such implementations is their microarchitectures.

By definition, architectural changes are visible to the programmer and require new software to be utilized. New architectural registers or SIMD instructions improve performance for programs that use them but will have no impact on legacy software. Microarchitectural changes are not visible to the programmer and can improve performance with no change in software.

Because microarchitectural changes maintain software compatibility, processor microarchitectures have changed much more quickly than architectures. Like almost all other changes in the semiconductor industry, Moore's law has driven this progress. The microarchitectural concepts in use today were not totally unknown to past designers; they were simply impractical. As Moore's law allows larger and larger transistor budgets, old microarchitectural ideas that were too complex to be implemented become possible.

Imagine writing a word processing program. The program should provide as much functionality and ease of use as possible, but must be limited to 10,000 lines of code. With this restriction, this word processor will be a very crude program indeed. If after two years, a new word processor is created using twice as many lines of code, there could be significant added functionality. If this pattern were to repeat, in 10 years a word processor written using over 300,000 lines of code might be dramatically more sophisticated. This does not mean that the earliest version was poorly designed; merely that it had different design constraints.

Growing transistor budgets allow microarchitects more resources to implement higher performance processors. An early design might include a single adder whereas a later design might include two, allowing more processing to occur in parallel. Inevitably other changes are needed to make efficient use of the ability to execute more instructions in parallel. The most successful way of performing more work in parallel has been through **pipelining**.

Pipelining

Most microarchitectural improvements have focused on exploiting **instruction level parallelism** (ILP) within programs. The architecture

Cycle	1	2	3	4	5	6	7	8
Instr 1	Fetch	Decode	Execute	Write				
Instr 2					Fetch	Decode	Execute	Write

Figure 5-1 Sequential processing.

defines how software should run and part of this is the expectation that programs will execute instructions one at a time. However, there are many instructions within programs that could be executed in parallel or at least overlapped. Microarchitectures that take advantage of this can provide higher performance, but to do this while providing software compatibility, the illusion of linear execution must be maintained. Pipelining provides higher performance by allowing execution of different instructions to overlap.

The earliest processors did not have sufficient transistors to support pipelining. They processed instructions serially one at a time exactly as the architecture defined. A very simple processor might break down each instruction into four steps.

1. *Fetch.* The next instruction is retrieved from memory.
2. *Decode.* The type of operation required is determined.
3. *Execute.* The operation is performed.
4. *Write.* The instruction results are stored.

All modern processors use clock signals to synchronize their operation both internally and when interacting with external components. The operation of a simple sequential processor allocating one clock cycle per instruction step would appear as shown in Fig. 5-1.

A pipelined processor improves performance by noting that separate parts of the processor are used to carry out each of instruction steps (see Fig. 5-2). With some added control logic, it is possible to begin the next instruction as soon as the last instruction has completed the first step.

We can imagine individual instructions as balls that must roll from one end of a pipe to another to complete. Processing them sequentially is like adding a new ball only after the last comes out. By allowing multiple balls to be in transit inside the pipe at the same time, the rate that the balls are loaded into the pipe is improved. This improvement happens even though the total time it takes one ball to roll the length of the pipe has not changed. In processor terms, the instruction bandwidth has improved even though instruction latency has remained the same.

Our simple sequential processor completed an instruction only every 4 cycles, but ideally a pipelined processor could complete an instruction

Cycle	1	2	3	4	5	6	7	8
Instr 1	Fetch	Decode	Execute	Write				
Instr 2		Fetch	Decode	Execute	Write			
Instr 3			Fetch	Decode	Execute	Write		
Instr 4				Fetch	Decode	Execute	Write	
Instr 5					Fetch	Decode	Execute	Write
Instr 6						Fetch	Decode	Execute
Instr 7							Fetch	Decode
Instr 8								Fetch

Figure 5-2 Pipelined processing.

every cycle. In reality, performance is significantly less than the ideal because of pipeline breaks. Ideal pipeline performance would be possible only if every instruction could be executed in exactly the same number of cycles and there were no dependencies between instructions. However, some instructions are inherently more complex than others and require more cycles of execution as a result. A divide operation requires more cycles of computation than an add operation. Floating-point operations require more cycles than integer operations. A load that misses in the cache requires more cycles than one that hits. Pipelining could still achieve ideal performance despite these variations in execution time if not for instruction dependencies.

Imagine that the first instruction is a divide, which stores its result in register 2, and that the next instruction is an add instruction, which reads register 2 as one of its inputs. This means that the add instruction cannot begin execution until the divide completes execution. This causes a pipeline break or stall as shown in Fig. 5-3.

Cycle	1	2	3	4	5	6	7	8
Div R2, R1	Fetch	Decode	Execute			Write		
Add R3, R2		Fetch	Decode	Wait		Execute	Write	
Instr 3			Fetch	Decode	Wait		Execute	Write
Instr 4				Fetch	Decode	Wait		Execute
Instr 5					Fetch	Decode	Wait	
Instr 6						Fetch	Decode	Wait

Figure 5-3 In-order pipeline data dependency.

Cycle	1	2	3	4	5	6	7	8
Div R2, R1	Fetch	Decode	Execute			Write		
Add R3, R2		Fetch	Decode	Wait		Execute	Write	
Branch			Fetch	Decode	Wait		Execute	Write
Instr 4								Fetch

Figure 5-4 In-order pipeline control dependency.

The causes of pipeline breaks are as follows:

Data dependencies

Control dependencies

Resource conflicts

Because of the **data dependency** on instruction 1, instruction 2 and all the following instructions are delayed. The pipeline will no longer complete an instruction every cycle because for some cycles it will be stalled while waiting for dependencies.

In addition to data dependencies, there are also control dependencies. The execution of a branch instruction determines which instruction should enter the pipe next. This means control dependencies cause breaks, stalling not just the execution of the next instruction but even the fetch step. If the third instruction were a branch, the pipeline might look like Fig. 5-4.

A third cause of pipeline breaks is resource conflicts. The microarchitecture of our processor may include only a single divider. If there are two divide instructions to be executed, even without any data or control dependencies, one may have to wait for the needed hardware to be available. Pipeline breaks limit how much speedup is achieved by pipelining.

The pipelines shown in Figs. 5-3 and 5-4 are called *in-order pipelines* because the order that the instructions reach execution is the same as the order they appear in the program. Performance is improved by allowing instructions to execute out of order, improving the average number of completed *instructions per cycle* (**IPC**).

Figure 5-5 shows an out-of-order pipeline, which starts as before with a divide instruction and an add instruction that uses the result of the divide. The fourth instruction is a multiply that uses the result of the add. These instructions have data dependencies that require them to be executed in order. However, the other instructions shown are subtracts, which are independent of the other instructions. Each one does not share its registers with any of the other instructions. This allows the first subtract to be executed before the add instruction even though it comes after it in the instruction flow.

Cycle	1	2	3	4	5	6	7	8
Div R2, R1	Fetch	Decode	Execute			Write		
Add R3, R2		Fetch	Decode	Wait		Execute	Write	
Sub R8, R7			Fetch	Decode	Execute	Write		
Mul R4, R3				Fetch	Decode	Wait	Execute	Write
Sub R10, R9					Fetch	Decode	Execute	Write
Sub R12, R11						Fetch	Decode	Execute

Figure 5-5 Out-of-order pipeline.

Out-of-order processors use schedulers, which scan a window of upcoming instructions for data dependencies. By comparing the operand registers of each instruction the scheduler determines which instructions must wait for results from others and which are ready to be executed. By scheduling instructions for execution as soon as possible, the average IPC is improved. Of course, a penalty in die area and design complexity must be paid to create the scheduler, and additional logic will be required to maintain the illusion of in-order execution.

Reordering of instructions improves performance by allowing the processor to work around data dependencies. It also helps reduce pipeline breaks due to resource conflicts. The scheduler can take into account not just the data an instruction needs but also any execution resources. If one divide instruction is waiting for another to complete, the processor can still make forward progress by executing other instructions that do not require the same hardware. Another solution to resource conflicts is to provide more resources. **Superscalar** processors improve IPC by adding resources to allow multiple instructions to be executed at the same step in the pipeline at the same time, as shown in Fig. 5-6.

Cycle	1	2	3	4	5	6	7	8
Div R2, R1	Fetch	Decode	Execute			Write		
Add R3, R2	Fetch	Decode	Wait			Execute	Write	
Sub R8, R7		Fetch	Decode	Execute	Write			
Mul R4, R3		Fetch	Decode	Wait			Execute	Write
Sub R10, R9			Fetch	Decode	Execute	Write		
Sub R12, R11			Fetch	Decode	Execute	Write		
Sub R14, R13				Fetch	Decode	Execute	Write	
Add R5, R4				Fetch	Decode	Wait		Execute

Figure 5-6 Superscalar pipeline.

A single issue out-of-order pipeline can at best achieve an IPC of 1, completing one instruction every cycle. A superscalar processor can achieve an IPC of greater than 1 by allowing multiple instructions to go through the pipeline in parallel. Superscalar designs are described by their issue width, the maximum number of instructions that can enter the pipeline simultaneously. Larger transistor budgets have made microarchitectures with issue widths of 2, 3, or more possible, but very wide issue designs have difficulty reaching their maximum theoretical performance.

Larger issue widths and longer pipelines mean that more independent instructions must be found by the scheduler to keep the pipeline full. A processor that is capable of an IPC of 3 may achieve an IPC of less than 1 because of numerous pipeline breaks. The added die area and complexity to build ever more sophisticated schedulers may not be justified by the performance improvements. This is the problem addressed by architectural solutions to expose more parallelism.

The EPIC architecture adds features to allow the compiler to perform most of the work of the scheduler. Encoding instructions with information about which can be executed in parallel dramatically simplifies the task of the scheduler. The compiler is also able to search a much larger window of instructions looking for independent instructions than would be possible in hardware. Speculative load and conditional move instructions allow more reordering by reducing control dependencies.

The **HyperThreading** architectural extensions simplify the scheduler's job by allowing the program to divide itself into separate independent threads. Except for special synchronizing instructions, the scheduler assumes any instruction in one thread is independent of any instruction in another thread. This allows the scheduler to fill pipeline breaks created by dependencies in one thread with instructions from the other thread.

In Fig. 5-7, the first instruction executed happens to be a very long one. If the next instruction depends upon the result, it must wait. However,

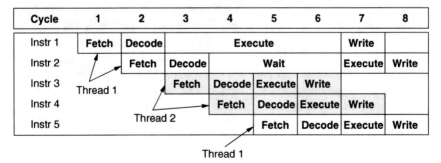

Figure 5-7 HyperThreading.

while that instruction is waiting, the processor can execute instructions from the second thread, which do not depend upon the first thread's results. The processor spends less time idle and more time performing useful work.

Out-of-order issue and superscalar issue are purely microarchitectural changes, which improve performance without any change in software. HyperThreading is an example of how architecture and microarchitecture working together can achieve even more performance at the cost of requiring new software.

Designing for Performance

What we really want from a fast computer is to quickly run the programs that we care about. Better performance simply means less program run time.

$$\text{Performance} \propto \frac{1}{\text{run time}} = \frac{\text{frequency} \times \text{instructions per cycle}}{\text{instruction count}}$$

To improve performance, we must increase frequency or average instructions per cycle (IPC) or reduce the number of instructions required. Choices in architecture will affect the IPC and instruction count. Choices in microarchitecture seek to improve frequency or IPC. Performance is improved by increasing either, but in general, changes that improve one make the other worse. The easiest way to improve frequency is by increasing the pipeline depth, dividing the execution of each instruction into smaller faster cycles.

The examples earlier in this chapter had a pipeline depth of 4, dividing instructions into 4 cycles. If instructions are balls rolling down a pipe, this means the time between adding new balls to the pipe is equal to 1/4 the time a ball takes to roll the length of the pipe. Adding new balls at this rate means there will always be 4 balls in the pipe at one time. If a pipeline depth of 8 had been chosen instead, the time between adding new balls would be 1/8 the time to roll the entire length and there would be 8 balls in the pipe at one time. Doubling the pipeline depth has doubled rate balls are added by cutting in half the time between new balls.

The time between adding new balls to the pipe is equivalent to a processor's cycle time, the inverse of processor frequency. Each instruction has a certain amount of logic delay determined by the type of instruction. Increasing the pipeline depth divides this computation among more cycles, allowing the time for each cycle to be less. Ideally, doubling the pipeline depth would allow twice the frequency. In reality, some amount of delay overhead is added with each new pipestage. This overhead limits the amount of frequency improved by increasing pipeline depth.

$$\text{Frequency} = \frac{1}{T_{\text{cycle}}} = \frac{1}{\dfrac{T_{\text{logic}}}{\text{depth}} + T_{\text{overhead}}} = \frac{\text{depth}}{T_{\text{logic}} + \text{depth} \times T_{\text{overhead}}}$$

The above equation shows how frequency is improved by dividing the logic delay of the instruction among a deeper pipeline. However, the rate of improvement will slow down as the amount of logic in each cycle approaches the overhead delay added with each new pipestage. Doubling frequency requires increasing pipeline depth by more than a factor of 2. Even if frequency is doubled in this fashion, performance will not double because IPC drops as pipeline depth increases.

A longer pipeline allows more instructions in the pipe at one time. More instructions in the pipe make data dependencies, control dependencies, and resource conflicts all more likely. Inevitably, increasing the pipeline depth increases the number of pipeline stalls per instruction and reduces the average instructions per cycle. Together the effects on frequency and IPC let us write an equation for how performance changes with pipeline depth.

$$\text{Performance} \propto \text{frequency} \times \text{IPC} = \frac{\text{depth}}{(T_{\text{logic}} + \text{depth} \times T_{\text{overhead}})(1 + \text{stalls per instruction})}$$

For the ideal case of no delay overhead and no added stalls, doubling pipeline depth will double performance. The real improvement depends upon how much circuit design minimizes the delay overhead per pipestage and how much microarchitectural improvements offset the reduction in IPC.

Of course, there is not really a single pipeline depth because different instructions require different amounts of computation. The processor cycle time will be set by the slowest pipestage. To prevent instructions requiring more computation from limiting processor frequency, they are designed to execute over more pipestages. An add instruction might require a total of 10 cycles whereas a divide might use a total of 40. Having instructions of different latencies increases resource conflicts. A short instruction can finish on the same cycle as a longer instruction started earlier and end up competing for access to write back results. There would be fewer conflicts if all instructions used the same pipeline depth, but this would unnecessarily penalize short instructions, which also tend to be common.

The most important rule in designing for high performance is to make the common case fast. There is no need to waste large amounts of design time and die area on functions that are seldom used. This idea is sometimes formally referred to as **Amdahl's law**, which estimates the overall speedup obtained by decreasing the time spent on some fraction of a job.[1]

Amdahl's law:

$$\text{Overall speedup} = \frac{\text{old time}}{\text{new time}} = \frac{1}{1-\text{fraction}} + \frac{\text{fraction}}{\text{speedup}}$$

If the fraction of execution time reduced is very small, then even a very large speedup for that piece will have very little impact on overall performance. The concept of Amdahl's law is used again and again in microarchitecture. Resources are often better spent on small improvements for the most common operations instead of large speedups for rare events. Many microarchitectural ideas slow down rare occurrences if that will speed up the common case.

Frequency is the same for all instructions, so Amdahl's law applies only to improvements in IPC. More functional units allow higher issue width. Better reordering algorithms reduce pipeline stalls. More specialized logic reduces the latency of computation. Larger caches can reduce the average latency of memory accesses. All of these microarchitectural changes improve IPC but at the cost of design complexity and die area. Fred Pollack compared the relative performance of different Intel processors and found that in the same fabrication technology, performance improved roughly with the square root of added die area. This relationship is often called *Pollack's rule*.[2]

Pollack's rule:

$$\text{Performance} \propto \sqrt{\text{die area}}$$

This rule of thumb means that a new microarchitecture that requires twice the die area will likely provide only a 40 percent improvement in performance. The large amounts of new transistors needed for higher performance microarchitectures have been made possible by Moore's law. Roadblocks to fabricating ever-smaller transistors may continue to be overcome, but continued increases in microarchitectural complexity

---

[1]Amdahl, "Validity of the single-processor approach."

[2]Pollack, "New Microarchitecture Challenges."

may not be practical. An important area of research is finding ways to beat Pollack's rule in order to continue to improve performance but with smaller increases in die area.

Measuring Performance

Processor performance can determine easily whether a computer product is a commercial success or a failure, and yet it is frustratingly hard to measure. Imagine the performance of several cars must be rated, giving each one a single number as a measure of its performance. Top speed could be used as a measure of performance, but consumers would likely complain that they don't often have the opportunity to drive at over a 100 miles per hour (mph). The lowest time accelerating from 0 to 60 mph might be a more meaningful measure of speed drivers would care about. Alternatively all the cars could be driven through an obstacle course and their times compared to see which was the fastest under more realistic conditions. Which car rated as the fastest could easily turn out differently depending on which measure was chosen, and the manufacturers of the other cars would inevitably complain that this rating system was unfair. Measuring processor performance has the same types of problems.

One of the earliest metrics of computer performance was determined by measuring *millions of instructions per second* (MIPS). The MIPS rating was simple and easy to understand. Unfortunately not all instructions require the same amount of computation. A million adds in one second are very different from a million branches. When DEC launched the VAX-11/780 computer in 1977, it was sold as a "1 MIPS" computer because it had similar performance to an IBM computer that was being marketed as a "1 MIPS" machine. The VAX computer was so popular comparisons to it were used to determine the MIPS of other computers. It was only in 1981 that tests showed the actual rate of instruction completion on the VAX machine to be about 0.5 MIPS.[3]

Had DEC been lying all along? In terms of the absolute rate of execution, the VAX computer did not run at 1 MIPS, but its more complicated instruction set let it accomplish the same work in fewer instructions. It could run programs in the same length of time while executing instructions at a slower rate, so it was a 1 MIPS machine in terms of relative performance. Eventually MIPS came to mean how much faster a computer was than the VAX-11/780. Because of the problems in comparing computers with different instruction sets, some in industry came to joke that MIPS really stood for "Meaningless Indicator of Processor Speed."

---

[3]Hennessy and Patterson, *Computer Architecture*, 72.

Trying to avoid the problem of different instruction sets, some computers are rated in terms of *millions of floating-point operations per second* (MFLOPS). This unfortunately is little better. Addition and division are both floating-point operations, but they require very different amounts of computation. A meaningful measure of MFLOPS requires assuming some distribution of adds, multiplies, and divides from a "typical" program. Choosing a typical program will also determine how much time is spent performing loads and branches and other instructions that are not counted. An imaginary program, which performed nothing but floating-point addition, would achieve far higher MFLOPS than any real application, but the temptation to quote this peak value is too much for almost any company to resist.

Because values like MIPS, MFLOPS, and IPC will all vary from one program to another, one of the most commonly used measures of performance is the processor frequency. It is simple and remains constant regardless of the program. Frequency is actually not a bad metric for comparing performance between processors with the same microarchitecture. Replacing one Pentium 4 with another at 20 percent higher clock frequency will improve performance about 20 percent. Having the same microarchitecture makes their IPC equal and having the same architecture makes their instruction counts equal. Best of all, frequency is something that is measured easily and not a subject of debate. However, frequency becomes a poor metric when comparing processors of different architectures or microarchitectures.

In some ways, measuring frequency is like measuring the rotations per minute of a motor. Running a motor faster will provide more power, but an eight-cylinder engine will do far more work than a four-cylinder engine at the same RPM. What matters is not just the number of rotations but the amount of work performed by each rotation. Two processors with different microarchitectures may give very different performance even at the same frequency. The best method for comparing performance is benchmarks. These metrics measure the time to run parts of real programs and average the results together to create a numeric measure of performance.

This is the equivalent to running cars through an obstacle course, but what are reasonable obstacles? Whereas some drivers might drive off road on a regular basis, others are more concerned with downtown traffic. Creating a computer benchmark requires picking pieces of programs believed to be representative of "typical" use. Some of the most widely used computer benchmarks come from the nonprofit Standard Performance Evaluation Corporation (SPEC). SPEC was created in the 1980s specifically to create better measures of computer performance. The SPECint2000 benchmark measures performance on 12-integer applications, and the

SPECfp2000 benchmarks measures performance on 14 floating-point applications.[4]

Figures 5-8 and 5-9 shows the SPECint2000 and SPECfp2000 results for various processors released in the years 2000 to 2004.[5] The graphs are drawn with processor frequency on the x-axis and benchmark performance per frequency on the y-axis. Performance per frequency gives us a relative measure of the IPC of a processor. The graphs show contours of constant benchmark performance. For each graph, performance increases when moving to the right or up. These two directions show the two basic ways of improving processor performance, through increasing frequency (toward the right of the graph) or improving IPC (toward the top of the graph).

The microarchitecture, the circuit design, and the manufacturing process determine frequency. Shrinking transistors or making changes to circuit design allow the frequency of a particular microarchitecture be steadily improved over time. This causes each of the processors in Figs. 5-8 and 5-9 to show a range of frequencies over the 4-year period shown.

The processor microarchitecture, compiler, and computer buses together determine the IPC for each program. A better compiler can order

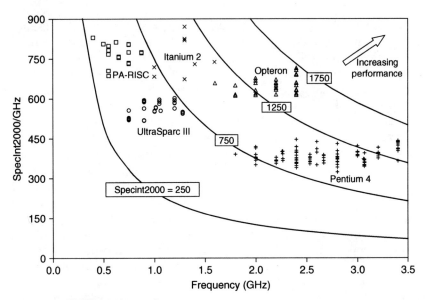

Figure 5-8 SPECint2000 results.

---

[4]Standard Performance Evaluation Corporation.
[5]ibid.

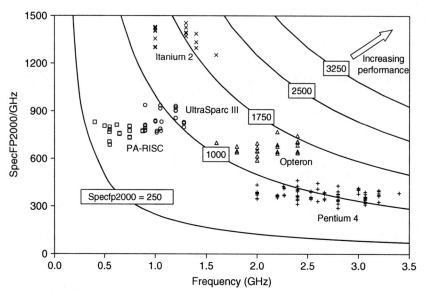

Figure 5-9 SPECfp2000 results.

instructions to be executed more efficiently, and higher performance buses can provide data to the processor more quickly, so that it spends fewer cycles waiting. Despite these variations the initial microarchitecture design of the processor is the most important factor affecting IPC. This is apparent in Figs. 5-8 and 5-9, where each processor tends to stay in a narrow vertical band of IPC compared to larger changes in frequency.

Intel's Pentium 4 has the longest pipeline of the processors shown in Figs. 5-8 and 5-9. By dividing its pipeline into very small steps this processor achieves very high frequencies. However, this comes at a price; its performance per cycle is low. The long pipeline causes many pipeline breaks and low IPC. At the other end of the spectrum the Itanium 2 achieves very high IPC in both integer and floating-point programs. Its EPIC architecture allows more reordering and parallelism to be exposed by the compiler; its microarchitecture is also designed to support simultaneous issue of many instructions. Although the Itanium 2 supports many parallel pipelines, the pipeline depth itself is relatively short, containing a large amount of logic in each pipestage. This results in the Itanium 2 being limited to lower frequencies than the Pentium 4. The Opteron processor takes a more balanced approach with frequencies and performance per cycle in between the Pentium 4 and Itanium 2.

These graphs show how different processor designs have chosen to focus on either frequency or IPC in the pursuit of higher performance. Eventually successful ideas from either approach end up being incorporated by designs from the opposite camp. The entire industry moves to higher performance by improving both frequency and IPC. Benchmarks are the best way of measuring processor performance, but they are far from perfect. The biggest problems are choosing benchmark programs, compiler optimizations, and system configurations. Different choices for any of these may suddenly change which computer scores as the fastest on a given benchmark.

Choosing realistic programs as the obstacle course is perhaps the most difficult problem because which programs are really important varies so much from one user to another. Someone who plays many 3D games will find the performance of floating-point intensive programs relevant, whereas someone who does not may care only about the performance of integer programs. An added complication is that the applications available change over time. The software takes advantage of improving processor performance and greater memory capacity by creating new applications that require more performance and memory. Benchmarks must regularly update the programs they use in order reflect current software. The SPEC integer benchmark program list was first created in 1988, but was updated in 1992, 1996, and 2000.

Processors with large caches tend to look better running older benchmarks. Processor cache memory has increased quickly over time but not nearly as dramatically as the size of typical applications. Usually a processor will have nowhere near enough cache to hold all the code or instructions of a contemporary program. However, it might have a large enough cache to hold all of a benchmark, especially if that benchmark is a few years old. This creates a strong temptation for some companies to quote older benchmarks long after they have ceased to be representative of current software.

In order to be run on processors of different architectures, benchmarks must be written in high-level programming languages, but this means that the compiler that translates the code for execution on a particular processor can have a very large impact on the measured performance. Some microarchitectures rely more upon the compiler for reordering while others are better able to compensate for a simpler compiler by reordering at run time. Even a simple compiler design typically has many different optimizations that can be selected. Some will expose more parallelism at the cost of increasing code size. This may improve or hurt performance depending upon the details of the microarchitecture.

Processors with complex hardware reordering look relatively better when making comparisons with simple compilers. Processors that need sophisticated compilers have helped drive improvements in compiler technology and these advances have been part of the steady progress in computer performance. However, sometimes undue attention is given to compiler optimizations that show dramatic improvements in performance for extremely specific sequences of instructions that happen to appear in important benchmarks. These optimizations may not help real applications at all or may even be detrimental for most programs. There have even been stories of compilers that turn on different optimizations depending on whether the program file being compiled has the same name as particular benchmark programs. To provide consistent results, each benchmark must choose which compiler optimizations are allowed by trying to pick those that could be used to achieve similar performance improvements for real programs.

Ultimately any benchmark measures not only processor performance but also computer performance. It is impossible for the processor to run any program at all without the support of the chipset and main memory, but these components are also important in the overall computer performance. A processor design that dedicates a large amount of die area to cache may suffer relatively little performance loss when used with a slower chip set or memory. Another processor that uses the same die area for more functional units could potentially achieve higher IPC, but it is more dependent upon the performance of the memory system. The computer system used for the benchmarking could end up determining which of these processors is rated as the fastest.

Huge sums of money are at stake when well-known benchmark results are reported. There is enormous pressure on every company to make its processors appear in the best possible light. To try and regain some control of how their products are measured, some companies create their own benchmarks. Inevitably every company's benchmarks seem to show their own processors with higher relative performance than the benchmarks created by any of their competitors. Unrealistic choices of benchmark programs, complier optimizations, and system configurations are all possible ways to "cheat" benchmarks. Truly fair comparisons of performance require a great deal of effort and continued vigilance to keep up with the changing world of hardware and software.

Microarchitectural Concepts

Modern processors have used pipelining and increased issue width to execute more instructions in parallel. Microarchitectures have evolved to deal with the data and control dependencies, which prevent pipelined processors from reaching their maximum theoretically performance. Reducing instruction latencies is the most straightforward way of easing dependencies. If results are available sooner, fewer independent instructions must be found to keep the processor busy while waiting. Microarchitectures can also be designed to distinguish between true dependencies, when one instruction must use the results of another, and false dependencies, which instruction reordering or better sharing of resources might eliminate. When dependencies can often but not always be eliminated, modern microarchitectures are designed to "guess" by predicting the most common behavior, spending extra time to get the correct result after a wrong guess. Some of the most important microarchitectural concepts of recent processors are:

- Cache memories
- Cache coherency
- Branch prediction
- Register renaming
- Microinstructions
- Reorder, replay, retire

All of these ideas seek to improve IPC and are discussed in more detail in the following sections.

Cache memory

Storing recently used values in cache memories to reduce average latency is an idea used over and over again in modern microarchitectures. Instructions, data, virtual memory translations, and branch addresses are all values commonly stored in caches in modern processors. Chapter 2 discussed how multiple levels of cache work together to create a memory hierarchy that has lower average latency than any single level of cache could achieve. Caches are effective at reducing average latency because programs tend to access data and instructions in regular patterns.

A program accessing a particular memory location is very likely to access the same memory location again in the near future. On any given day we are much more likely to look in a drawer that we also opened the day before than to look in a drawer that has been closed for months. If we used a particular knife to make lunch, we are far more likely to use the same knife to make dinner than some utensil in a drawer that hasn't been opened for months. Computers act in the same way, being more likely to access memory locations recently accessed. This is called *temporal locality*; similar accesses tend to be close together in time.

By holding recently accessed values in caches, processors provide reduced latency when the same value is needed repeatedly. In addition, nearby values are likely to be needed in the near future. A typical program accessing memory location 100 is extremely likely to need location 101 as well. This grouping of accesses in similar locations is called *spatial locality*.

If I read a page from the middle of a book, I probably will read the next page as well. If I use a tool from my toolbox, it is likely that I will also need another tool from the same box. If program behavior was random, then caches could provide no performance improvement, but together temporal and spatial locality make it possible for many different types of caches to achieve very high hit rates. Of course, any cache will have some misses where the needed data is not found in the cache, and different programs will exhibit varying amounts of locality giving different miss rates. Looking at the different causes of cache misses can help in finding ways of improving performance.

Types of cache misses are as follows:

Capacity miss. Cache is not large enough to store all the needed values.

Conflict miss. Needed value was replaced during another recent miss.

Cold miss. Memory location is being accessed for the first time.

A capacity miss occurs when the cache is not large enough to hold all the needed values. A program that serially accesses 1 MB of memory will miss when starting the same sweep over again if run on a processor with a cache smaller than 1 MB. The performance of some programs varies dramatically when run on processors with caches slightly larger or slightly smaller than the block of data the program is working upon. Conflict misses occur when there is sufficient capacity to store all the needed values, but the cache has made a poor choice in which values should be replaced as new values are loaded. If the cache design chooses the same cache location to store two different needed values, these values may conflict with one another and cause a high miss rate even if the total amount of data being accessed is small.

A cache is said to be "cold" when a new program first begins to run. Because the program has just started, the cache has not yet built up a store of values likely to be needed. Hit rates will be very low for the first accesses. Each of these types of misses is affected by the choice of cache size, associativity, and line size.

The cache size is simply the amount of data it can store. Increasing cache size is the easiest way to improve hit rates by reducing capacity misses. This performance improvement comes at the cost of larger die size. A large cache will also increase the cache delay, because the more memory cells in a cache, the longer it takes to access. This means increasing the cache size will usually also require increasing the number of clock cycles allocated for a cache access or reducing the processor clock frequency. Either of these will cause a loss in performance that may easily offset the improvement gained by a better hit rate. Dividing a large cache into multiple levels allows a microarchitecture to balance size and latency to provide the best overall performance.

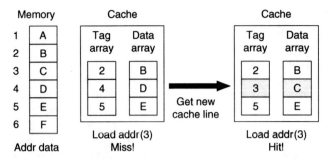

Figure 5-10 Cache data and tag arrays.

Each cache is actually made up of two separate arrays, the data array and the tag array. The data array contains the actual values being stored. When describing the size of the cache it is the data array size that is quoted. The tag array contains the addresses of the values currently stored in the data array. The tag array is effectively an index for the contents of the data array.

Each portion of the data array that has its own tag is called a *cache line*. When a cache miss causes new data to be fetched, the entire cache line is replaced and the matching tag array entry is updated. In Fig. 5-10, a load from address 3 misses in the cache initially. This triggers the data from that address to be loaded, replacing the value stored in the cache for address 4.

This could cause a conflict miss if the data from address 4 is needed again in the near future. In order to reduce conflict misses and make better use of limited size, some caches allow a particular memory location to be stored in more than one location in the cache. In these designs, when new data is loaded, the cache chooses which value is replaced based on which values have been used most recently. The number of locations in which a value can be stored is called the *associativity of the cache* (Fig. 5-11).

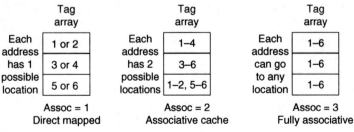

Figure 5-11 Cache associativity.

A cache that stores each value in only one location has an associativity of 1 and is said to be direct mapped. A cache that can store in two locations has an associativity of 2, and a cache that can store a value in any location is said to be fully associative. Caches that are more associative are less likely to replace values that will be needed in the future, but greater associativity can cause increased cache delay and power, since an associative cache must look for data in multiple places.

Some cold misses are inevitable any time a new program is run. The first miss on a location will be what causes it to be loaded, but increasing cache line size will reduce the number of cold misses. When a byte of memory is accessed, it is very likely that nearby bytes will also be needed. By fetching larger blocks of data with each miss, it is more likely that a needed piece of data will already have been brought into the cache, even before it is to be used for the first time.

Large lines will reduce the number of cold misses but increase the number of conflict misses. The same size cache using larger lines will have a smaller number of lines. This makes conflicts between pieces of data being mapped to the same line more likely. The byte immediately after one just accessed is very likely to be needed, but the next one less so and the one after that even less. As line size is increased beyond some optimum point, the cache is loading more and more data that is less likely to be needed and may be replacing data more likely to be used. Another limit on line size is that large line sizes produce more bus traffic. Performance may be hurt if the processor has to wait to access the bus while a very large cache line is being loaded.

In addition to the main data and instruction caches, a modern processor actually contains many other caches with different specialized functions. One example is the *translation lookaside buffer* (**TLB**). This is a cache that stores virtual memory page numbers in the tag array and physical memory page numbers in the data array. Whenever a memory access occurs, the TLB is checked to see if it contains the needed virtual to physical translation. If so, the access proceeds. If not, the needed translation is fetched from main memory and is stored in the TLB. Without a TLB, every load or store would require two memory accesses, one access to read the translation and another to perform the operation. Virtual memory is an important architectural feature, but it is microarchitectural features such as the TLB that allow it to be implemented with reasonable performance.

Some misses are inevitable for any cache. If we could implement an infinitely large, fully associative cache, it still would suffer from cold misses. Having to choose a finite size adds capacity misses and making the array less than fully associative will add conflict misses. A line size must be chosen, and any cache could be implemented as a hierarchy of multiple levels of cache. These choices will have a large impact on the die size and performance of any microprocessor.

Cache coherency

Caches create special problems when multiple devices are sharing main memory. Even in a computer system with a single CPU, other components can access main memory. Graphics cards, peripherals, and other hardware may read and write memory. When other devices read memory it is important they get the latest value, which may exist only in the processor cache. When other devices write memory it is important that the data stored in the processor cache is kept current. In multiprocessor systems, there are multiple processors all sharing memory, each with their own caches. The ability to keep all these caches synchronized is called **cache coherency**.

The first requirement for supporting cache coherency is a bus protocol that supports snooping. Snooping is when an agent on the bus monitors traffic to check whether it affects data in its own cache. A typical bus protocol might support four phases per transaction, which are as follows:

1. Request
2. Snoop
3. Response
4. Data

During the request phase the agent performing a read or write drives request and address pins to indicate what data it will be accessing. The snoop phase gives any agent sharing the bus the chance to check whether its own cache is affected. The address from the request phase must be compared with all the cache tags to determine if the data being accessed is in the cache. If so, the snooping processor can provide the data itself. If not, during the response phase the intended receiver can signal that it is ready to send or receive data. The responder might signal that it must defer the transaction. It does not have the data needed for a read or is not ready to accept a write. The responder will create a new transaction at a later time to satisfy this request. If the responder can handle the transaction, then during the data phase, the actual data is exchanged over the bus. Even when processors are idle and not executing any instructions, they must continue to snoop the bus to keep all the caches synchronized.

In addition to snooping the bus, each processor must be able to keep track of which lines in its cache are shared with other processors. It would be inefficient to have to perform a bus transaction every single time the cache is written just in case some other processor shares the data. Instead, flags bits are associated with each line that track the ownership of the data. The most common protocols have four ownership states.

148 Chapter Five

Modified (M). Processor owns line and copy is different than main memory.

Exclusive (E). Processor owns line and copy is the same as main memory.

Shared (S). Processor shares line with others.

Invalid (I). Line is invalid.

These coherency schemes are called **MESI** protocols after the supported states. A modified line has been written by the processor, so that it now holds the only copy of the latest data. The processor must provide this data if another agents try to read this line. An exclusive line is held only in one cache but still matches the copy in main memory. A processor can write this line and change its state to modified, without notifying other agents. A shared line means the same data is held in main memory and more than one cache. Any writes to this line must be broadcast on the bus to allow other processors to update their caches. An invalid line is typically one that has been written by another processor making the lines in other caches out of date. Some of these transitions are shown in the example in Fig. 5-12.

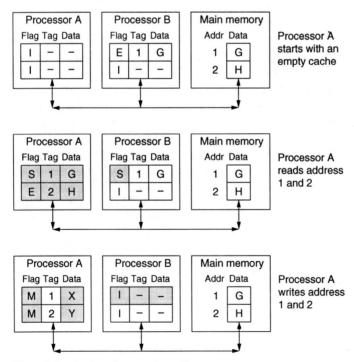

Figure 5-12 Cache coherency transitions.

Figure 5-12 shows two processors, each with their own cache, at three different moments in time. Normally there would also be a Northbridge chip handling communication with memory but for simplicity this has been left out. Each cache line has a flag showing that line's MESI state, a tag holding the address of the data stored, and the data itself. To start out, processor A's cache is empty with all lines invalid, and processor B's cache exclusively owns the line from address 1. When processor A reads the lines from address 1 and 2, processor B snoops the bus and sees the request. Processor B ignores the request for line 2, which is not in its cache, but it does have line 1. The cache line state must be updated to share because now processor A's cache will have it as well. When processor A writes both cache lines, it writes line 2 without a bus transaction. Because it is the exclusive owner, it does not need to communicate that this write has happened. However, line 1 is shared, which means processor A must signal that this line has been written. Processor B snoops this write transaction and marks its own copy invalid. Processor A updates its copy to modified.

Only through this careful bookkeeping and communication can caches be safely used to improve performance without causing logical errors.

Branch prediction

One type of specialized cache used in modern microarchitectures is a branch prediction cache. Branches create a special problem for pipelined and out-of-order processors. Because they can alter the control flow, all the instructions after them depend upon their result. This **control dependency** affects not just the execution of later instructions, but whether they should be fetched at all. For many programs, 20 percent or more of the instructions are branches[6]. No pipelined processor could hope to achieve any reasonable speedup without some mechanism for dealing with branch control dependencies. The most common method is **branch prediction** (Fig. 5-13). The processor simply guesses which instruction should be fetched next.

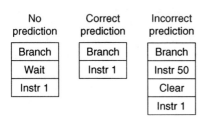

Figure 5-13 Branch prediction.

---

[6]Hennessy and Patterson, *Computer Architecture*, 109.

Without branch prediction a pipeline break is created, since the processor must wait until the branch has been resolved before fetching the proper instruction. With branch prediction, the processor guesses whether the branch will be taken or not, and based on that prediction it continues fetching instructions without a break. If the prediction is correct, time has been saved and performance is improved. If the prediction is incorrect, several instructions will have incorrectly entered the pipeline before the mistake is discovered. It will take longer to begin executing the correct instruction than if no prediction had been made and the processor had merely waited to fully resolve the branch. These lost cycles are known as the branch mispredict penalty. For branch prediction to improve performance, it must be accurate enough for the correct predictions to save more cycles than are lost to incorrect predictions. As the pipeline gets longer, the number of cycles from prediction to discovering an incorrect prediction increases. This increases the branch mispredict penalty and makes prediction accuracy more important.

Some architectures allow the compiler to include a suggested prediction as part of the encoding for each branch instruction. For architectures that do not include this, the simplest type of hardware prediction is static prediction, which causes each branch to be predicted the same way every time. On average, backward branches (those going to earlier instructions) tend to be taken and forward branches (those going to later instructions) tend not to be taken. For branches with relative targets, this makes deciding to guess taken or not taken a simple matter of looking at the sign bit of the target: for a negative target address guess taken, and for a positive target address guess not taken.

Static prediction has the advantage of being extremely simple but may have very poor results for some branches. Prediction accuracy is greatly improved by using dynamic prediction, which uses past branch behavior to predict future behavior. One of the most commonly used dynamic prediction schemes is called *2-bit prediction*.[7]

The processor contains a cache memory holding 2-bit counters for each of the recently accessed branches. This cache is called a *branch prediction buffer* or *branch history table*. Each counter holds the value 0, 1, 2, or 3. Whenever a branch is taken, its 2-bit counter is incremented if it is below the maximum value of 3. Whenever a branch is not taken, its 2-bit counter is decremented if it is above the minimum value of 0. When a branch is fetched, its address is used to lookup the value of the counter for that branch. If the value is a 0 or 1, the branch is assumed to be not taken. If the counter is a 2 or 3, the branch is assumed to be taken. See Fig. 5-14.

---

[7] Smith, "Study of Branch Prediction".

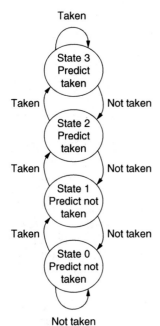

Figure 5-14 2-bit branch prediction.

For the SPEC89 benchmarks, a 4096 entry 2-bit branch prediction cache gives a prediction accuracy of 82 to 99 percent.[8] The repetitive behavior of most branches allows for this relatively simple dynamic scheme to achieve much better results than any static scheme. To improve accuracy even further, a two-level prediction scheme can be used.[9]

In two-level prediction, simple repetitive patterns are accurately predicted by maintaining multiple-branch prediction caches. A shift register is used to record the behavior of the most recent branches. This global branch history vector determines which of the prediction caches will be accessed when making a prediction or updating a branch's counter. Imagine a branch that alternates between taken and not taken. A 2-bit scheme might cause the branch counter's value to alternate between 1 and 2, and the prediction to be wrong every time. In the simplest two-level scheme, there would be two prediction caches, one accessed when the last branch was taken and one when the last branch was not taken. For each branch, there are two 2-bit counters; for this alternating branch one would quickly saturate at 0 whereas the other would go to 3. This would lead to correct predictions for this simple pattern.

---

[8]Hennessy and Patterson, *Computer Architecture*, 199.
[9]Yeh and Patt, "Two-Level Adaptive Branch Prediction."

A longer global branch history vector allows detection of more complex patterns, or a history vector can be maintained for each branch rather than a single vector recording the behavior of all branches. There are many possible variations and their success varies greatly from one application to another. This makes it difficult for a microarchitecture designer to choose between them. One solution is to implement two different prediction algorithms in hardware and allow the processor to dynamically select between them.[10]

In addition to whatever hardware is supporting each prediction scheme, another cache of 2-bit counters is maintained tracking which scheme is more successful. If method A is correct and method B is incorrect, the counter is decremented. If method A is incorrect and method B is correct, the counter is incremented. If both methods are correct or incorrect, no change is made. The branch prediction is then made using method A for branches whose counter is 0 or 1 and using method B for branches whose counter is 2 or 3. These hybrid methods achieve the highest prediction accuracies at the cost of the most hardware support.

For branches that are predicted taken, the next instruction should be fetched from the target address of the branch. For relative branches, this means the target offset must be added to the instruction pointer. To allow fetching to begin sooner, many branch prediction caches store not only the branch history but also the branch target address. When this is done the branch prediction cache is called a *branch target buffer* (BTB). Indirect branches are a special problem because their target varies during run time. The branch may be correctly predicted but with the wrong target address. To deal with this, some BTBs allow indirect branches to have multiple entries with different target addresses. Past branch behavior is used to pick which of the target addresses is most likely.

Register renaming

Because branches control the flow of the program, pipeline breaks created by branches are called *control dependencies*. Pipeline breaks caused by one instruction having to wait for the result of another are called *data dependencies*. True data dependencies are caused by the passage of data from one instruction to another, but in addition there are false data dependencies created by the reuse of registers in programs. These false dependencies are eliminated and performance improved by **register renaming**.

---

[10]Hilgendorf et al., "Evaluation of branch-prediction methods."

Each computer architecture defines some finite number of registers that instructions use to read and write results. To keep instruction size and code size to a minimum, older architectures tended to define a relatively small number of registers. This made sense at the time, especially since manufacturing processes were not capable of creating processors with large numbers of registers. As a result of having very few architectural registers to choose from, programs tend to reuse registers after only a few instructions. This reuse creates false dependencies. True dependencies are when a register is read by one instruction after being written by another. This is a *read-after-write* (RAW) dependency. False dependencies are *write-after-read* (WAR) or *write-after-write* (WAW) dependencies. A write-after-read dependency causes one instruction to have to wait to write a register until another instruction has read the value. In a write-after-write dependency, two instructions conflict with each other by both trying to write the same register. In code without branches WAW dependencies would not happen because at least one instruction would always read a result after it was written, but with branches or interrupts, programs may write the same register more than once without any reads.

Register renaming removes false dependencies by mapping the architectural registers of the program code into physical registers on the processor. The number of physical registers can be greater than the number of architectural registers, allowing registers to be reused less often. Eliminating false dependencies makes possible more instruction reordering and therefore more performance. An on-die *register alias table* (**RAT**) can be used to record which physical register each architectural register is currently mapped to. Figure 5-15 shows an example code sequence before and after register renaming.

In Fig. 5-15, the only true data dependency is between the first two instructions. The add instruction's result is an input to the multiply instruction. These two instructions must be executed in the program

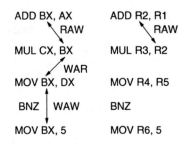

Figure 5-15 Register renaming.

order. After these two instructions go through renaming, architectural registers AX, BX, and CX have been mapped to physical registers R1, R2, and R3. There is a WAR dependency between the multiply and move. The move instruction is written to write to the same architectural register that is to be read by the multiply. This dependency is removed by mapping the architectural register BX to a different physical register for the move instruction. After a branch a second move also writes to register BX. The dependency is removed by mapping BX to yet another physical register. After renaming only the single true dependency remains.

This mapping from architectural to physical registers is very similar to the mapping of virtual to physical memory addresses performed by virtual memory. Virtual memory allows multiple programs to run in parallel without interfering with each other by mapping their virtual memory addresses to separate physical addresses. Register renaming allows multiple instructions to run in parallel without interfering with each other by mapping their architectural registers to separate physical registers. In both cases, more parallelism is allowed while the results of each program are unaffected.

Architectures that define a large number of architectural registers have less need of hardware register renaming since the compiler can avoid most false dependencies. However, because control flow of programs varies at run time, false dependencies still appear and even processors with these architectures can benefit from register renaming.

Microinstructions and microcode

We can imagine a processor pipeline being a physical pipe with each instruction like a ball rolling down the pipe. If some balls roll more slowly down the pipe, other balls will stack up behind it. The pipeline works best when all the balls travel the pipeline in the same length of time. As a result, pipelined processors try to break down complicated instructions into simpler steps, like replacing a single slow ball with several fast ones.

RISC architectures achieve this by allowing only simple instructions of relatively uniform complexity. Processors that support CISC architectures achieve the same affect by using hardware to translate their complex machine language instructions into smaller steps. Before translation the machine language instructions are called **macroinstructions**, and the smaller steps after translation are called **microinstructions**. These microinstructions typically bare a striking resemblance to RISC instructions. The following example shows a simple macroinstruction being translated into three microinstructions.

Macroinstruction:

ADD Mem(AX), BX   # Increment value at memory location AX by value BX

Microinstructions:

LD R3, Mem(R1)   # Load value at memory location R1 into R3
ADD R4, R3, R2   # Add R2 and R3 and place result in R4
STD R4, Mem(R1)   # Store R4 at memory location R1

The macroinstruction adds a value to a location in memory. In microinstructions, which are also called *uops*, this is performed in three separate steps: getting the value from memory, performing the addition, and storing the result in memory. The processor can more quickly and efficiently execute programs as uops because the instructions are more uniform in complexity. When trying to execute CISC macroinstructions directly, each instruction might access memory once, twice, or not at all. This makes scheduling and keeping the pipeline full more difficult. As uops, each instruction either accesses memory once or performs a calculation but never both. By focusing on optimizing a smaller number of simpler more uniform instructions, processor performance can be improved. Also, the form and types of uops can be changed with each new processor design to suit the microarchitecture while maintaining software compatibility by consuming the same macroinstructions.

Most CISC architecture macroinstructions can be translated into four or fewer uops. These translations are performed by decode logic on the processor. However, some macroinstructions could require dozens of uops. The translations for these macroinstructions are typically stored in a *read-only memory* (ROM) built into the processor called the *microcode*. The microcode ROM contains programs written in uops for executing complex macroinstructions. In addition, the microcode contains programs to handle special events like resetting the processor and handling interrupts and exceptions.

In many ways, the microcode performs the same functions for the processor as the BIOS ROM does for the system. The BIOS ROM stores device drivers, which allow the same software to use different computer hardware, and also contains the code run during reset to initialize the computer. The microcode ROM allows the same macroinstructions to be executed differently on processors with different microarchitectures and contains the code run during reset to initialize the processor. Both the BIOS and microcode are providing hardware abstraction and initialization.

Each time the computer is reset, the processor first runs the microcode reset routine. After this is done, initializing the processor, the first program it runs is the BIOS ROM reset routine. The BIOS reset program has the job of starting the operating system. The *operating system* (OS) then loads applications for the user. Once running, the application runs by sending macroinstructions to the microprocessor for execution. The instructions can send specific requests for service to the operating system by calling device drivers. The OS will in turn execute its own routines on the processor or call device drivers stored in the BIOS.

Figure 5-16 shows this interaction of computer software and hardware. Compiling an application for particular computer architecture converts it from a high-level programming language to macroinstructions that are understood by the processor. In addition, the application will be written assuming a particular operating system and set of services provided by that OS. Recompiling an application converts it to the macroinstructions of a different architecture, but parts of the application may need to be rewritten by hand when running under a different OS.

Accessing the computer hardware through the OS does allow the application to be written without knowing the specific details of how the OS manages the system hardware. Likewise, the OS relies upon the BIOS and can be written without specific details of how the BIOS accesses specific physical components on the motherboard. Many modern operating systems use their own performance optimized device drivers rather than rely upon the BIOS, but the BIOS routines are still required for system start-up when OS must be read from the hard drive.

In the end, the application, OS, and BIOS are all just programs, which run as machine language instructions on the processor. Once on the processor the macroinstructions can be translated to uops using basic logic circuits or by reading from a microcode ROM. The uops execute in the functional units on the processor, and some will produce results

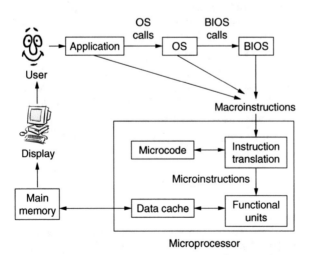

Figure 5-16 Computer software and hardware.

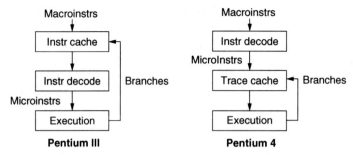

Figure 5-17 Trace cache.

that will be stored in the data cache. These values are forwarded on to main memory from where they can be displayed to the user.

The microarchitecture of the Pentium 4 limits the performance impact of having to translate macroinstructions to uops by storing uops in a trace cache. See Fig. 5-17.

The Pentium III microarchitecture stores macroinstructions in its instruction cache. These are then translated before execution. If one of the instructions is a branch, the new program path must be fetched from the instruction cache and translated before execution can continue. In the Pentium 4, the instructions translated to uops before being stored in the cache. The instruction cache is called a **trace cache** because it no longer stores instructions as they exist in main memory, but instead stores decoded uop translations. When a branch is executed the already translated uops are immediately fetched from the trace cache. The instruction decode delay is no longer part of the branch mispredict penalty and performance is improved. The disadvantage is that the decoded instructions require more storage and the trace cache must be larger in size to achieve the same hit rate as an instruction cache.

In general, the use of uops allows only the translation circuits on the processor to see a macroinstruction. The rest of the processor operates only on uops, which can be changed from one processor design to the next as suits the microarchitecture while maintaining software compatibility. The use of uops has allowed processors supporting CISC architectures to use all of the performance enhancing features that RISC processors do and has made the distinction between CISC and RISC processors less important.

Reorder, retire, and replay

Any processor that allows **out-of-order execution** must have some mechanism for putting instructions back in order after execution. Software is written assuming that instructions will be executed in the order specified in the program. To produce the expected behavior, interrupts and exceptions must be handled at the proper times. Branch prediction means that some instructions will be executed that an in-order processor would never have fetched. There must be a way of discarding the improper results produced by these instructions. All the performance enhancing microarchitecture features work to improve performance while maintaining software compatibility. The true out-of-order nature of the processor must be concealed from the program, and the illusion of in-order execution maintained. The most common way to accomplish this is to use a *reorder buffer* (**ROB**).

A reorder buffer takes out-of-order instructions that have completed execution and puts them back into the original program order. When instructions are being fetched and are still in order each one is allocated an entry in the ROB in the order the instructions are supposed to be executed. After the instructions are executed out of order, the ROB is updated to track which have been completed. The results produced by any instruction are guaranteed to be correct only if all earlier instructions have completed successfully. Only at this point are the results committed and the instruction truly finished. This is called *retiring an instruction*.

Before retirement any result produced by an instruction is speculative. New instructions will use the latest speculative result, but the most recent committed results must also be maintained in case the speculative results must be discarded. Only when an instruction retires are the committed results updated with that instruction's result.

Figure 5-18 shows an example of the ROB and the RAT at three different moments in time as a set of three instructions is retired. The code

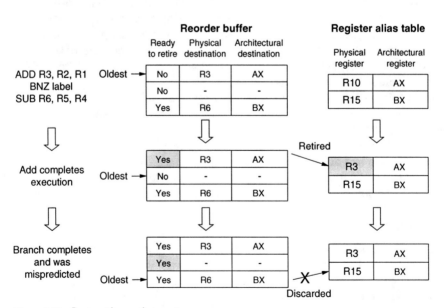

Figure 5-18 Instruction retirement.

shown on the left specifies an add instruction to be followed by a branch and then a subtract instruction if the branch is not taken. The ROB maintains a pointer to the oldest entry not yet retired, which starts at the entry for the add. To start with, only the subtract is ready to retire, having been executed out of order before the other two instructions. The ROB also stores the physical destination register for the instruction and the architectural register that physical register had been mapped to. In this example, the add instruction is to write to register 3, which has been mapped to architectural register AX, and the subtract is to write to register 6, which has been mapped to architectural BX. The branch does not write to a register. At the right the RAT maintains a list of the physical registers holding the most recent committed in-order results for each architectural register.

Imagine the add instruction now completes execution. As the oldest instruction, it is allowed to retire. The ROB oldest pointer is incremented to point to the branch's entry. The RAT is updated to show that the most recent committed results for AX are stored in register 3. Upon execution of the branch, it is discovered to have been mispredicted. The branch is now the oldest entry in the ROB, so it is allowed to retire. However, the subtract should never have been executed at all. When it retires its results are discarded because the early retired branch was mispredicted. The RAT is not updated and the latest committed results for BX are still in register 15. The registers used by the subtract will be recycled for use by other instructions; the result the subtract produced will be overwritten without ever having been read.

If instructions were like balls rolling through a pipe, then the in-order portion of the processor would have the job of carefully labeling each ball with a number that indicates the order they should be retired. The out-of-order execution of the processor is like the balls being thrown into the pipe in random order. The ROB has the job of gathering up the balls that come out of the pipe, and using their labeled numbers to put them back into order and to discard any balls that should never have come through the pipe in the first place.

The example in Fig. 5-18 shows instructions retiring successfully and an instruction being discarded. Another possibility is that an instruction will need to be replayed. Some processors allow execution of instructions whose data may not be available yet. The most common case is scheduling instructions that depend upon a load. Because the load may or may not hit in the cache, its latency is not known. All the dependent instructions could be stalled until a hit or miss is determined but this causes some delay. Performance can be improved by assuming the load will hit (the most common case), and scheduling instructions before this is known for certain.

Taking into account the worst possible latency of each type of instruction would make filling the pipeline much more difficult. Instead, the processor can assume that each instruction will execute in the shortest possible number of cycles. After execution the instruction is checked to see this guess was correct. Scheduling assuming data will be available is sometimes called data speculation. It is very similar to branch prediction. Both methods make predictions during scheduling and reduce delay when correct, at the cost of increased delay when incorrect. When branch prediction is wrong instructions are incorrectly executed and must be discarded. When data speculation is wrong, the correct instructions are executed but with the wrong data. These instructions cannot be discarded; instead, they must be executed again, this time with the correct data. Sending instructions back into the pipe is called *replay*. If the events that cause replay are rare enough, overall performance is improved.

Life of an Instruction

The basic steps any microprocessor instruction goes through have changed little since the first pipelined processors. An instruction must be fetched. It must be decoded to determine what type of instruction it is. The instruction is executed and the results stored. What has changed is that steps have been added to try and improve performance, such as register renaming and out-of-order scheduling. The total number of cycles in the pipeline has increased to allow higher clock frequency. As an example of how a processor microarchitecture works, this section describes in detail what actions occur during each step of the original Pentium 4 pipeline.

The Pentium 4 actually has two separate pipelines, the front-end pipeline, which translates macroinstructions into uops, and the execution pipeline, which executes uops.

The front-end pipeline has the responsibility for fetching macroinstructions from memory and decoding them to keep a trace cache filled with uops. The execution pipeline works only with uops and is responsible for scheduling, executing, and retiring these instructions. Table 5-1

TABLE 5-1 Pentium & Pipeline

Front-end pipeline	Microinstruction pipeline (20)
Instruction prefetch	Microbranch prediction (2)
L2 cache read	Microinstruction fetch (2)
Instruction decode	Drive (1)
Branch prediction	Allocation (1)
Trace cache write	Register rename (2)
	Load instruction queue (1)
	Schedule (3)
	Dispatch (2)
	Register file read (2)
	Execute (1)
	Calculate flags (1)
	Retirement (1)
	Drive (1)

shows the number of clock cycles allocated for each step in the execution pipeline for a total of 20 cycles. This is the best case pipeline length with many instructions taking much longer. Most importantly 20 cycles is the branch mispredict penalty. It is the minimum number of cycles required to fetch and execute a branch, to determine if the branch prediction was correct, and then to start fetching from the correct address if the prediction was wrong. Intel has not provided details about the number of cycles in the front-end pipeline. The following sections go through each of the steps in these two pipelines.

Documentation of the details of each pipeline stage of the Pentium 4 is not always complete. The following description is meant to be a reasonable estimation of the processor's operation based on what has been publicly reported by Intel and others.[11, 12, 13]

Instruction prefetch

When a computer is turned off, all the instructions of all the software are retained on the hard drive or some other nonvolatile storage. Before any program is run its instructions must first be loaded into main memory and then read by the processor. The operating system performs the job of loading applications into memory. The OS treats the instructions of other programs like data and executes its own instructions on the processor to copy them into main memory. The OS then hands control of the processor over to the new program by setting the processor's *instruction pointer* (IP) to the memory address of the program's first instruction. This is when the processor begins running a new program. The processor has decoded none of these instructions yet, so they are all macroinstructions. The instruction prefetcher has the task of providing a steady stream of macroinstructions to be decoded into uops.

Keeping the processor busy requires loading instructions long before they are actually reached by the program's execution. The prefetcher starts at the instruction pointer address and begins requesting 32 bytes at a time from the L2 cache. The instruction pointer address is actually a virtual address, which is submitted to the *instruction translation lookaside buffer* (ITLB) to be converted into a physical address before reading from the cache. A miss in the ITLB causes an extra memory read to load the ITLB with the needed page translation before continuing. When looking up the physical address of the needed page, it may be discovered that the page is not currently in memory. This causes the processor to signal a page fault exception and hand control back over to the operating system to execute the instructions needed to load the page before returning control.

---

[11]"IA-32 Architecture Reference Manual."

[12]Hinton et al., "Microarchitecture of the Pentium 4."

[13]Shanley, *The Unabridged Pentium 4.*

Once the page is in memory and the address translation is complete, the processor issues a read address to the L2 cache and begins reading instructions.

L2 cache read

The Pentium 4 L2 cache is a unified cache, meaning that it stores both instructions and data. Reads looking for either are treated the same way. Part of the read address bits is used to select a particular line in the data and tag array. The most significant address bits are then compared with value read from the tag array. A match indicates a cache hit and no match indicates a cache miss. The Pentium 4 L2 cache is actually 8-way associative which means that 8 separate lines are read from the tag array and data array. Comparing all 8 tag lines to the requested address determines which if any of the 8 data lines is the correct one.

If the needed data is not in the cache, the bus controller is signaled to request the data from main memory. If the data is found in the cache, at this point the instructions are treated as only a stream of numbers. The processor does not know what the instructions are or even how many have been retrieved. This will be determined during the next step, instruction decode.

Instruction decode

During the instruction decode step the actual instructions retrieved from the L2 cache are determined. Because the number of bytes used to encode a macroinstruction varies for the x86 architecture, the first task is to determine how many macroinstructions were read and where their starting points are. Each instruction's bits provide an opcode that determines the operation as well as bits encoding the operands.

Each macroinstruction is decoded into up to 4 uops. If the translation requires more than 4 uops, a place keeper instruction is created instead, which will cause the full translation to be read from the microcode ROM at the proper time. If one of the macroinstructions being decoded is a branch, then it is sent for branch prediction.

Branch prediction

A prediction of taken or not taken is made for every macroinstruction branch. After being found by the decoder the address of the branch is used to look up a counter showing its past behavior and its target address in the branch target buffer (BTB). A hit enables the BTB to make a prediction and provide the target address. A miss causes a new entry to be allocated for the branch in the BTB, and the branch to go through static prediction. In general, backward branches are predicted as taken, and forward branches are predicted as not taken.

On a prediction of taken, a new instruction address is provided to the instruction prefetch, so that instructions from the most likely program path are fetched even before the branch has executed.

Trace cache write

The decoded uops, including any uop branches, are written into the trace cache. The contents of the trace cache are not the same as main memory because the instructions have been decoded and because they are not necessarily stored in the same order. Because branch prediction is used to direct the instruction prefetch, the order that macroinstructions are decoded and written into the trace cache is the expected order of execution, not necessarily the order the macroinstructions appear in memory.

Writing into the trace cache finishes the front-end pipeline. Uops will wait in the trace cache until they are read to enter the execution pipeline. All the steps until this point have just been preparing to start the performance-critical portion of the pipeline. After being loaded into the trace cache, a uop may remain there unread for sometime. It is part of the code of the program currently being run, but program execution has not reached it yet. The trace cache can achieve very high hit rates, so that most of the time the processor performance is not affected by the latency of the front-end pipeline. It is important that the front-end pipeline has enough bandwidth to keep the trace cache filled.

Microbranch prediction

To begin the execution pipeline, the processor must determine which uop should enter the pipeline next. The processor really maintains two instructions pointers. One holds the address of the next macroinstruction to be read from the L2 cache by the instruction prefetch. The other holds the address of the next uop to be read from the trace cache. If the last uop was not a branch, the uop pointer is simply incremented to point to the next group of uops in the trace cache. If the last uop fetched was a branch, its address is sent to a trace cache BTB for prediction. The trace cache BTB performs the same function as the front-end BTB, but it is used to predict uop branches rather than macroinstruction branches. The predictions of the front-end BTB steer the instruction prefetch to read needed macroinstructions from the L2 cache. The predictions of the trace cache BTB steer the microinstruction fetch to read needed uops from the trace cache.

Uop fetch and drive

Using the address determined by microbranch prediction, the trace cache is read. If there is a trace cache miss, the needed address is sent to the L2 cache. The data read from the L2 cache then flows through the front-end pipeline to be decoded and loaded into the trace cache. If there is a trace cache hit, the trace cache delivers up to 3 uops per cycle. Some uops read from the trace cache are actually pointers to uop routines stored in the microcode ROM. In this case, the ROM is read and it begins feeding uops into the execution pipeline instead. This allows macroinstructions that must be translated into large numbers of uops to not take up space in the trace cache.

The uops must then travel across the die to the next step in the execution pipeline. The Pentium 4 is unusual in that its pipeline allows for two "drive" cycles where no computation is performed but data is simply traveling from one part of the die to another. This allows the processor to achieve very high frequencies while still providing some flexibility in where the different blocks of the execution pipeline are physically placed on the die. Designers attempt to create a floorplan where blocks that communicate often are placed close together, but inevitably every block cannot be right next to every other block with which it might communicate. The presence of drive cycles in the pipeline shows how transistor speeds have increased to the point where now simple wire delay is an important factor in determining a processor's frequency.

Allocation

The first stop after being fetched from the trace cache is the allocation step. At this point the uops still reflect the original program order (at least if branch predictions have been correct). It is important to record this order before the uops enter the out-of-order portion of the pipeline. Each uop is allocated an entry in the reorder buffer (ROB). These entries are allocated in program order and will be retained by each uop until it is retired or discarded. The Pentium 4 ROB has 126 entries, which means at one time there are at most 126 uops in the execution pipeline. If the ROB is full, the allocation step must stall the uops it has and wait for space to become available before allowing the uops to proceed.

To demonstrate some of the actions of the following steps, we will follow a single example uop through the pipeline, shown in Fig. 5-19.

At this point in the pipeline, the uop has a ROB entry and the information encoded into it before it was stored in the trace cache. This

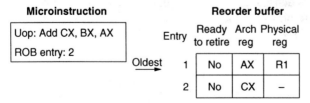

Figure 5-19 Uop at allocation.

information includes an operation, in this case an add instruction and architectural registers for two sources and destination.

Register rename

At the register rename step the register alias table (RAT) is read to determine which physical registers hold the uops source data and which will be used to store its result. There are actually two RAT tables. One is a speculative table containing the most recent mappings. The other is the retirement RAT, which ignores the mappings of any uops that have not yet retired. If uops that have been speculatively executed need to be discarded, this is accomplished by returning to the known good values in the retirement RAT.

Our example uop now has physical registers for its sources and destination and these mappings are held in the speculative RAT, as shown in Fig. 5-20.

Load instruction queue

Uops are then loaded into one of two instruction queues. The memory queue holds loads and stores and the general queue holds all other uops. The order of uops is maintained within each queue, but there is no ordering enforced between the queues. If progress on memory operations is slowed by cache misses, a separate queue allows other instructions to continue to make progress. Uops stay in the queue waiting their turn until there is space in a scheduler.

Schedule and dispatch

When the oldest uop is read from the memory queue, it is loaded into the memory scheduler. Uops read from the general queue are fed into three other schedulers, depending upon their instruction type. In every cycle, each scheduler compares the source registers of its uops with a scoreboard showing which registers will soon have valid data. The schedulers then dispatch for execution the oldest uops whose sources are

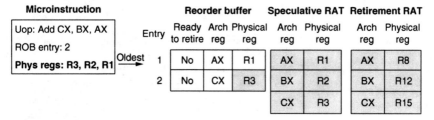

Figure 5-20 Uop at rename.

ready. This means that the uops are no longer in the original program order. Uops can dispatch before older uops if their sources are ready first. When a uop is dispatched, its destination register and minimum latency are used to update the scoreboard showing which registers have ready data. This means that dependent uops may be scheduled too soon if a uop takes longer than expected, for instance if a load misses in the cache. Uops that are scheduled too soon will have to be replayed, going through dispatch again to receive their correct source data. The Pentium 4 can dispatch a maximum of 6 uops in one cycle.

Register file read

The only values that are used in computations are those stored in the register files. There is one register file for integer values and another for floating-point values. Superscalar processors use register files with multiple read and write ports, allowing multiple uops to read out their source data or write back their results at the same time.

Figure 5-21 shows an example uop that has the source data it needs to perform its computation.

Execute and calculate flags

All the steps up until this point have just been to get the uops to the proper functional units on the die with the data they need to actually perform their operation. There are three separate parts of the processor responsible for the actual execution of uops. The *integer execution unit* (IEU) performs all the integer operations and branches. Although integer arithmetic uops are performed in half of a cycle, most instructions take longer. The *floating-point unit* (FPU) performs all the floating-point and SIMD operations. The *memory execution unit* (MEU) performs loads and stores.

The MEU includes the level 1 data cache, which is accessed by all load and store instructions. A miss in the level 1 data cache triggers an access to the L2 cache. The MEU also contains the *data translation lookaside buffer* (DTLB), which performs virtual to physical address translations for loads and stores.

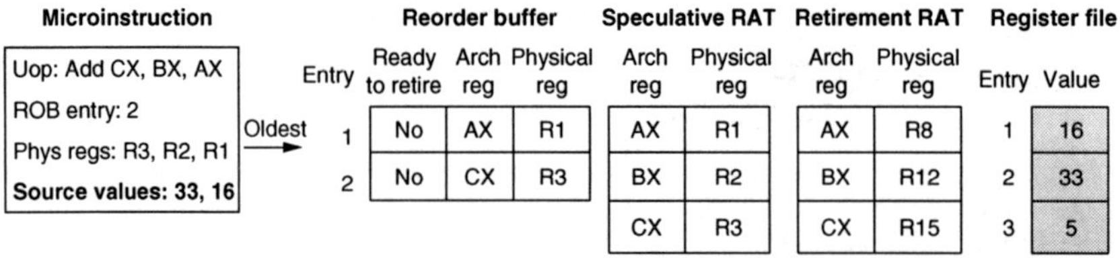

Microinstruction			Reorder buffer			Speculative RAT		Retirement RAT		Register file	
Uop: Add CX, BX, AX		Entry	Ready to retire	Arch reg	Physical reg	Arch reg	Physical reg	Arch reg	Physical reg	Entry	Value
ROB entry: 2	Oldest	1	No	AX	R1	AX	R1	AX	R8	1	16
Phys regs: R3, R2, R1		2	No	CX	R3	BX	R2	BX	R12	2	33
Source values: 33, 16						CX	R3	CX	R15	3	5

Figure 5-21 Uop at register file read.

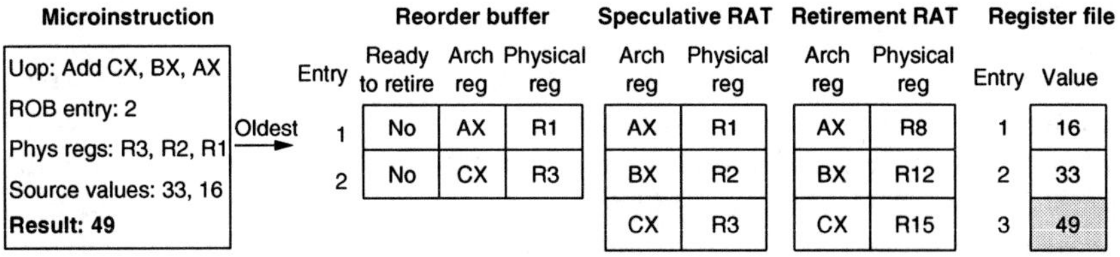

Figure 5-22 Uop at execution.

Microinstruction		Entry	Reorder buffer			Speculative RAT		Retirement RAT		Register file	
Uop: Add CX, BX, AX			Ready to retire	Arch reg	Physical reg	Arch reg	Physical reg	Arch reg	Physical reg	Entry	Value
ROB entry: 2	Oldest	1	No	AX	R1	AX	R1	AX	R8	1	16
Phys regs: R3, R2, R1		2	No	CX	R3	BX	R2	BX	R12	2	33
Source values: 33, 16						CX	R3	CX	R15	3	49
Result: 49											

When uop execution is complete, the result (if any) is written back into one of the register files along with flag values. Flag values store information about the result such as whether it was 0, negative, or an overflow. Any of the flag values can be a condition for a later branch uop.

At this point the add uop has finally performed its computation and written the result into the register file, as shown in Fig. 5-22.

Retirement and drive

Having completed execution and generated a result, the processor must now determine whether the correct instruction was executed with the correct data. If the uop was a branch, its actual evaluation is compared to the original prediction. The front-end BTB and trace cache BTB are both updated with the branch's latest behavior. If the branch was mispredicted, the trace cache BTB is signaled to begin reading uops from the correct address. All uops with entries in the ROB after the mispredicted branch should not have been executed at all and will have their results discarded.

Even if branch prediction was correct, the uop may have operated on incorrect data. If an earlier instruction had longer expected latency, then the uops that depended upon it may have been scheduled too soon. This ROB checks for this and if necessary uops are sent back to the dispatch step to be replayed.

If branch prediction was correct and no special events occurred, then the execution was successful and the uop is retired. Upon retirement, the uop's results are committed to the current correct architectural state by updating the retirement RAT and all the resources allocated to the instruction are released. The Pentium 4 can retire 3 uops per cycle.

The add instruction can now retire with satisfaction at a job well done (Fig. 5-23).

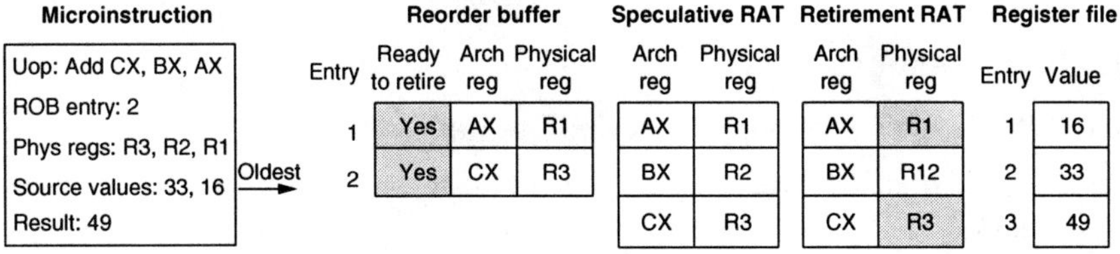

Microinstruction		Entry	Reorder buffer			Speculative RAT		Retirement RAT		Register file	
Uop: Add CX, BX, AX			Ready to retire	Arch reg	Physical reg	Arch reg	Physical reg	Arch reg	Physical reg	Entry	Value
ROB entry: 2		1	Yes	AX	R1	AX	R1	AX	R1	1	16
Phys regs: R3, R2, R1	Oldest	2	Yes	CX	R3	BX	R2	BX	R12	2	33
Source values: 33, 16						CX	R3	CX	R3	3	49
Result: 49											

Figure 5-23 Uop at retirement.

Conclusion

Designing a processor microarchitecture involves trade-offs of IPC, frequency, die area, and design complexity. Microarchitectural design must balance these choices based on the target market. There are many complicated algorithms that could be implemented in hardware to add performance but are not used because of the design time and die area they would take to implement. Each square millimeter of die area required by a microarchitectural feature will either add to the overall cost or take area away from some other feature. A larger branch target buffer may mean a smaller trace cache. The design team must find the features that provide the greatest performance improvement per area for the programs that are most important for their product. Some of the most important design choices will include:

- Pipeline length
- Instruction issue width
- Methods to resolve control dependencies
- Methods to resolve data dependencies
- Memory hierarchy

These and other microarchitectural choices will define most of the high-level details of the design. The following chapters will focus on the job of implementing the design.

Key Concepts and Terms

Amdahl's law	Microinstructions
Branch prediction	Out-of-order execution
Cache coherency	Pipelining
Control dependency	RAT
Data dependency	Register renaming
HyperThreading	ROB
IPC	Superscalar
MESI	TLB
Microcode	Trace cache

Review Questions

1. What are the causes of pipeline breaks?
2. What effect will increasing the pipeline length likely have on frequency, IPC, and performance?

3. How does HyperThreading improve performance?
4. Explain Amdahl's law in words.
5. What are some commonly used measures of processor performance? What are their drawbacks?
6. What are the three causes of cache misses? How can each be reduced?
7. Why is cache coherency necessary? What are the four states of MESI cache coherency protocols?
8. How can processors avoid control dependencies?
9. How can processors avoid data dependencies?
10. What are the trade-offs in choosing an instruction cache or a trace cache?
11. [Discussion] How will data dependencies, control dependencies, and resource conflicts affect processors trying to achieve high frequency through deep pipelines? How will wide issue superscalar processors be affected?
12. [Discussion] Assuming the gap between processing speed and main memory latency continues to grow, computer performance may become dominated by memory access time. Which microarchitectural features would improve performance in this case? Which would not? How might a computer with a processor 100 times faster than today, but memory only twice as fast be designed differently than current computers?

Bibliography

Amdahl, Gene. "Validity of the single-processor approach to achieving large scale computing capabilities." *AFIPS Conference Proceedings*, Atlantic City, NJ: 1967, pp. 483–485.

Hennessy, John and David Patterson. *Computer Architecture: A Quantitative Approach.* 3d ed., San Francisco, CA: Morgan Kaufmann, 2003.

Hilgendorf, Rolf, Gerald Heim, and Wolfgang Rosenstiel. "Evaluation of Branch-Prediction Methods on Traces from Commercial Applications." *IBM Journal of Research and Development*, July 1999, pp. 579–593.

Hinton, Glenn, et al. "The Microarchitecture of the Pentium 4 Processor." *Intel Technology Journal*, Q1 2001, pp. 1–13.

"IA-32 Intel Architecture Optimization Reference Manual." Intel Press, Order #248966-011, http://www.intel.com/design/pentium4/manuals/index_new.htm.

Johnson, Mike. *SuperScalar MicroProcessor Design.* Englewood Cliffs, NJ: Prentice-Hall, 1991. [Johnson is a senior fellow at AMD and was an important part of the K5 and K7 design teams. This book goes through all the design choices for supporting superscalar and out-of-order execution with tons of simulated data on effects these choices have on performance.]

Marr, Deborah et al. "Hyper-Threading Technology Architecture and Microarchitecture." *Intel Technology Journal*, Q1 2002, pp. 4–15.

Pollack, Fred. "New Microarchitecture Challenges in the Coming Generations of CMOS Process Technologies." Micro32 Keynote Address, Haifa, Israel: November 1999.

Shanley, Tom. *The Unabridged Pentium 4*. Boston, MA: Addison-Wesley, 2004. [This massive 1600-page book is by far the most definitive description of the Pentium 4 available anywhere.]

Smith, James.. "A Study of Branch Prediction Strategies." *Proceedings of the 8th Annual Symposium on Computer Architecture*, Minneapolis, MN: 1981, pp. 135–148.

Standard Performance Evaluation Corporation. http://www.spec.org. [All the SpecInt2000 and SpecFP2000 results shown in this chapter are available for download from the SPEC website.]

Yeh, Tse-Yu and Yale Patt. "Two-Level Adaptive Branch Prediction." *Proceedings of the 24th Annual International Symposium on Computer Microarchitecture*, Albuquerque, New Mexico: 1991, pp. 51–61.

Chapter

6

Logic Design

Overview

This chapter discusses the process of converting a microarchitectural design into a *hardware description language* (HDL) model. Validation of the logic design model is described as well as design automation flows, which allow HDL to be converted into layout. By hand logic minimization is demonstrated for combinational and sequential circuits.

Objectives

Upon completion of this chapter the reader will be able to:

1. Be familiar with the different levels of HDL abstraction.
2. Describe the trade-offs of different design automation flows.
3. Describe the goals and difficulties of pre-silicon validation.
4. Be familiar with the symbols and behavior of common logic gates.
5. Perform logic minimization of combinational or sequential circuits using Karnaugh maps (K-maps).

Introduction

Microarchitectural choices will have the greatest impact on the processor's performance and die area, but these choices are almost never black and white. Microarchitecture must trade off performance with die area and complexity. The result of the microarchitecture design is typically block diagrams showing the interactions of the processor's different components and a written specification describing their algorithms. Some simple simulations or hand calculations may be performed to estimate performance, but from this set of documents there is no easy way to determine whether the processor will actually produce the correct result for different programs. The task of logic design is to turn a microarchitectural specification into a logical model that can be tested for correctness.

Very early computer designs used few enough transistors in their components that circuit design could be performed directly from the design specification. The march of Moore's law has led to the use of vastly more transistors and more complexity, making it critical to simulate the correctness of the logic before attempting to create a transistor-level design. Indeed, modern microprocessors could not be successfully designed without microprocessors to simulate their designs.

These simulation models are created using HDL. These are computer programming languages specifically intended to simulate computer hardware. The processor design is divided into smaller and smaller logical blocks until the functionality of one block is simple enough to allow HDL for that block to be written easily by a single person. HDL code for these different blocks is then combined to create models of large pieces of the processor microarchitecture and eventually a model of the behavior of the entire processor. In addition, an HDL model of the processor's environment must be created to allow the processor model to be stimulated with the proper inputs and to check the processor outputs for correctness.

HDL models can be written to be relatively abstract or extremely detailed. Ultimately, a structural model must be created with sufficient detail to be implemented easily in transistors. Design automation tools allow high-level HDL code to be automatically converted into structural HDL models and eventually layout. The ability of logic synthesis tools to do this has been critical to keeping design times under control; all modern processors rely upon this automation for at least part of their design. However, because hand tuning provides performance benefits, the structural models of at least parts of many processors are still created by hand.

The logical description of a modern microprocessor can require a million of lines of HDL code, turning a hardware design problem into a significant task of software engineering. All the problems faced when creating any large piece of software must be addressed. When will interfaces between different parts of the design be defined? How will changes in one block impact the rest of the design? How will code revisions being turned in simultaneously by dozens of programmers be handled? What functionality will be available at different stages of the project?

Hardware simulations have the added problem of distinguishing between logical and simulation errors. If the HDL model of the processor produces the wrong result, the problem may be a logical bug in the design. If not corrected, this bug would cause the hardware to produce the same error. However, the problem could be with just the simulation of the processor. Perhaps the simulation does not accurately model what the behavior of the real hardware would be or is trying to simulate a set of conditions that could never happen in the real world. The job of finding, analyzing, and fixing these bugs is called *design verification* or *validation*; it requires the skills of not only a hardware designer but of a software designer as well.

Hardware Description Language

The complexity of modern microprocessors requires their logic design to be extensively tested before they are manufactured. **Hardware description language (HDL)** models are created to allow this simulation. Compared with going directly to circuit design, HDL modeling dramatically reduces design time and logic bugs. Specifying the logic at a higher level of abstraction is far easier for the designer, and simplicity reduces the chance of logic errors. Unlike a text specification the HDL model can be tested, allowing many logic bugs to be found and fixed before any later design steps are attempted. HDL languages are also designed to be independent of the manufacturing process, so that logic designs are moved easily from one manufacturing generation to the next.

The most commonly used forms of HDL are Verilog®and VHDL (Very High-Level HDL). HDL code is similar to general-purpose high-level programming languages like C and Fortran, but they include the concept of concurrency. Traditional programming languages assume that each program instruction will be executed serially one after the other. As we have seen in earlier discussions of microarchitecture, the processor works hard to maintain the illusion that this is how it really operates, but in reality there are typically many things happening simultaneously in hardware. Unlike other program languages, HDL allows logic designs to model hardware that performs multiple operations at the same time. To show the difference, Fig. 6-1 compares two pieces of code, one in C and one in Verilog.

The C code executes serially. The variable Z is calculated as the sum of X and Y and then printed. After the value of X has been changed, variable Z is printed again. In C, changing the value of variable X has no effect on Z. In the Verilog code, an "always" block is declared to be executed any time the variable X or Y change. The block sets Z to be the sum of X and Y. An "initial" block then executes instructions serially in the same way that C does.

In Verilog, changing the value of X causes the always block to be executed again, and the value of Z to be updated. This makes output show a different value for Z after only the variable X has been changed. The behavior of the always block in Verilog represents the way the combinational logic of an adder would behave in hardware. The logic always produces the sum of two values, and the output of the circuit will change anytime the inputs to the circuit change.

Figure 6-1

C Code	Verilog
X = 2;	always @(X or Y)
Y = 3;	Z = X + Y;
Z = X + Y;	initial begin
printf ("Z = %d\n", Z);	X = 2;
X = 7;	Y = 3;
printf ("Z = %d\n", Z);	$display ("Z = %d\n", Z);
Output	X = 7;
Z = 5	$display ("Z = %d\n", Z);
Z = 5	end
	Output
	Z = 5
	Z = 10

Figure 6-1 C vs. Verilog.

HDL code is easier to write and easier to read at higher the levels of abstraction. More abstraction also reduces simulation time, but a high-level simulation can miss subtle logic bugs and be more difficult to convert to a circuit implementation. HDL code can be written at three general levels of abstraction.

1. Behavior level
2. Register transfer level
3. Structural level

A behavior level model is the most abstract. It would include all the important events in the execution of the processor, but the exact timing of these events would not be specified. The **register transfer level (RTL)** simulates the processor clock and the specific events that happen at each cycle. An RTL model should be an accurate simulation of the state of the processor at each cycle boundary. The RTL level does not model the relative timing of different events within a cycle. The structural level shows the detailed logic gates to be used within each cycle. Figure 6-2 shows examples of behavioral, register transfer, and structural level Verilog.

All three of the examples in Fig. 6-2 are written in Verilog, and all define a counter that increments from 0 to 2 before returning to 0. The only difference between them is the level of abstraction. The behavioral code looks much like programs created with general-purpose programming languages. A "task" has been defined, which could be called from elsewhere each time the counter is to be incremented. There is no processor

Behavioral	Register transfer	Structural

integer Count;
 	
task IncCount;
begin
  if (Count == 2)
    Count = 0;
  else
    Count = Count + 1;
end
	reg [1:0] Count;
	
	always @(posedge Clock)
	begin
	  if (Count == 2)
	    Count = 0;
	  else
	    Count = Count + 1;
	end
		module MyCounter (Clock, Count);
		input Clock;
		reg [1:0] Count;
		output [1:0] Count;
		
		nor (NS0, Count[1], Count[0]);
		MSFF (Count[1], Count[0], Clock);
		MSFF (Count[0], NS0, Clock);
		
		endmodule;

Figure 6-2 Behavioral, register transfer, and structural level Verilog.

clock and the count itself is declared as a generic integer. The register transfer level uses an "always" block to specify that the count is incremented at the rising edge of each new clock. Also, the count is now defined as a 2-bit register. The structural model creates a counter module. Within this module other gates and modules defined elsewhere are instantiated to create the logic desired.

In reality, all of these coding styles could be mixed together. The first version of the HDL model might be written at the behavioral level with different sections being gradually changed to lower levels of abstraction as the details of the design are decided. Alternatively the different levels of abstraction could be maintained as separate independent models. This has the advantage of allowing fast simulations to verify logical correctness on the highest level of abstraction. The lower-level models only need to be shown to be logically equivalent to the next higher level of abstraction. The disadvantage of this approach is the added effort of maintaining multiple models and keeping them in synch. In either case, a structural representation must eventually be created before the actual chip is built. This representation can be created by hand, or logic synthesis tools can create it automatically from code at a higher level of abstraction.

Design automation

As processor designs have steadily increased in complexity, processor design teams have also grown. For high-performance designs, modern design teams may employ as many as a thousand people. To allow continued increases in complexity while preventing design teams from growing even further and product design times from become unacceptably long, modern engineers must rely upon design automation.

Logic synthesis is the process of converting from a relatively abstract HDL model of the desired behavior to a structural model that can be realized in hardware. In the past, this step was always performed by hand and was one of the most difficult and error-prone steps in chip design. Today many automated tools exist, which perform these steps automatically. Some tools allow behavioral models as inputs, but most require an RTL model. The three basic choices of automation for changing an RTL model into layout are shown in Fig. 6-3.

The highest level of design automation is **automated logic synthesis**. Synthesis tools automatically convert from RTL to layout. Processor blocks to be designed using synthesis do not require a circuit designer or mask designer. Logic designers write the RTL for the behavior they wish, and synthesis CAD tools automatically generate layout. A library of cells and cell layout must be created first, but once these cells are created, any logic designer can create layout directly from RTL.

Synthesis makes it easy to make changes in a design. All that is required is updating the RTL and rerunning the synthesis tool. Converting to a new manufacturing technology is also vastly simplified. A new cell library must be created, but once available the same RTL can be resynthesized to create a new implementation. The disadvantage of this level of automation is that for some types of circuits the result produced by automated synthesis will not be as small in area or as fast in frequency as a more manual design could be.

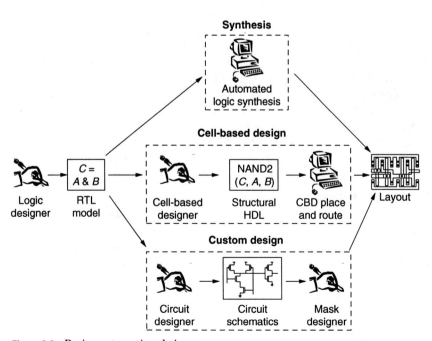

Figure 6-3 Design automation choices.

Cell-based design (CBD) flows make use of some automation but allow human designers more input. In these types of designs, structural RTL is created by hand using only cells from an already prepared library. This library may be the same as or separate from the synthesis library. The designer gains more control over the design by choosing the cells to be used rather than leaving it up to automation. This may allow the designer to create a smaller and faster implementation. Extra time and effort are required to generate structural RTL, but automated tools can still place and route together the cells from the library to automatically generate layout. This approach requires more effort than automated synthesis but still makes use of a reusable cell library.

Custom designs use the least amount of automation. A circuit designer creates by hand a transistor-level circuit schematic, which implements the desired logic, and a mask designer creates layout by hand to implement the desired circuit. The circuit schematic has the same basic information as structural RTL but is not limited to use preexisting cells. Instead, the circuit designer may create any combination of transistors needed, including those that cannot be modeled in RTL. This allows for the highest-performance circuit but also requires the most care to prevent failures due to human error in the design process. A mask designer working by hand can also create layout that combines the different gates of a block more compactly than an automated assembly of library cells that were created without the knowledge of how they would be used. However, this means the layout created will typically be used in this one block alone.

It is possible to design an entire processor using only one of these three design flows, but many modern processors are a combination of all three. Different portions of the die use different levels of automation as needed. Synthesis is best suited to complex "random" logic, like instruction decode, without regular patterns of which a human designer could take advantage. These are areas that human designers have the most difficultly with and in some cases the result of synthesis will be better than any design done by hand. Synthesis is also typically used where area or speed are not critical or there are likely to be last minute changes to the logic, since synthesis provides the fastest time from a logic change to new layout.

CBD might be used in functional units like an adder where there are regular patterns of which a human designer can take advantage. These functional units are also more likely to limit the processor frequency. Custom design is best used on memory arrays like caches. In these blocks, small circuits are repeated thousands or millions of times. Very small improvements in these circuits have a large impact on the overall chip, so these are areas worth optimizing by hand. Custom design may also be used to improve the speed of the slowest circuit paths. If these circuits limit the overall frequency of the processor, extra design time may be worth the effort.

TABLE 6-1   Steps between RTL and Layout

	Synthesis	Cell-based design (CBD)			Custom
	Full synthesis	Autoplaced cells	Autorouted cells	Cell-based custom	Full custom
Automatic logic minimization	Yes	No	No	No	No
Automatic placement	Yes	Yes	No	No	No
Automatic routing	Yes	Yes	Yes	No	No
Reusable cell library	Yes	Yes	Yes	Yes	No

In summary, synthesis provides the fastest design time but the least optimized result. By performing all the circuit and layout design by hand, custom design produces the most optimal result but requires the most time. CBD provides a middle ground, with better speed and area than synthesis in less time than full custom design.

Even within a single logic block, levels of automation may be mixed together. A designer may choose some cells by hand but allow synthesis tools to pick the rest. Each of the steps between RTL and layout can be automated or performed by hand (see Table 6-1).

To create layout from RTL, first the logic gates to be used must be chosen. This logic minimization is the heart of automated synthesis tools. The final layout also requires each of the cells to be placed relative to the others and wires routed to connect them. Cell-based design flows may perform these operations automatically or allow the user to control them. What synthesis and CBD have in common is working from a reusable cell library. As automation tools become more sophisticated, full custom design is becoming less common, but more use of synthesis and CBD requires larger and more sophisticated cell libraries. The individual cells that make up the library are typically created using custom design, so the work of custom design is not disappearing but instead shifting from creating large functional blocks to cell libraries in support of synthesis and CBD. In the end, automation will never completely replace custom circuit design and layout.

Pre-silicon validation

Manufacturing a new processor design is expensive. Masks must be created to transfer the layout of the different material layers onto the wafer. This cost does not change even if only a few test prototypes are needed. Manufacturing new chips also takes time. Changing from producing one revision of the design to the next takes weeks or months. This makes it critically important that the design be tested as thoroughly as possible before any chips are actually made. This is the job of **pre-silicon validation.**

Imagine you are the chief validation engineer for a processor project. The team has been working for years and is finally ready to begin producing the first test chips. An HDL model has been created and matching circuits and layout produced for the entire design. Now you are asked whether the entire project should move forward with the creation of the first actual silicon chips or continue to modify the design. The essential question is whether the current design is logically correct or not. If test chips are produced, will they actually work? The wrong answer to this question will cost the project time and money.

Pre-silicon validation attempts to answer this question by first checking the HDL model for correctness and then checking to make sure the implementation will match the behavior of the model. These two tasks are called **design verification** and **implementation verification**. Design verification relies on using the HDL model to simulate the behavior of the processor running different programs. This behavior is compared to the microarchitectural specification of what the behavior should be. Usually there are also compatibility checks where the "correct" behavior is simply what the previous generation processor did.

Without HDL models processors with today's complexity could never be validated and probably could never be made to function properly at all. However, HDL models have a serious limitation. They are very, very slow. The HDL model of a 3-GHz processor might simulate instructions at the rate of a physical processor running at 1 Hz.[1] This means that years of pre-silicon simulation are required to test the behavior of just a few minutes of actual operation. The design team could purchase more or faster computers to perform the simulations, but this would make only a tiny difference in the vast gap between software simulation and hardware. Perhaps with twice as many computers that are each twice as fast, the validation team could in years of simulation duplicate the behavior of half an hour of operation instead of just a few minutes. This still comes nowhere near testing all the software programs that might be considered important.

To improve the speed of simulation, some projects use hardware emulation. *Field programmable gate arrays* (FPGAs) contain logic and memory elements that can be programmed to provide a particular behavior. By programming FPGAs to imitate the behavior of an RTL or structural model, the behavior of the processor might be simulated in hardware at a few megahertz. This might be thousands of times slower than the actual processor will run but millions of times faster than pure software simulation. FPGAs were used extensively in the pre-silicon validation of AMD's K6 processor.[2]

---

[1]Bentley, "Validating the Intel Pentium 4," 245.

[2]Shriver and Smith, *Anatomy of a Microprocessor*, 28.

Whether hardware or software emulation is used, any simulation will be dramatically slower than the physical processor; the number of possible combinations of instructions and operands is far too vast to test completely even with full-speed processors. One hundred PCs running HDL simulations at 1 Hz for a year could simulate more than 10^9 cycles. One hundred PCs testing 1 GHz prototypes for a year could simulate more than 10^{18} cycles, but the total number of possible inputs operands to a single double-precision floating-point divide is more than 10^{38}. The infamous Pentium FDIV bug is an example of how difficult finding logic errors is. In 1994, after Intel had already shipped more than a million Pentium processors, it was discovered that the processor design contained a logic bug that caused the result of the floating-point divide instruction to be incorrect for some combinations of inputs. The chance of hitting one of the combinations that triggered the bug by applying random inputs was estimated at 1 in 9 billion.[3]

Testing instructions at random or even testing random pieces of real programs will never be able to thoroughly validate the design. The number of processor cycles that can be simulated is limited by the time required. The true art of design verification is not simulating lots of code, but choosing code that is likely to expose bugs in the HDL model. Although validation engineers do not perform design themselves, they must become intimately familiar with the details of the design in order to find its weak points. Ultimately these engineers may have a better understanding of the overall logical operation of the processor than anyone else on the design team.

To measure the progress of design verification, most validation teams create measures of test coverage. The number of different key microarchitectural states and events triggered by each test simulation are measured. New tests are written specifically to try out uncovered areas. Running the same types of tests over and over gives the illusion of a well-tested design without actually improving the chance of finding new bugs. Real validation progress is measured by tracking improvements in test coverage rather than simply the total number of cycles simulated.

Design verification tests only the behavior of the HDL model. The other half of pre-silicon validation is implementation verification, which makes sure that the processor's circuits and layout will faithfully reproduce the behavior of the HDL. In the past, test vectors were simulated on the HDL model and a netlist that described the circuit implementation, in order to perform implementation verification. The behavior of HDL and implementation were compared to make sure they were identical. The weakness in this approach is that behavior is proved to be the same only for the test vectors tried. There is always the chance that there are differences in behavior for other untested combinations of inputs.

---

[3]"Statistical Analysis of Floating Point Flaw."

This was how the Pentium FDIV bug escaped detection. It was actually caused by a flaw in the circuit design that was missed during implementation verification. The HDL model always yielded the correct result, but the implementation did not exactly match the behavior of the HDL. Behavior was identical almost all the time, but not always. To avoid these types of escapes, implementation verification is today more commonly done by formal equivalence checking. The HDL and circuit netlist are both converted to sets of boolean equations and these are proven to be logically equivalent. This requires far more sophisticated CAD tools but avoids the problem of having to choose test vectors by proving that the two models will give the same result for all possible combinations of inputs.

Logic bugs found early in the logic design process cost almost nothing to fix. The HDL is simply changed to eliminate the bug. Bugs found when the chip implementation is almost complete can become the critical path for completion of the design and delay the first prototypes. After prototypes have been produced the cost of bugs is still higher, possibly requiring steppings (revisions) of the design. The amount of coverage by pre-silicon validation is one of the critical factors in determining when to tapeout a chip design and manufacture the very first stepping. The first stepping of a new processor almost always has bugs. This is because pre-silicon validation can simulate only a limited number of programs and circuit simulations are only an approximation of real behavior. It is the job of post-silicon validation to test completed chips in order to find and fix any bugs not found by pre-silicon validation. Finding bugs during post-silicon validation is even more critical since finding significant flaws after shipping has begun could lead to disastrous recalls.

Even though some bugs are expected in any first stepping, if the problems are too serious, it may be difficult to make progress with post-silicon validation. On the other hand a design that has no logic bugs on very first stepping may be a sign that tapeout should have been scheduled sooner. The goal is to begin shipping as quickly as possible. The overall schedule may be hurt by spending too much time in pre-silicon validation trying to make the very first stepping perfect. The shortest time to market is achieved by scheduling tapeout as soon as the design is of high enough quality to allow effective post-silicon validation. Some bugs found before the first design tapeout may even be left to be fixed in later steppings if they will not pose a significant barrier to post-silicon work.

As processors have grown in complexity, pre-silicon validation has become a more important factor in determining the overall required design time. The number of logic bugs to be found and fixed has tended to grow linearly with the number of lines of HDL code,[4] and finding these bugs is not getting any easier. Improved design automation tools allow late changes to the HDL to be converted easily to layout, but these last minute changes still

---

[4]Bentley, "High Level Validation Microprocessors."

require validation. The date when HDL must be frozen can be easily limited not by the time to implementation, but by the loss of design verification coverage and the time to retest each HDL change.

To make design verification practical for future processors, it seems likely more abstract models of processor behavior will be required. These could be behavioral HDL models or new *architectural description languages* (ADL). These models would try to describe the microarchitecture at a very high level but still allow meaningful design verification. Traditional RTL models will probably still be necessary to bridge the gap between an ADL model and the implementation, since it is unlikely that implementation verification tools would be able to compare a behavioral model directly to layout. There would clearly be significant added effort in maintaining separate ADL and RTL models throughout the life of a project, but the ADL model could run quickly in simulation and be frozen early in the design. Late changes to the RTL would be validated by comparing behavior to the ADL model. This makes implementation verification a two-tier process. First RTL behavior would be compared to the ADL model, and then layout behavior would be compared to RTL. The steady growth of processor design complexity and pre-silicon validation effort make more efficient validation methods one of the most important needs of future processor designs.

Logic Minimization

When performed by hand the process of converting from an RTL model to a structural model is often called *logic minimization* rather than *logic synthesis*. Synthesis tools use libraries of cells to convert an RTL model all the way to layout. This is fast and may provide the best solution for some cases, but sometimes a better result can be reached by hand. Even when automation is to be used, it is important to understand the mechanisms behind automated synthesis. Different RTL models with the same behavior can yield very different results when run through automated tools. It helps in understanding this to be familiar with the steps of by hand logic minimization.

The next sections make use of logic gates without discussing their circuit implementations. Chapter 7 will show how these different logic gates are realized with actual transistors, and the real world considerations of delay, noise, and power that come with real electrical circuits. In this chapter, we consider only the logical behavior of idealized gates.

Combinational logic

The simplest types of logic are combinational logic circuits that contain no memory. These circuits do not preserve any internal state, so the outputs are functions of only the inputs. An adder is a good example of combinational logic since the sum depends only on the two numbers being added, not on what add instructions might have been performed earlier.

Combinational logic with any behavior can be built up out of simple logic gates, the most common of which are **NOT, AND, OR**, and **XOR** gates (see Fig. 6-4).

The output of a NOT gate (also called an inverter) is true only if its input is false. This gate inverts the value of its input. The output of an AND gate is true only if all its inputs are true. The output of an OR gate is true if any of its inputs is true. The output of *exclusive-OR* (XOR) gate is true if an odd number of inputs are true. The AND, OR, and XOR gates each have complementary versions (**NAND, NOR**, and **XNOR**) that are the logical opposites of the original functions. They are equivalent to adding an inverter to the output of each gate.

We can imagine how these gates might be used to monitor a keyboard. An AND gate might trigger a capital letter if a shift key and a letter are pressed simultaneously. Because there are two shift keys, an OR gate might create a shift signal that is true if the left shift key or the right shift key is pressed. Your mouse might use an XOR to trigger some action if the left button or right button is pressed, but not both simultaneously. Another AND gate might cause a different action when the left and right mouse buttons are pressed together. Computer keyboards and mice are in reality much more complicated than this, but the idea of building behaviors out of logic gates is the same.

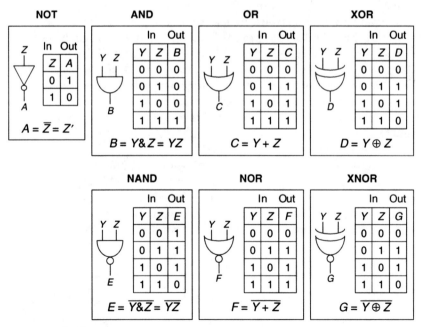

Figure 6-4 Logic gates.

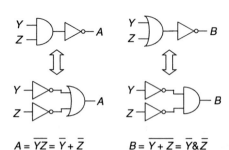

$A = \overline{YZ} = \overline{Y} + \overline{Z}$ $B = \overline{Y+Z} = \overline{Y}\&\overline{Z}$

Figure 6-5 DeMorgan's theorem.

All the two-input functions shown in Fig. 6-4 can be defined for any number of inputs. It is possible to perform logic design using only some of the types of gates, but it is often far more intuitive to use all the types. In fact, **DeMorgan's theorem** shows that using inverters, AND logic can be changed to OR logic or vice versa (Fig. 6-5).

DeMorgan's theorem points out that inverting any function is accomplished by inverting all the terms of the function while changing ANDs to ORs and ORs to ANDs. This means that any function could be created using only NAND gates or NOR gates. While performing logic design it is typically easiest to use more types than this, but during circuit design DeMorgan's theorem is often used to pick the most convenient type of gate. The equation given in Fig. 6-6 shows the effect of repeated application of DeMorgan's theorem.

Implementing any complex logical function begins with writing out the truth table, the table listing the desired outputs for each possible combination of inputs. The goal of logic design is to find the logic gates and connections needed to implement the function. ANDing together different combinations of the inputs and their complements can create functions that are true for only one possible combination of inputs. These are called *fundamental products*. By ORing together the needed fundamental products, any desired function can be realized. A logical function written in this fashion is called a *canonical sum*.

Implementing a logical function as a canonical sum is very straightforward but may also be very wasteful. It is often possible to simplify the canonical sum and implement the same function using fewer and simpler logic gates. The canonical sum of the example function in Fig. 6-7 would require 4 three-input gates and 1 four-input gate to implement.

$$\overline{V(WX+YZ)} = \overline{V} + \overline{(WX+YZ)} = \overline{V} + \overline{WX}\,\&\,\overline{YZ} = \overline{V} + (\overline{W}+\overline{X})(\overline{Y}+\overline{Z})$$

Figure 6-6 Applying DeMorgan's theorem.

X	Y	Z	Fundamental Product
0	0	0	$\bar{X}\bar{Y}\bar{Z}$
0	0	1	$\bar{X}\bar{Y}Z$
0	1	0	$\bar{X}Y\bar{Z}$
0	1	1	$\bar{X}YZ$
1	0	0	$X\bar{Y}\bar{Z}$
1	0	1	$X\bar{Y}Z$
1	1	0	$XY\bar{Z}$
1	1	1	XYZ

$A = \bar{X}Y\bar{Z} + \bar{X}YZ + X\bar{Y}\bar{Z} + X\bar{Y}Z$

X	Y	Z	A
0	0	0	1
0	0	1	0
0	1	0	1
0	1	1	0
1	0	0	1
1	0	1	1
1	1	0	0
1	1	1	0

Figure 6-7 Canonical sum.

Boolean algebra shows that this function could actually be implemented with only 3 two-input gates. (See Fig. 6-8).

The simplest sum-of-products form of a function is called the **minimal sum**. This is the form with the fewest and simplest terms. One of the most important tasks of logic design is performing logic minimization to find the minimal sum. Unfortunately this is not always straightforward. Applying boolean algebra to try and simplify an equation can lead to different results, depending on in what order transformations are applied. In this fashion, there is not a clear of way determining whether a new form of the function is really the simplest possible. The most common way of finding minimal sums by hand is using Karnaugh maps.

Karnaugh maps

Karnaugh maps are a graphical way of minimizing logic by allowing us to easily see terms that can be combined to simplify the function.[5] The maps work by using the concept of "neighboring" fundamental products. Two fundamental products are considered neighbors if they differ by only one input. Any neighboring fundamental products can be combined into a single term. See Fig. 6-9.

Just looking at equations or a truth table, it is difficult to determine which terms are neighbors. Karnaugh maps are tables drawn in such a way that the results of logical neighbors are placed physically next to each other.

$A = \bar{X}Y\bar{Z} + \bar{X}YZ + X\bar{Y}\bar{Z} + X\bar{Y}Z$

$A = \bar{X}Z(\bar{Y} + Y) + X\bar{Y}(\bar{Z} + Z)$

$A = \bar{X}Z(1) + X\bar{Y}(1)$

$A = \bar{X}Z + X\bar{Y}$

Figure 6-8 Logic minimization.

---

[5]Karnaugh, "Map Method for Synthesis."

$\overline{X}\overline{Y}\overline{Z} + \overline{X}Y\overline{Z}$ Neighboring products
⇩   Differ only by Y input
$\overline{X}\overline{Z}$ Can be combined

           Not neighboring products
$\overline{X}\overline{Y}\overline{Z} + XY\overline{Z}$ Differ by X and Y inputs
           Cannot be combined

Figure 6-9 Neighboring products.

In Karnaugh maps, the cells to the left, right, above, and below any cell are holding the function's value for neighboring fundamental products. Cells that are diagonally separated are not neighbors. For example in the two-variable map (Fig. 6-10), fundamental product F_0 is a neighbor of the products of F_1 and F_2, but not F_3. Being able to graphically determine logical neighbors allows the minimal sum to be found more easily than repeated equation transformations.

To find the minimal sum, first the Karnaugh map is filled in with the function's values from the truth table. Then all the 1's in the Karnaugh map must be circled as part of a set of 1's. All the 1's in a set must be neighbors and the number of 1's in a set must be a power of 2. The goal is to include all the 1's using the smallest number of sets and the largest sets possible. Each set corresponds to a term in the minimal sum, the one created by combining all the neighboring products in the set. This term is the one that includes only the input values, which are the same for all the members of the set. By examining each set, the terms of the minimal sum can be written.

In the first example in Fig. 6-11, the two 1's in the map can be included in a single set of two cells. The value of input Y is different for the cells in the set, which means the term corresponding to this set does not include Y. The value of input Z is 0 for all the cells in the set, so the term corresponding to this set will include $\overline{Z}$. Because this is the only set needed and Y and Z are the only input variables, the complete minimal sum is $\overline{Z}$.

In the second example in Fig. 6-11, the three 1's in the map require two sets of two cells. A set of three is not allowed (since three is not a power of 2). The three 1's could be covered by one set of two cells and

Y	Z	F
0	0	F_0
0	1	F_1
1	0	F_2
1	1	F_3

            Y
           0   1
Z   0 | F_0 | F_2 |
    1 | F_1 | F_3 |

Figure 6-10 Two-variable Karnaugh map.

Y	Z	A
0	0	1
0	1	)
1	0	
1	1	

$A = \overline{Z}$

Y	Z	B
0	0	1
0	1	1
1	0	
1	1	

$B = \overline{Y} + Z$

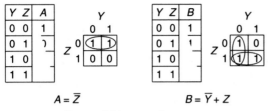

Figure 6-11 Two-variable examples.

one set of one cell, but this would not give the simplest sum. To find the minimal sum with the fewest and simplest terms, we should use the fewest number of sets and the largest sets possible. For the vertical set, the input Y is 0 for both cells, meaning this set corresponds to the term $\overline{Y}$. For the horizontal set, the input Z is 1 for both cells giving the term Z.

For two input functions, it might be possible to write the minimal sum by inspection. Karnaugh maps are valuable when minimizing functions of three variables or more.

In a three-variable Karnaugh map, two cells in the same row and in the leftmost and rightmost columns are considered neighbors (Fig. 6-12). For example, F_0 is a neighbor of F_4 and F_1 is a neighbor of F_5. Functions are minimized as before by including all the 1's in the fewest number of largest possible sets.

In the first example in Fig. 6-13, two sets of two cells are sufficient to cover all the 1's. Picking these two sets allows us to quickly find the same minimal function that was found by boolean transformations in Fig. 6-8. In the second example, one set of two cells is chosen including cells F_0 and F_4. The final 1 in the map has no neighboring cells that are 1's, so it must be included in a set by itself.

In a four-variable Karnaugh maps, just like three-variable maps, two cells in the same row and in the leftmost and rightmost columns are considered neighbors. In addition, two cells in the same column and in the

X	Y	Z	F
0	0	0	F_0
0	0	1	F_1
0	1	0	F_2
0	1	1	F_3
1	0	0	F_4
1	0	1	F_5
1	1	0	F_6
1	1	1	F_7

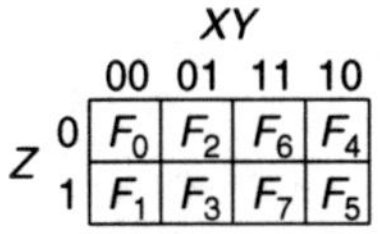

Figure 6-12 Three-variable Karnaugh map.

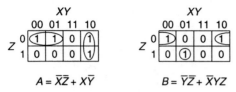

Figure 6-13 Three-variable examples.

top and bottom rows are considered neighbors (Fig. 6-14). For example, F_0 is a neighbor of F_8 and F_4 is a neighbor of F_6.

In the first example in Fig. 6-15, all the 1's are covered by one set of two cells and two sets of four cells. Sets of eight cells are possible by combining two neighboring sets of four. The second example uses two sets of four. One set includes all the center cells, whereas the other contains all four corners. Because cells at the edges of the map are considered neighbors of cells on the opposite side, it is possible to create a set of four out of all the corner cells.

Karnaugh maps can be used with five variable functions by creating 2 four-variable maps, which differ by just one input. Cells in the exact same position in the two maps are considered neighbors. For six variable functions, 4 four-variable maps are needed. Arranged in two rows and columns, cells are neighbors with cells in the same position in the

W	X	Y	Z	F
0	0	0	0	F_0
0	0	0	1	F_1
0	0	1	0	F_2
0	0	1	1	F_3
0	1	0	0	F_4
0	1	0	1	F_5
0	1	1	0	F_6
0	1	1	1	F_7
1	0	0	0	F_8
1	0	0	1	F_9
1	0	1	0	F_{10}
1	0	1	1	F_{11}
1	1	0	0	F_{12}
1	1	0	1	F_{13}
1	1	1	0	F_{14}
1	1	1	1	F_{15}

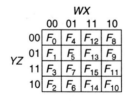

Figure 6-14 Four-variable Karnaugh map.

Logic Design

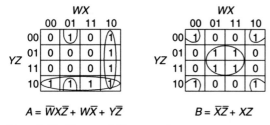

Figure 6-15 Four-variable example.

maps in the same row or column. For functions of more than six variables, tabular methods are required. In summary, the general steps in reducing a logic equation by hand are:

1. Write truth table.
2. Fill in Karnaugh map.
3. Cover all 1's with fewest number of largest sets.
4. Write minimal sum.

There may not be a unique minimal sum. Some functions have more than one solution with the same number of terms and inputs per term.

A good example of the process is deriving the logic for a binary add. Each digit of a binary full adder must input three binary values: the two numbers being added and a carry bit. The output will be a sum bit for that digit and a carry out to go to the next digit. For binary addition, the sum bit will be true if one or three of the inputs is true. There will be a carry out if two or more of the inputs are true. Figure 6-16 shows the truth table, Karnaugh maps, and minimal sums for a 1-bit full adder.

A	B	C	C_{out}	Sum
0	0	0	0	0
0	0	1	0	1
0	1	0	0	1
0	1	1	1	0
1	0	0	0	1
1	0	1	1	0
1	1	0	1	0
1	1	1	1	1

$$C_{out} = AB + BC + AC$$

$$Sum = \overline{A}\overline{B}C + \overline{A}B\overline{C} + ABC + A\overline{B}\overline{C}$$

Figure 6-16 1-bit full adder logic.

Sum = $\bar{A}\bar{B}C + \bar{A}B\bar{C} + ABC + A\bar{B}\bar{C}$

Sum = $\bar{A}(B \oplus C) + A\overline{(B \oplus C)}$

Sum = $A \oplus B \oplus C$

Figure 6-17 Simplifying with XORs.

Unfortunately no real simplification was possible for the sum output. Each of the four fundamental products has no neighbors, which means that Karnaugh maps cannot help to reduce this function. A tactic that helps in these cases is looking for XOR functions in the minimal sum. In general, XORs tend to produce fundamental products with few or no neighbors. Functions that cannot be reduced easily by Karnaugh maps are often simplified using XORs. This process is shown for the sum function in Fig. 6-17.

Using the minimal sum for the C_{out} from Fig. 6-16 and the expression for Sum from Fig. 6-17, we can now draw the logic gate representation of a full adder. Figure 6-18 shows three full adders connected to form a 3-bit binary adder.

This type of adder is called a *ripple carry adder* because the carry output of one full adder becomes an input of the next full adder. By connecting full adders in this fashion we can create an adder for any number of bits. Ripple carry adders are typically used only for small adders because the delay of this adder increases linearly with the number of bits but for adding small numbers this simple implementation may be the best choice.

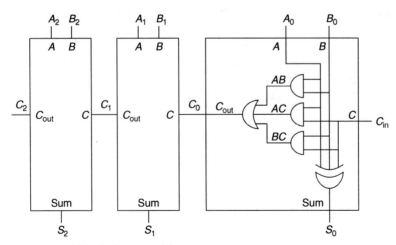

Figure 6-18 3-bit ripple carry adder.

Most importantly this example shows how an abstract concept such as addition can be written down as a truth table and ultimately implemented using logic gates.

Many automated tools are available today to perform logic minimization; it is increasingly rare to see engineers using Karnaugh maps to minimize functions by hand. However, for some simple problems this may be the easiest solution; for more complex problems where automation is used, it is important to understand the mechanism behind logic minimization.

Sequential logic

Logic gates allow microprocessors to perform computations on data, but to run programs a processor must also store data. Sequential logic circuits include memory elements, allowing them to retain knowledge of what has happened before. In combinational logic circuits, the outputs are functions only of the inputs. In sequential logic circuits, the outputs are also functions of the current memory state stored by the circuit. Combinational logic is sufficient to add numbers, but it could not, for example, control an elevator. Whether an elevator goes up or down depends not just on what button is currently being pressed, but also on which buttons were pressed in the past. Pressing the button for a lower floor will cause a different behavior in an idle elevator than one that is currently going up. Sequential logic can create control circuits that react differently to the same inputs depending upon the state of the machine. They can also implement memory arrays that store values to be read later. In fact, an entire microprocessor can be thought of as a single very large sequential logic circuit.

A common element for storing data within sequential logic is a **flip-flop**. A flip-flop is a memory element that captures the data at its input at the rise of a clock signal and holds it until the next rise of clock. Any changes in the data input between rising edges of the clock signal are ignored.

Figure 6-19 shows the voltage waveforms of a flip-flop's inputs and outputs. At the first rise of clock, the input is high, so the flip-flop output changes to high. Transitions on the inputs before the second rise of clock are ignored. Because the input is high at the second rise of clock, the flip-flop output does not change. The input is low for the third rise of clock, which causes a change in the output. How flip-flops are implemented with transistors is discussed in Chap. 7.

Pipelined processors are filled with combinational logic that must perform the same action for each instruction as it reaches that stage of the pipeline. These circuits are commonly separated by flip-flops. The flip-flops capture the inputs to the logic at the beginning of each processor

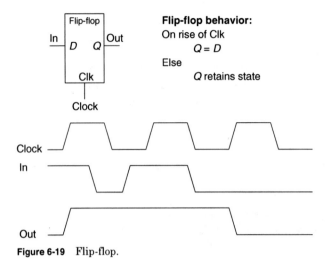

Figure 6-19 Flip-flop.

clock cycle and hold them until the next instruction's data is ready to enter this pipestage. In effect, the flip-flops act like stoplights controlling the flow of data from one pipestage to the next. See Fig. 6-20. At the beginning of each cycle, all the flip-flops "turn green," beginning a new cycle of logic. Immediately afterward, they turn red to prevent the arrival of any new data from disrupting the current computation. At the beginning of the next cycle, the process begins again.

Designing sequential logic begins with deciding how many unique states the circuit has. This will determine the number of flip-flops or other memory elements required. For example, imagine a circuit which counts up from 0 to 2, increasing by 1 each cycle, before starting over at 0. This counter would have three states and therefore require two flip-flops to implement. To choose the logic gates needed, we write a truth table with the present state of the flip-flops as inputs and their next state as outputs. We then follow the same steps used to minimize purely combinational logic.

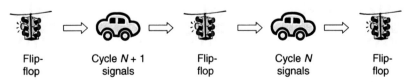

Figure 6-20 Flip-flops separate pipestages.

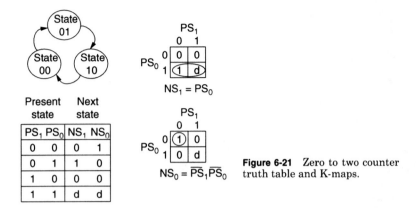

Figure 6-21 Zero to two counter truth table and K-maps.

There are four possible states for the values of two flip-flops, but in this case we have only made use of three of them. Because the state of "11" is never reached, it doesn't matter what the next state would be after that one. In the truth table in Fig. 6-21, these values are marked not as 0's or 1's but with a "d" standing for "don't care." When minimizing logic, "don't cares" are treated as 1's or 0's, whichever will lead to the most convenient result. When minimizing the equation for NS_1, the "don't care" is treated as a 1. The equation for NS_0 is simpler, when treating its "don't care" as a 0. Based on these equations for the next state based on the value of the present state, an implementation of a zero to two counter is shown in Fig. 6-22.

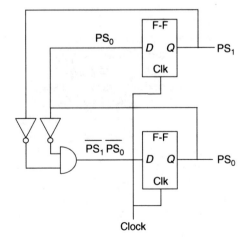

Figure 6-22 Zero to two counter implementation.

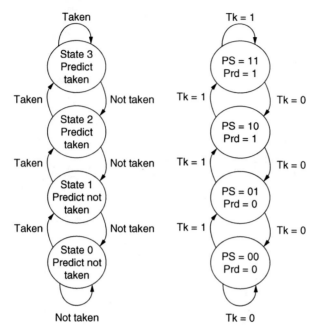

Figure 6-23 Branch prediction state diagram.

Sometimes the important result of sequential logic is its current state, but often in addition to the next state a sequential logic circuit must create an output that depends upon the present state. For example, in implementing a circuit to perform branch prediction, the important result is not the past behavior of a branch, but what the current prediction will be.

Chapter 5 discussed a branch prediction algorithm in which a state value stored in 2 bits was incremented each time a branch was taken and decremented each time it was not taken. The branch is to be predicted taken for state values of 2 or 3 and predicted not taken for state values of 0 or 1. Figure 6-23 shows the original state diagram describing the prediction scheme as well as a new state diagram where specific bit values have been chosen for each of the states as well as inputs and outputs. The input "Tk" is set to 1 if the branch was taken, and the output "Prd" is set to 1 if the branch should be predicted taken.

The state diagram in Fig. 6-23 contains eight arrows indicating transitions to a new state. Each one of these is represented by a row of a truth table, as shown in Fig. 6-24. The first row of the truth table shows that when the present state is "00" and the *taken signal* (Tk) is a 0, the next state will also be "00." The second row shows if the taken signal is

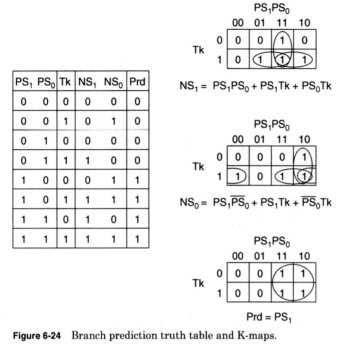

PS₁	PS₀	Tk	NS₁	NS₀	Prd
0	0	0	0	0	0
0	0	1	0	1	0
0	1	0	0	0	0
0	1	1	1	0	0
1	0	0	0	1	1
1	0	1	1	1	1
1	1	0	1	0	1
1	1	1	1	1	1

Figure 6-24 Branch prediction truth table and K-maps.

a 1, the state will change from "00" to "01." The next state values for each are derived from the state diagram, and the prediction signal (Prd) is set to a 1 when the present state is either "10" or "11." From the truth table, minimal sum equations are found, as also shown in Fig. 6-24. In the final implementation, shown in Fig. 6-25, a "branch" signal is assumed to be a 1 when a branch has occurred and the "Tk" signal is valid. This is signal is ANDed with the clock, so that the state will be updated only after a branch has occurred. In summary, the steps for creating a sequential circuit are:

1. Draw state diagram.
2. Assign values to states, inputs, and outputs.
3. Write truth table with present state as inputs and next state as outputs.
4. Fill in Karnaugh maps.
5. Cover all 1's with fewest number of largest sets.
6. Write minimal sums.

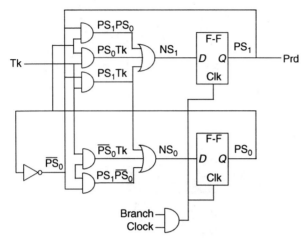

Figure 6-25 Branch prediction implementation.

Conclusion

At its heart, logic design is this process of taking an idea for a microarchitectural feature and changing it into an implementation of logic gates ready for the next step of circuit design. Logic design starts with a microarchitectural specification, which may be no more than a text document. It transforms this idea for a processor into HDL models that can be validated and implemented. Pre-silicon validation is critical to make sure operation of the first prototype chips is as close to the original specification as possible, but inevitably some bugs are found only in post-silicon validation. Making pre-silicon validation faster and more thorough is one of the most important needs of future processor designs.

Automation tools allow logic design to be the last step before manufacturing for some designs or at least parts of some designs. These automated flows allow design time to be reduced at some sacrifice in area and speed. It is possible to support many different levels of automation to allow the appropriate level of hand control to be chosen separately for each part of the design. When optimizing combinational logic by hand, Karnaugh maps are commonly used, and sequential circuits are designed in similar fashion by treating the current state of the circuit as just another set of inputs.

Logic design does not consider all the details of the real transistors that will be used to make the logic circuit, but in converting a specification into a model that can be simulated and tested for correctness, logic design makes the critical leap from idea to implementation.

Logic Design

Key Concepts and Terms

AND, OR, XOR	Hardware description language (HDL)
Automated logic synthesis	Implementation verification
Cell-based design (CBD)	Karnaugh maps
Custom design	Minimal sum
DeMorgan's theorem	NAND, NOR, XNOR
Design verification	Pre-silicon validation
Flip-flop	Register transfer level (RTL)

Review Questions

1. Describe the differences between HDL and traditional programming languages.
2. What is the primary purpose for creating an HDL model?
3. What are the differences between different HDL levels of abstraction?
4. Describe the two tasks of pre-silicon validation?
5. Why do most processor prototypes have logic bugs?
6. Describe how the impact of logic bugs changes during the course of a design project?
7. Describe the three different flows from RTL to layout and their trade-offs.
8. Find the truth table, K-map, and minimal sum for the equation:

$$A = \overline{X}\,\overline{Y}\,\overline{Z} + \overline{X}\,\overline{Y}Z + X\overline{Y}Z$$

9. [Bonus] Find the truth table, K-map, and minimal sums for a controller of a stop light, which on each clock cycle changes from green to yellow, yellow to red, or red to green if the "GO" signal is false. If the GO signal is true, the light changes as normal unless it is green. If the GO signal is true and the light is green, the light stays green.
10. [Discussion] When implementing each of the different microarchitectural features described in Chap. 5, which type of design automation would be most appropriate and why? How could going from a microprocessor concept to layout be further automated?

Bibliography

Bentley, Bob. "High Level Validation of Next-Generation Microprocessors." *7th IEEE International High-Level Design Validation and Test Workshop*, Cannes, France: October 2002, pp. 31–35.

Bentley, Bob. "Validating the Intel Pentium 4 Microprocessor." *Design Automation Conference*, Las Vegas, NV: June 2001, pp. 244–248.

Johnson, Everett and Mohammad Karim. *Digital Design: A Pragmatic Approach*. Prindle, Boston, MA: Weber & Schmidt Publishing, 1987. [Covers much of the same material as McCluskey's *Logic Design Principles*. Not quite as detailed but a somewhat easier to read.]

Karnaugh, Maurice. "The Map Method for Synthesis of Combinational Logic Circuits." *Trans. AIEE*. vol. 72, 1953, p. 593. [The original paper describing K-maps. When this paper was written, the first silicon transistor had not yet been created. Although technology drives so many aspects of processor design, some things are timeless.]

McCluskey, Edward. *Logic Design Principles with Emphasis on Testable SemiCustom Circuits*. Englewood Cliffs, NJ: Prentice-Hall, 1986. [McCluskey is a giant in logic design. His book is not an easy read, but it covers with enormous precision all the details of logic minimization for any type of circuit.]

Palnitkar, Samir. *Verilog HDL: A Guide to Digital Design and Synthesis*. Mountain View, CA: SunSoft Press, 1996. [Good programmer's reference for Verilog along with useful tips on coding for automated synthesis.]

Petzold, Charles. *Code: The Hidden Language of Computer Hardware and Software*. Redmond, WA: Microsoft Press, 1999. [A very gentle introduction to the concepts of encoding numbers (as binary), logic design, and encoding instructions to write software. Does a good job of assuming no previous technical knowledge and getting all the way from using a flashlight to send Morse code up to high-level programming languages.]

Shriver, Bruce and Bennett Smith. *The Anatomy of a High-Performance Microprocessor: A Systems Perspective*. Los Alamitos, CA: IEEE Computer Society Press, 1998. [A detailed description of the microarchitecture of the AMD K6 and its interaction with the system.]

"Statistical Analysis of Floating Point Flaw in the Pentium Processor." Intel Corporation, 1994, http://support.intel.com/support/processors/pentium/fdiv/wp/. [A paper by Intel, giving details on the cause and likelihood of encountering the infamous FDIV "flaw."]

Sternheim, Eliezer, Rajvir Singh, and Yatin Trivedi. *Digital Design with Verilog HDL*. Cupertino, CA: Automata Publishing, 1990.

Chapter 7
Circuit Design

Overview

This chapter covers how logic design is converted to a transistor implementation. The chapter describes the behavior of MOSFETs and how to use them to create logic gates and sequentials as well as the electrical checks of circuit timing, noise, and power.

Objectives

Upon completion of this chapter the reader will be able to:

1. Describe the regions of operation of a MOSFET.
2. Describe the differences between NMOS and PMOS devices.
3. Draw circuit diagrams for CMOS logic gates.
4. Understand fanout and how it affects circuit timing.
5. Understand P-to-N ratio and how it affects noise immunity.
6. Draw circuit diagrams for a latch and a flip-flop and understand their operation.
7. Describe different sources of circuit noise and ways of reducing them.
8. Describe different types of power consumption and ways of reducing them.

Introduction

Circuit design is the task of creating electrical circuits that implement the functionality determined by logic design. Logic design and the steps before largely ignore the details of the behavior of the transistors that will ultimately make up the microprocessor, but in the end the processor is primarily just a collection of transistors connected by wires. No matter how sophisticated or complicated the microarchitecture or logic design of the processor, its functions must be built using transistors and wires.

Although *bipolar junction transistors* (BJTs) were invented first, today all microprocessors use *metal oxide semiconductor field effect transistors* (**MOSFETs**). Modern microprocessors contain many large arrays of memory elements and MOSFETs are ideal for memory storage as well as performing logic. In addition, MOSFET logic can operate at lower voltages than required by BJTs. As reduced device sizes have required lower voltages, this has become extremely important and MOSFETs have come to dominate digital hardware design.

To make the job of designing a processor with hundreds of millions of transistors manageable, circuit design must be broken down into very small elements. This chapter describes how MOSFETs are used to build logic gates and sequential elements, which are connected together to form the desired logic. For logic design, transistors can be thought of as simple switches that are either on and conducting or off and not conducting. However, circuit design must consider not just getting the correct logical answer but also the trade-offs of the circuit implementation. Different circuits provide the same answer but with very different speed, cost, or power. Nonideal behavior of transistors must be considered as well. There will be electrical noise on signal wires, leakage currents from off transistors, and constraints on not only the maximum delay (**maxdelay**) allowed but also the minimum delay (**mindelay**). The circuit designer must take all these factors into account while still delivering a logically correct implementation. This means circuit designers must understand the details of not only the behavior of the logic they are to implement, but also the behavior of the transistors themselves.

MOSFET Behavior

As described in Chap. 1, each MOSFET has three terminals called the **gate, source**, and **drain**. When the transistor is on, current flows from the source to the drain, and when the transistor is off, no current flows. MOSFETs are made in two basic types using N-type and P-type dopants. **NMOS** transistors use N-type dopant for the source and drain separated by P-type. This creates a MOSFET that will turn on when the gate voltage is high. The high voltage draws the negative carriers from the source and drain under the gate. **PMOS** transistors use P-type dopant for the source and drain separated by N-type. This creates a MOSFET that will turn on when the gate voltage is low. The low voltage draws the positive carriers from the source and drain under the gate. See Fig. 7-1.

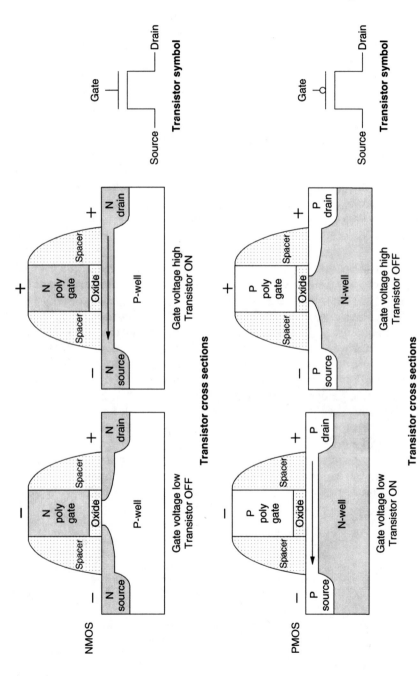

Figure 7-1 NMOS and PMOS transistors.

In the 1970s, processors were made using either just NMOS or just PMOS devices, but starting in the 1980s chips were built using both. A manufacturing process that creates N-type and P-type MOSFETs is called *complementary MOS* (**CMOS**). Having NMOS and PMOS devices allows logic gates to be designed where whenever one device turns on another turns off. This way the output of the gate is always being pulled to the supply voltage or ground but never both. These two voltages are used to represent the 1's and 0's of binary numbers. Any instruction or piece of data can be encoded as a number, and any number can be represented as a binary number using only two voltages.

The main reason why computers use binary is to keep circuit design simple. It would be possible to design a processor that used more than two voltage levels to represent each digit of a number. Ten voltage levels per digit would allow numbers to be stored in decimal notation as we are used to writing them. This would allow a smaller number of digits to be used to represent each number, but the circuits processing those digits would become enormously more complicated. Distinguishing between 10 input voltage levels and producing the correct value of 10 possible output voltages would require treating the transistors as analog devices. Many more transistors would be required to process each digit, and these circuits would become much more susceptible to electrical noise. Even though binary circuits require using more digits, they perform computations faster at lower cost and vastly simplify the job of circuit design.

Binary circuits let circuit designers think of transistors as being either only on or off when designing logic gates, but to estimate the speed or power of these gates we must be familiar with how the transistor's current varies with voltage. A MOSFET can be in three regions of operation depending on the differences in voltage at the three terminals. These voltage differences are written, as shown in Fig. 7-2.

The **threshold voltage** (V_t) at which the transistor will turn on is determined by the gate oxide thickness and amount of dopant in the well between the source and the drain. For an NMOS, if the gate voltage is

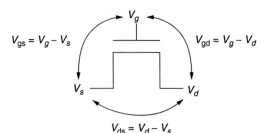

Figure 7-2 MOSFET voltage differences.

TABLE 7-1 MOSFET Current Equations

Region	NMOS	PMOS	Current equation
Cutoff	$V_{gs} < V_t$ and $V_{gd} < V_t$	$V_{gs} > V_t$ and $V_{gd} > V_t$	$I \approx 0$
Saturation	$V_{gs} > V_t$ and $V_{gd} < V_t$	$V_{gs} < V_t$ and $V_{gd} > V_t$	$I \approx \frac{1}{2}\beta(V_{gs} - V_t)^2$
Linear	$V_{gs} > V_t$ and $V_{gd} > V_t$	$V_{gs} < V_t$ and $V_{gd} < V_t$	$I \approx \beta\left(V_{gs} - V_t - \frac{V_{ds}}{2}\right)V_{ds}$

not a threshold voltage above either the source or the drain, the transistor will be off. This is called the **cutoff** region, and there will be almost no current flow. If the gate voltage is a threshold voltage above the source but not the drain, the transistor is said to be in **saturation**. In this case, the channel formed beneath the gate will not reach all the way to the drain. Current will flow but how much will be determined only by the voltages at the source and gate. If the gate voltage is a threshold voltage above both the source and drain, the transistor is in the **linear** region. The channel will reach all the way from source to drain, and the current flowing will be a function of the voltages on all three terminals. PMOS devices function in the same way but their threshold voltages are negative and the gate voltage must be a threshold voltage below the source or drain to form a channel (Table 7-1).

The transistor dimensions and whether it is N-type or P-type determine the β value of the transistor. Figure 7-3 shows the four important dimensions for every transistor.

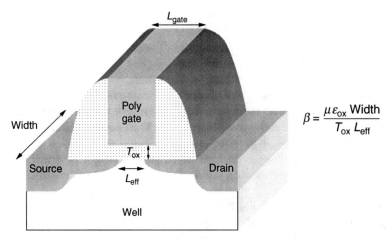

$$\beta = \frac{\mu \varepsilon_{ox} \text{ Width}}{T_{ox} L_{eff}}$$

Figure 7-3 Transistor β.

The length of the transistor is the spacing between the source and drain. This can be specified as L_{gate}, the actual width of the poly gate, or as L_{eff}, the effective separation between the source and drain dopants under the gate. L_{gate} is a physical parameter that can be measured with an electron microscope, but it is L_{eff} that will actually determine how much current the transistor produces. Because the dopants of the source and drain diffuse through the silicon underneath the gate, L_{eff} is always smaller than L_{gate}. When we speak of Moore's law driving the semiconductor industry forward by making transistors smaller, we are primarily taking about reducing the minimum transistor length.

As the transistor length is reduced, the thickness of the gate oxide (T_{ox}) must also be reduced for the gate to have good control of the channel. If T_{ox} is not scaled, there could be significant current even in the cutoff region. Moving the gate physically closer to the channel makes sure the transistor can be effectively turned off. The width of the transistor is determined by the size of the source and drain regions. A wider channel allows more current to flow in parallel between the source and drain.

The factor ε_{ox} measures the electric permittivity of the gate oxide material. This will determine how strongly the electric field of the gate penetrates the gate oxide and how strong it will be when it reaches the silicon channel. Finally, the factor μ is a measure of the mobility of the charge carriers. The drift velocity of the carriers across the channel is estimated as their mobility times the electric field between the source and drain. Together these five factors, the transistor length, width, oxide thickness, oxide permittivity, and carrier mobility, determine the β value of a transistor.

Of these five, the circuit designer has only control of the transistor's length and width. Most transistors on a die will be drawn at the minimum allowed channel length in order to produce the most current and provide the fastest switching. Some sequential circuits will make use of transistors as "keepers," which only hold the voltage level of a node when it is not being switched by other transistors. These keeper devices may be intentionally drawn with nonminimum lengths to keep them from interfering with the normal switching of the node.

The circuit designer increases a transistor's switching current by increasing the width, but this also increases the capacitive load of the gate. Increasing the width will make the drain of this transistor switch faster, but the transistor driving the gate will switch more slowly. Also, a wider transistor requires more die area and more power. Much of a circuit designer's job is carefully choosing the widths of transistors to balance speed, die area, and power.

The manufacturing process determines the oxide thickness, oxide permittivity, and charge mobility. Modern processes use gate oxides less than 10 molecules thick, so further scaling of the oxide may be impractical.

Instead, research is focusing on new high-K materials that will increase the oxide permittivity. Charge mobility is improved by strained silicon processes, which either push the silicon atoms in PMOS channels closer together or pull the atoms in NMOS channels apart. For all silicon MOSFETs, the mobility of holes in PMOS devices is only about half the mobility of electrons in NMOS devices. The movement of holes is caused by the movement of electrons in the opposite direction, but because many different electrons are involved in moving a single hole, the process is much less efficient than NMOS devices where each charge carrier is a single free electron. To make up for the difference in mobility, circuit designers routinely make the width of PMOS devices double the width of NMOS devices.

Another important difference between NMOS and PMOS devices is their ability to pull their drain terminals all the way up to the supply voltage (usually written as V_{dd}) or down to the ground voltage (usually written as V_{ss}). Because NMOS devices will switch off if both source and drain are within a threshold voltage of the gate, an NMOS with its gate tied to the supply voltage cannot pull its drain to the supply voltage. The resulting waveform when this is tried is shown in Fig. 7-4.

In Fig. 7-4, the gate of an NMOS is held at the supply voltage whereas V_{in} is driven first high and then low. The terminal V_{out} tracks with V_{in} up to one threshold voltage (V_t) from the supply voltage, but then the NMOS goes into cutoff, and V_{out} stops rising. On the other hand when

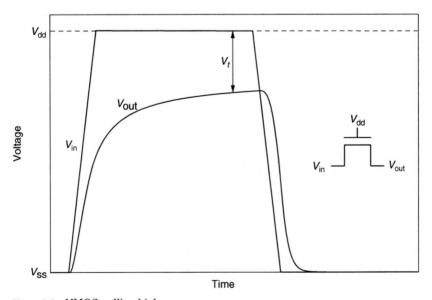

Figure 7-4 NMOS pulling high.

V_{in} goes low, V_{out} is able to follow all the way to the ground voltage, and current continues to flow in the NMOS until V_{in} and V_{out} are equal.

PMOS devices have the opposite problem. Because PMOS devices will switch off if both source and drain are only a threshold voltage above the gate, a PMOS with its gate tied to ground cannot pull its drain to ground. The resulting waveform when this is tried is shown in Fig. 7-5.

The inability of NMOS devices to pull high and PMOS devices to pull low is why almost all logic gates use both types of devices in order to make sure the output of the gate can be switched through the full range of voltages.

The MOSFET current equations are only approximations. Current in cutoff is not exactly zero. Saturation current does vary slightly with the drain voltage. The equations assume that charge mobility is constant at any electric field, but this is only true over a limited range. The current equations presented in this section are a useful way of thinking about how transistors behave and how their current varies with voltage, but in reality modern designers never solve these equations by hand in their day-to-day work. One reason is because modern designs use so many transistors that solving current equations by hand would be exceedingly tedious. An even better reason is that these equations are only very crude approximations of real transistor behavior. To accurately model transistor currents, we must rely upon computer simulations using much more complicated equations, which are still only approximations.

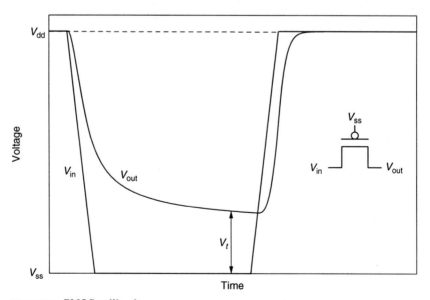

Figure 7-5 PMOS pulling low.

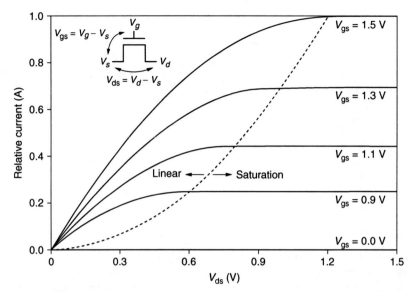

Figure 7-6 NMOS current vs. voltage.

The basic current equations are worth learning not for the purpose of calculation, but to guide one's understanding of how transistors operate and to get an intuitive feel for how a circuit will operate. We see the qualitative nature of this behavior in the graph of ideal NMOS current in Fig. 7-6.

When the difference between the gate and source voltage (V_{gs}) is zero the device is in cutoff, and the current is effectively zero at any drain voltage. For values of V_{gs} above the threshold voltage, current increases linearly with the drain voltage up to a point. This is the linear region. At a large enough difference between the drain and source voltage (V_{ds}) the current stops increasing. This is the saturation region. The higher the value of V_{gs}, the higher the slope of current with V_{ds} in the linear region, and the higher the current value will be in saturation.

Although the precise modeling of current behavior is important for detailed analysis, the general behavior of almost all digital circuits can be understood with only a solid grasp of the basic current regions.

CMOS Logic Gates

To implement the logic of the HDL model, the circuit designer has just three basic elements to work with: NMOS transistors, PMOS transistors, and the wires that connect them. By connecting NMOS and PMOS transistors in different ways, simple logic gates are created where electricity

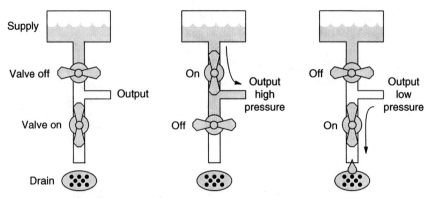

Figure 7-7 Water circuit.

flows only where and when needed. These gates are used together to produce more and more complicated behaviors until any logical function can be implemented. If we imagine electricity flowing like water, then transistors would be the valves that control its flow, as in the imaginary water "circuit" shown in Fig. 7-7.

In this circuit, the output starts at low pressure with no water in it. When the top valve is turned on and the bottom is turned off, current flows from the supply to the output bringing it to high pressure. If the top valve is then turned off and the bottom valve turned on, the water drains from the output bringing it back to low pressure. Electrical current flows through transistors in the same way, changing the voltage or electrical "pressure" of the wires. The water circuit in Fig. 7-7 is analogous to an electrical inverter shown in Fig. 7-8.

The supply is a high voltage and the electric ground acts to drain away current. A single input wire is connected to the gates of a PMOS and an NMOS transistor. When the voltage on the input is low (a logical 0), the PMOS transistor turns on and the NMOS turns off. Current flows

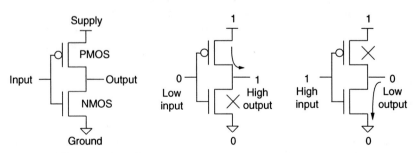

Figure 7-8 Inverter operation.

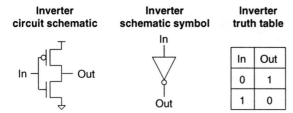

Figure 7-9 Inverter symbol and truth table.

through the PMOS to bring the output to a high voltage. When the voltage on the input is high (a logical 1), the PMOS transistor turns off and the NMOS turns on. The NMOS drains away the charge on the output returning it to a low voltage. Because the voltage on the output always changes to the opposite of the voltage on the input, this circuit is called an **inverter**. It inverts the binary signal on its input.

For commonly used logic gates, circuit designers use symbols as shorthand for describing how the transistors are to be connected. The symbol for an inverter is a triangle with a circle at the tip. The truth table for a logic gate shows what the output will be for every possible combination of inputs. See Fig. 7-9. Because an inverter has only one input, its truth table need only show two possible cases. More complicated logic gates and behaviors can be created with more transistors.

The second most commonly used logic gate is a **NAND** gate. The simplest NAND has two inputs and provides a 0 output only if both inputs are 1. A two-input NAND circuit and its operation are shown in Fig. 7-10.

The NAND uses two P-transistors and two N-transistors. The P-transistors are connected in parallel from the output (labeled X in the

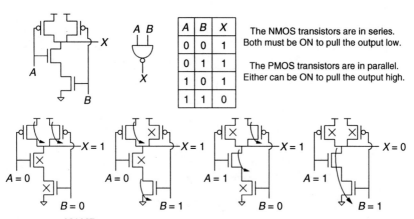

Figure 7-10 NAND gate.

schematic) to the supply line. If either turns on, the output will be drawn to a 1. The N-transistors are connected in series from the output to the ground line. Only if both turn on, the output will be drawn to a 0. The two inputs, A and B, are each connected to the gate of one P-transistor and one N-transistor.

Starting at the bottom left of Fig. 7-10, the NAND circuit is shown with both inputs at 0. Both N-devices will be off and both P-devices will be on, pulling the output to a 1. The middle two cases show that if only one of the inputs is 0, one of the N-devices will be on, but it will be in series with the other N-device that is off. A 0 signal is blocked from the output by the off N-device, and the single on P-device still pulls the output to a 1. At the bottom right, the final case shows that only if both inputs are high and both N-devices are on, the output is pulled to a 0.

Similar to the NAND is the **NOR** logic gate. The two-input NOR and its operation are shown in Fig. 7-11.

The NOR provides a 0 output only if either input is a 1. The NOR also uses two P-transistors and two N-transistors. The P-transistors are connected in series from the output to the supply line. If both turn on, the output will be drawn to a 1. The N-transistors are connected in parallel from the output to the ground line. If either turns on, the output will be drawn to a 0. The two inputs are each connected to the gate of one P-transistor and one N-transistor.

Shown at the bottom left of Fig. 7-11, if both inputs are 0, both N-devices will be off and both P-devices will be on, pulling the output to a 1. In the two center cases, if only one of the inputs is 0, one of the P-devices will be on, but it will be in series with the other P-device, which is off. A 1 signal

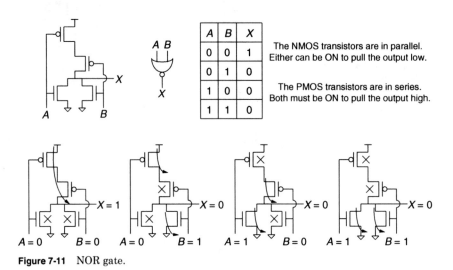

Figure 7-11 NOR gate.

is blocked from the output by the off P-device and the signal on N-device pulls the output to a 0. The final case at the bottom right shows both inputs high and both N-devices pulling the output to a 0.

CMOS logic gates always include one network of transistors to pull the output down and another to pull the output up. The networks are designed, so that they are never both be on at the same time. NMOS transistors are used for the pulldown network because they are effective at driving low voltages. PMOS transistors are used for the pullup network because they are effective at driving high voltages. The result is that CMOS logic gates are always inverting. A rising input will turn on an NMOS and turn off a PMOS, which will either cause the gate output to go low or stay the same. A falling input will turn on a PMOS and turn off an NMOS, which will either cause the gate output to go high or stay the same. A rising input never causes a rising output, and a falling input never causes a falling output. To make noninverting logic like an AND, CMOS logic uses an inverting gate like an NAND followed by an inverter.

Using the same pattern of NMOS pulldown networks and PMOS pullup networks, we can construct three-input CMOS gates (Fig. 7-12).

The NMOS and PMOS networks for each gate in Fig. 7-12 are mirror images of each other. When the NMOS transistors receiving two inputs are in series, the PMOS transistors are in parallel and vice versa. There is no limit to the number of inputs a CMOS gate can be designed for, but in practice gates with more than four or five inputs are rarely used. As the number of inputs grows, more transistors (either NMOS or PMOS) must be connected in series. The current from two series transistors is roughly cut in half and from three in series is roughly cut to a third. This means that gates with more inputs will draw less current when switching and therefore will switch more slowly.

Also, gates with more inputs are more susceptible to noise. It only takes a small amount of noise on the inputs of a large number of transistors in

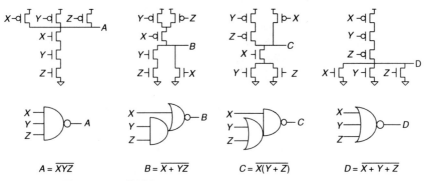

Figure 7-12 Three-input CMOS gates.

parallel to make a gate switch when it is not supposed to. For performance and noise immunity, it is better to implement logic requiring a large number of inputs using multiple stages of small gates rather than a single very complex CMOS gate.

Transistor sizing

Something on which circuit designers spend a great deal of time is choosing the best size for each of the transistors in their circuits. Most transistors will be drawn at the minimum allowed length in order to provide the most current when switching, but transistor widths must be chosen to balance speed, area, power, and resistance to noise.

How much noise it takes to improperly switch a gate is determined in part by the relative sizing of its PMOS and NMOS devices. The ratio of PMOS width and NMOS width is the **P-to-N ratio** of the gate, which determines at what input voltage the current from the N-pulldown network matches the current from the P-pullup network. Usually this ratio is set so that the input must have switched halfway between ground and the supply voltage before the output will switch halfway. The effect of P-to-N ratios is seen by plotting the output voltage of a CMOS gate as a function of its input voltage. This is called the **transfer curve**.

Figure 7-13 shows the transfer curve of an inverter for three different P-to-N ratios. For any ratio as the input goes from low to high, the output will go from high to low. Because the mobility of a PMOS device is about half that on an NMOS device, a P-to-N ratio of 2 makes the output fall to half the supply at about the point the input has risen to half the supply. An inverter with this ratio must have noise on its input close to half

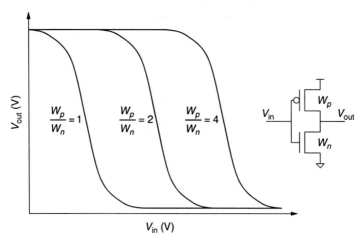

Figure 7-13 Inverter transfer curve.

the supply voltage before a significant amount of noise is propagated to its output.

A P-to-N ratio of 1 will make the NMOS relatively stronger than the PMOS. The output of the gate will be drawn low at an input voltage below half the supply. A P-to-N ratio of 4 will make sure that the PMOS is stronger than the NMOS, and the output of the gate will not go low until an input voltage is above half the supply. The low ratio inverter is more susceptible to a low signal with noise pulling it up, and the high ratio inverter is more susceptible to a high signal with noise pulling it down. When sizing more complicated gates, transistors in series must be upsized if pullup and pulldown networks are to be made equal in strength.

Figure 7-14 shows typical relative sizings for an inverter, NAND, and NOR gates. The NAND has equal strength pulling high and low for a P-to-N ratio of 1 because the NMOS devices are stacked in series. The NOR requires a ratio of 4 because the PMOS devices are stacked. The large PMOS devices required by NORs make circuit designers favor using DeMorgan's theorem to change to NANDs wherever possible. Different ratios can be used if it is desirable to have a gate switch more quickly in one direction, but this will always come at the cost of greater sensitivity to noise and switching in the opposite direction more slowly.

In addition to the ratio between PMOS and NMOS devices within a gate, the ratio of device sizes between gates must be considered. As discussed in Chap. 1, the delay of a gate is determined by the capacitive load on its output, the range of voltage it must switch, and the current it draws.

$$T_{delay} = \frac{C_{load} \times V_{dd}}{I}$$

Each gate will drive at least one other gate, and the capacitance of the receiver's input is directly proportional to the total width (W_r) of its devices. There will be some additional base capacitance on the driver's

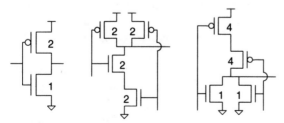

Figure 7-14 Typical CMOS P-to-N ratios.

output due to the total width of the drain diffusion (W_d) of the driver's own devices.

$$C_{load} = C_{gate} \times W_r + C_{base} \times W_d$$

The current draw of the driver gate is proportional to the width of its devices (W_d) and the current per width ($IperW$) provided by the process. The delay of a gate is decreased by reducing the size of the receiver, to reduce the capacitive load, or by increasing the size of the driver, to provide more current.

$$T_{delay} = \frac{C_{load} \times V_{dd}}{I} = \frac{(C_{gate} \times W_r + C_{base} \times W_d)V_{dd}}{W_d \times IperW}$$

$$= \left(\frac{C_{gate} \times V_{dd}}{IperW}\right)\left(\frac{W_r}{W_d}\right) + \frac{C_{base} \times V_{dd}}{IperW}$$

The delay of every gate is determined by parameters that are functions of the process technology and the ratio of the widths of the receiver's devices and the driver's devices. This ratio is called the **fanout** of the gate. A gate driving three copies of itself, which are all sized the same, will have a fanout of 3. A gate driving a single copy of itself with devices three times as wide will also have a fanout of 3. In both these cases, the delay will be the same. The delay of any CMOS gate can be written as some delay per fanout (T_{fan}) plus some minimum base delay.

$$T_{delay} = T_{fan} \times \text{fanout} + T_{base}$$

$$\text{Fanout} = \frac{W_r}{W_d}$$

The ratio of the size of a driver and the size of its receiver determines delay, not just the size of the driver itself. The width of transistors should not be blindly increased to try and make a circuit run faster. Every time the width of the devices in one gate is increased, the delay of that gate is decreased but the delay of that gate's drivers increase. We see an example of this trade-off in Fig. 7-15.

In Fig. 7-15, the same path of three inverters is shown three times with different sizings to show how choosing transistor widths impacts overall delay. We assume that the sizes of the first and third inverters are fixed, and that for this technology T_{fan} and T_{base} for an inverter are both 1 unit of delay. In the top example, the first and second inverters have the

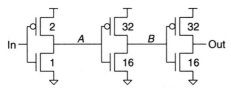

Delay (In → A) = 2   Delay (A → B) = 17   **Total = 19**

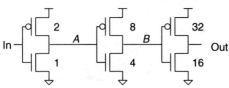

Delay (In → A) = 17   Delay (A → B) = 2   **Total = 19**

Delay (In → A) = 5   Delay (A → B) = 5   **Total = 10**

Figure 7-15 Sizing for minimum delay.

same sizes. The fanout is 1 and therefore the delay of the first inverter is 2. The second inverter is driving devices 16 times larger. Its fanout is 16, and therefore the delay is 17. The total delay through both inverters is 19.

To speed up this circuit, a circuit designer might increase the width of the transistors in the second inverter. This is shown in the middle example. Here the second inverter has been made the same size as the third inverter, so that its delay is only 2. However, now the delay of the first inverter is 17, so the total delay is still 19.

To get the minimum delay, we must gradually increase the inverter sizes, as is shown in the last example. In the last example, the size of the second inverter has been chosen to make the fanout for both stages 4. This makes both delays 5 and the total delay 10. In general, to minimize overall delay, the designer must try to keep the fanout and delay at each gate roughly equal. Anytime the delay at a gate's input is very different from the delay on its output, resizing can reduce total delay.

Assuming that gates have zero delay at zero fanout, it can be shown that the optimal fanout for minimizing delay is the irrational number

e ($\approx$ 2.718).[1] Unfortunately real gates have some delay even with no load at all. This makes the optimum fanout a function of the process technology, which will determine the ratio of the delay per fanout (T_{fan}) and the delay at zero fanout (T_{base}). Typical values for the optimal fanout of an inverter are between 3 and 4. Because different types of logic gates will have different values of T_{fan} and T_{base}, their optimal fanout will be different. In general, as the number of gate inputs increases, the optimal fanout of the gate goes down. The more complex the gate, the greater its delay, and the less it should be loaded when optimizing for delay. The final sizing for minimum delay will be a function of all the different types of logic gates in the path as well as the loading due to interconnects between the gates. Although CAD tools can find the optimal sizing for simple paths through iteration, there is no analytical solution. An easy way to spot inexperienced designers is to look for engineers making one gate exactly 2.718 times the size of another.

Even if there were a single value for optimal fanout, it would be very poor circuit design to always size for this value. Sizing for optimal fanout is trying to minimize delay. However, some paths in any circuit block will have more logic gates than others and therefore more delay. Only the very slowest paths will limit the operating frequency, and only these paths should be sized to minimize delay. Other paths can use smaller transistors to reduce the die area and power needed. Device sizing is one of a circuit designer's most powerful tools, but there is not a single perfect sizing or single perfect logic gate.

Sequentials

Different logic gates can be connected to form any desired combinational logic function, but logic gates cannot store values. This is the job of sequentials. The simplest possible sequential is made of a single transistor, as shown in Fig. 7-16.

When the clock signal is high, the NMOS transistor is on and the voltage on the data signal will pass through to the out node. When the clock

Clock	Data	Out
0	0	Old data
0	1	Old data
1	0	0
1	1	1

Figure 7-16 1T memory cell.

---

[1]Glasser and Dobberpuhl, *Analysis of VLSI Circuits*, 259.

signal is low, the transistor will be off, and the capacitance of the out node will hold the charge stored there. This is the memory cell used by DRAM chips to form the main memory store of computers. This cell has the advantage of using the minimum possible number of transistors and therefore the minimum possible die area, but it has a number of disadvantages as well.

We have already noted that NMOS transistors cannot drive a node all the way up to the supply voltage. This means that when writing a 1 into the single transistor memory cell (**1T memory cell**), the voltage stored on the out node will actually be one threshold voltage down from the voltage on the clock signal. Also, charge will tend to leak away from the out node over time. With nothing to replace the charge, the data stored will eventually be lost. DRAM chips deal with this problem by rewriting all their data every 15 ms. Because of the signal loss when storing a high voltage and the need for refreshing the cell, microprocessors do not use the 1T memory cell for their sequentials. Instead, this circuit is altered to form a more robust storage circuit, the **latch**. The changes needed to form a latch from a 1T cell are shown in Fig. 7-17.

The first step to make a more robust memory cell is to make sure that it can fully charge the storage node to either the supply voltage or ground. To do this, the circuit must use an NMOS and a PMOS transistor together. The second circuit in Fig. 7-17 shows a CMOS pass gate. In the circuit, NMOS and PMOS transistors are connected in parallel. The controlling signal goes directly to one gate and through an inverter to the other. This means the NMOS and PMOS will always either be both on or both off. When they are on, together they draw the output node to any voltage, and when they are off, the capacitance on the output will hold the last charge put there.

To avoid needing to refresh the stored value, the next step shown in Fig. 7-17 adds a keeper. A pair of back-to-back inverters is added, which take the voltage stored at the output and drive it back onto the output. This loop will maintain the stored value as long as the circuit has power. The feedback inverter driving the storage node is marked with a W to indicate that this must be a "weak" gate. The width and length of the transistors in this inverter must be chosen, so that when the pass gate is open a new value can be forced onto the storage node even though the keeper is trying to maintain the previous value. This type of latch is called a *jam latch* because the new value must be forced in against the current of the keeper.

To make sure that the gate driving in through the pass gate has sufficient current draw to overcome the keeper, it can be built into the latch circuit. Placing an inverter on the input of the pass gate allows any gate to drive inputs to this latch, without worrying about the relative sizes of the driving gate and the keeper in the latch. The final circuit in Fig. 7-17 also adds an inverter on the output. This prevents the need to

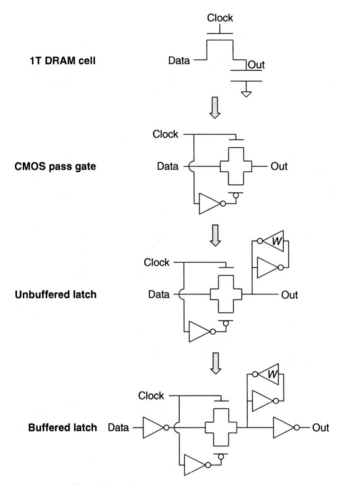

Figure 7-17 Latch circuit.

route the storage node through a potentially long wire to some other gate, which could make the storage node susceptible to noise. Adding an output inverter allows the storage node to control only the inverter that drives the true output of the latch.

There are actually two basic types of latches, as shown in Fig. 7-18. In an N-latch, the clock signal goes first to the NMOS device in the pass gate and through an inverter to the PMOS device. This means an N-latch will be open when the clock is high. In a P-latch, the clock signal goes first to the PMOS device in the pass gate and through an inverter to the NMOS device. This means a P-latch will be open when the clock is low.

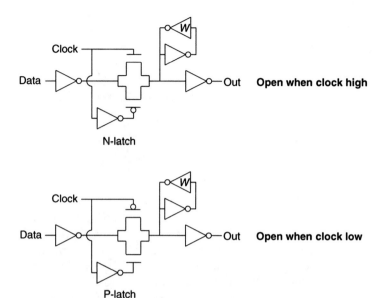

Figure 7-18 N-latch and P-latch.

When a latch is open, new data flows directly through to the output, but when the clock changes and the latch closes, it will catch whatever data is there at the time and hold it until the latch opens again. New data arriving while the latch is closed must wait for the clock signal to change before it can pass through.

Latches are called *sequentials* because not only do they retain previously captured values, they also enforce a sequence between the different data signals. In a pipelined processor, the data from multiple different instructions is flowing through the processor circuits at the same time. Because some paths through the circuits are faster than others, it is possible for later issued instructions to overtake instructions issued earlier. If this were to happen, it would be impossible to distinguish the correct data for each instruction. Sequentials prevent this by enforcing a separation in time between the data of different instructions. Chapter 6 discussed how **flip-flops** are commonly used to separate the pipestages of a pipelined processor. A circuit implementation of a flip-flop can be created by combining two latches.

The flip-flop shown in Fig. 7-19 is a P-latch immediately followed by an N-latch. When the clock is low, the P-latch will be open but the N-latch will be closed. When the clock is high, the P-latch will be closed but the N-latch will be open. There is never a time when both latches are open at the same time. This means the clock signal must go low and then go

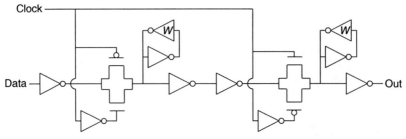

Figure 7-19 Flip-flop.

high for data to pass through a flip-flop. When the clock is low, data will pass through the P-latch and wait at the input to the N-latch. As the clock goes high, the P-latch will close, capturing the data value at that moment, and the N-latch will open to pass that value to the output. One latch "flips" while the other "flops." This behavior makes a flip-flop an "edge-triggered" sequential. It captures and passes through its input value on each rising clock edge. Transitions at the input to the flip-flop before or after the rising edge of clock are ignored.

Sequentials are an essential part of any processor design. They allow data values to be stored and later retrieved. Also, they enforce the synchronization of all the processor functions by passing data values only according to the transitions of a clocking signal.

Circuit Checks

Once the needed logic gates and sequentials have been created, the circuit designer combines them to implement the logical behavior specified by the HDL model. A number of automated checks will need to be performed to make sure the circuit implementation will behave as desired. The first was already mentioned in Chap. 6: logical equivalence checking. The actual circuit implementation is too complex to simulate its behavior running real applications. Instead, CAD tools convert the behavior of the circuit into a set of boolean equations. The same is done with the HDL model, and these equations are compared to test whether the circuit faithfully reproduces the behavior of the HDL model. Whether the HDL model itself is correct is a separate question, which is explored but never definitively answered by pre-silicon validation.

However, even if the circuit implementation is logically correct, there are a number of potential electrical problems that must be checked. The timing of the circuit could lead to poor frequency or incorrect behavior. Electrical noise could make a logically correct circuit still produce the wrong answer. The power required to operate a circuit might be too high for the desired application. All of these checks require modeling the true electrical behavior of the transistors rather than simply treating them as being on or off. To quickly and accurately model these behaviors for large circuits, modern designers rely heavily upon CAD tools.

Timing

The area on which circuit designers typically spend the most effort is circuit timing. Chapter 6 described how flip-flops act like stoplights between the different stages of a processor pipeline. The same clock signal is routed to all the flip-flops, and on its rising edge all the flip-flops "turn green" and allow the signal at their input to pass through. The signals rush through the logic of each pipestage toward the next flip-flop. For proper operation, all the signals must have reached the next flip-flop before the clock switches high again to mark the beginning of the next cycle. The slowest possible path between flip-flops will determine the maximum processor frequency. It is therefore important to balance the logic within each pipestage to make them all have roughly the same delay. One fast pipestage is wasteful since the clock rate will always be limited by whichever pipestage is slowest. The circuit designers work toward the same clock frequency target, and circuits that are too slow are said to have a maxdelay violation or a speedpath.

Figure 7-20 shows an example of the timing check that must be performed for maxdelay violations. The rise of clock signal *A* (Clk*A*) causes the first flip-flop to drive its data value onto signal *A* (Sig*A*). This causes transitions in the logic gates as the particular combinational logic of this pipestage is carried out. This logic may cause the input to the second

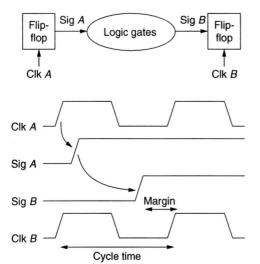

Figure 7-20 Maxdelay timing.

flip-flop, signal B (Sig$B$), to switch. This must happen before the rise of clock signal B (Clk$B$) 1 cycle after the rise of Clk$A$. The amount the Sig$B$ is earlier than needed is the maxdelay timing margin.

In addition to requiring a maximum delay for each pipestage, there is also a need for a minimum delay. A very fast pipestage might allow its signals to pass through and reach the next flip-flop before it has closed. This would allow the signals to pass through two flip-flops in 1 cycle and would end with them in the wrong pipestage. Like cars crashing into one another, the proper data from one pipestage would be lost when it was overtaken from behind by the signals from the previous pipestage. The main function of flip-flops is to stop speeding signals from entering the next pipestage too soon, keeping the signals of each pipestage separate. A circuit that has so little delay that its data could jump ahead to the wrong pipestage is said to have a mindelay violation or a race.

Figure 7-21 shows an example of the timing check that must be performed for mindelay violations. The second rise of Clk$A$ causes another transition of Sig$A$. Variation in the arrival time of the clock signal can cause this rise of Clk$A$ to happen before the rise of Clk$B$. This uncertainty in clock timing is called **clock skew**. Ideally, each transition of the global clock signal would reach every flip-flop on the die at the exact same moment, but in reality there is some variation in the arrival time at different points on the die. This can be caused by different amounts of capacitance on different parts of the global clock signal or by manufacturing variation in the gates driving the clock signal. The second transition of

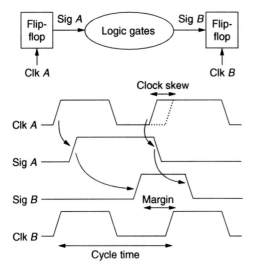

Figure 7-21 Mindelay timing.

SigA can cause a second transition of SigB, which must occur after flip-flop capturing SigB is closed. The amount SigB is later than needed is the mindelay timing margin.

A mindelay violation is also called a *race* because a single clock transition causes two things to happen. One clock edge makes the input to the flip-flop switch and causes the flip-flop to capture data. One event must happen before the other for the circuit to function properly, and it is the race between these two events that must be checked. This makes mindelay violations particularly worrisome because changing the clock frequency will have no effect on them. Slowing down the clock and allowing more time for signals to propagate through the logic of the pipestage can fix a maxdelay violation. This will mean selling the processor at a lower frequency for less money, but the processor is still functional. A mindelay violation is independent of the clock frequency and usually forces a processor to be discarded.

Because there are many possible paths through a circuit, a single pipestage can have maxdelay and mindelay violations at the same time. Much of a circuit designer's effort is spent balancing the timing of circuits to meet the project's frequency target, without creating mindelay violations.

Figure 7-22 shows what the output of a particular pipestage might look like over many cycles. Each new rise of ClkA can cause a transition high or low, and each transition happens after varying amounts of delay. There is only a narrow window of time when the output SigB is guaranteed not to change. This is the **valid window**. The slowest path through the logic determines the lead of a valid window, and the fastest

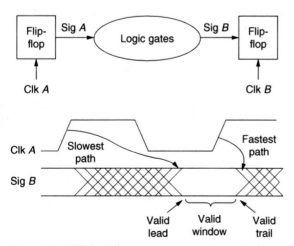

Figure 7-22 Valid windows.

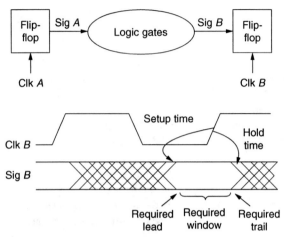

Figure 7-23 Required windows.

path determines the trail. In this window of time, SigB will always have the value from one particular cycle. Before or after the valid window it is not certain which cycle's data is currently at SigB.

The sequential must try to capture the pipestage's output during the valid window, but the sequential itself has some finite delay. In order to be captured properly, the signal must arrive some amount of time before the rising clock edge and be held for some length of time afterward. This is the **required window** of the sequential, as shown in Fig. 7-23.

The time an input must arrive before the clock edge to be captured is the setup time of the sequential. The time an input must be held after the clock edge is the hold time. Together they form the minimum required window of the sequential. However, uncertainty in the clock timing makes the required window grow larger, as shown in Fig. 7-24.

Because clock skew causes the same clock edge to arrive at different times at different sequentials, both sides of the required window must

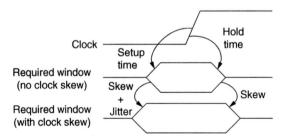

Figure 7-24 Required windows and clock uncertainty.

add margin for clock skew. In addition, there will be some variation in the cycle time of the clock. A 3-GHz processor will attempt to generate a clock signal at exactly 3 GHz, but the clock generation itself is not perfect. Variations in process, voltage, and temperature cause the clock frequency to vary, and this variation in cycle time is called **clock jitter**. When checking the timing of maxdelay paths from one cycle to the next, margin for clock jitter must also be included. Because uncertainty in clock timing increases the requirements on both maxdelay and mindelay, it must be very carefully controlled for there to be any hope of meeting timing.

To always capture the correct data, the valid window at the sequential's input must completely overlap the required window. Figure 7-25 shows this comparison. If the leading edge of the valid window comes after the leading edge of the required window, this is a maxdelay failure. If the trailing edge of the valid window comes before the trailing edge of the required window, this is a mindelay violation. Designing for circuit timing is fundamentally the task of balancing the fastest and slowest paths through the logic and choosing device sizes to make sure each pipestage's valid window overlaps its required window.

The overhead of sequential setup and hold times as well as clock uncertainty is an important limit to how much frequency can be gained by increasing pipeline depth. By increasing the pipeline depth, the amount of logic in each pipestage is reduced. However, the maximum frequency is still limited by the sequential setup time as well as clock skew and jitter. This overhead per pipestage does not change as the number of pipestages increases.

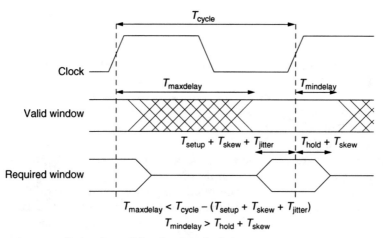

Figure 7-25 Comparing valid and required windows.

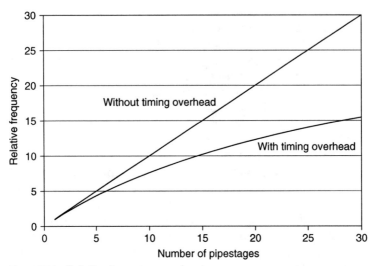
Figure 7-26 Relative frequency vs. pipestages.

Figure 7-26 shows how relative frequency can change as pipestages are added. If there were no timing overhead, increasing from a single stage to a pipeline of 30 stages would allow a frequency 30 times higher but with timing overhead the real achieved speedup could easily be only half of the ideal. Each new pipestage reduces the processor's performance per cycle, so as the improvement in relative frequency slows, it becomes less and less likely that adding pipestages will actually increase overall performance. The trend of processors improving performance by increasing pipeline depth cannot continue indefinitely.

Noise

In an ideal digital circuit, every wire would always be at the supply voltage or at ground except for brief moments when they were switching from one to the other. In designing digital logic, we always consider the behavior when each node is a high voltage or a low voltage, but an important part of circuit design is what happens when voltage signals are not clearly high or low.

Any real world signal will have some amount of electrical noise, which will make the voltage deviate from its ideal value. Part of what makes digital circuits useful for high-speed computation is that they tend to be very tolerant of noise. We can adjust the P-to-N ratio of CMOS gates, so only an input voltage very far from its ideal value could incorrectly switch the gate. We can measure how tolerant a gate is of noise by finding the unity gain points on its voltage transfer curve.

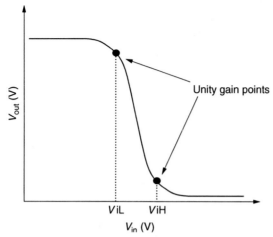

Figure 7-27 Unity gain points.

Figure 7-27 marks the points on a transfer curve where the slope is exactly −1. These points are significant because they mark the voltages where the gate's behavior will change from suppressing noise to amplifying it. Any input voltage above ground will cause the output voltage to be somewhat less than the full-supply voltage, but noise less than the voltage input low (ViL) unity gain point will be reduced. The amount the output drops below the supply voltage will be less than the amount the input is above ground. However, any noise added above ViL will be amplified. This part of the transfer curve has a steeper slope, and the change in output voltage will be larger than the change in input voltage. There is a similar unity gain point for input voltages that should be high but have dropped below ground. Noise above the voltage input high (ViH) point will be reduced, whereas noise below that point will increase. How far the unity gain points are from the ideal ground and supply voltages is a good measure of how sensitive a gate will be to noise.

The most common source of electrical noise is **cross talk**. Every time a wire changes voltage its change in electrical field will change the voltage of all the wires around it. Almost everyone has experienced cross talk first hand at one time or another on the telephone. If not properly shielded, telephone wires running next to one another induce noise in each other. We hear bits of another conversation when the changing voltage of their line affects the voltage on our line.

Figure 7-28 shows how cross talk affects digital circuits. Signal A is routed on a wire that runs in between signals B and C. If signals B and C switch from a high voltage to a low voltage, they will tend to draw signal A down with them. If signal A is being driven high by a gate, it will quickly recover to its proper voltage, but momentarily it will dip

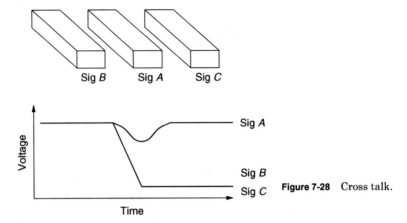

Figure 7-28 Cross talk.

below the supply voltage. The maximum amount of noise induced by cross talk is proportional to the percentage of the total capacitance of signal A to nets that are switching. If 30 percent of signal A's capacitance is to signals B and C, then the noise droop could be as much as 30 percent of the total supply voltage. Cross talk noise is minimized by decreasing capacitance to lines that switch or by increasing capacitance to lines that do not switch, such as power or ground lines.

Another source of noise is variation in the voltage on the power and ground lines themselves. In simulating circuits, we often treat the supply and ground voltages as ideal voltage sources that never vary, but these voltages must be delivered to each gate through wires just like data signals. They are susceptible to cross talk noise just like data signals as well as voltage noise due to large current draws.

Figure 7-29 shows an example of how **supply noise** can affect digital circuits. An inverter is driving signal A, which is received by a NOR

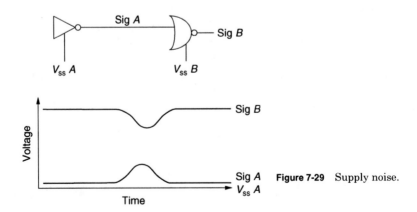

Figure 7-29 Supply noise.

gate some distance away. The inverter has no noise on its input, and there is no cross talk affecting signal A, but a large current draw by another gate near the inverter causes the ground voltage at the inverter ($V_{ss}A$) to be pulled up. The inverter continues to drive its output to ground, but now its ground voltage is higher than the ground voltage at the receiver. The NOR gate will begin to switch low because its input is now at a higher voltage than the NOR gate's local ground line. Providing very low resistance paths for the supply and ground voltages to all gates minimizes supply noise.

Paths with very little timing margin cause another type of noise. When inputs to sequentials switch just before or after the sequential closes, the storage node of the sequential will be disturbed. Even with all other sources of noise eliminated, just having very little timing margin can put nodes at intermediate voltages between the supply and ground.

Figure 7-30 shows example waveforms for an N-latch. Setup noise is caused when the data input switches just before the latch closes. Because the latch closes before the storage node Q has switched all the way to the

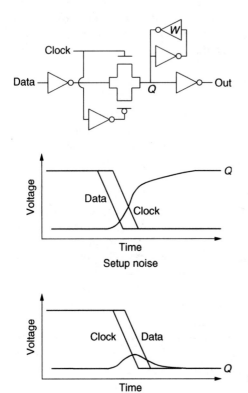

Figure 7-30 Setup and hold noise.

supply, it is left to the keeper gate to pull it up the rest of the way. Because the keeper is sized to be weak, this could take a full clock cycle or more.

Hold noise is caused when the data input switches just after the latch closes. Because the latch is not completely closed, the storage node Q is disturbed from its ideal value; once again it is left to the keeper gate to restore the proper value. The amount of setup and hold noise that can be tolerated must be considered when performing maxdelay and mindelay timing checks.

In the late 1970s, transistors had been scaled down to small enough sizes to be affected by another source of noise, radiation. Ceramics and lead solder in chip packaging often contain some radioactive isotopes. When they decay, these atoms release alpha particles, which can penetrate the silicon chip in the package. The level of radiation is trivial, but the tiny transistors of modern semiconductors act as extremely sensitive detectors. As alpha particles travel through the silicon, they collide with electrons in the valence band, knocking them loose from their silicon atoms. Tens of thousands of electrons and holes can be created in this fashion and if all the electrons are collected by one piece of diffusion, they can be enough to switch the voltage on that node. Memory bits are the most susceptible because they have the least capacitance and therefore require the smallest amount of added charge to flip. These errors are called *soft errors* because although the current data being held is lost, the memory cell is not permanently damaged.

Alpha particles are charged, which prevents them from traveling very far through any solid material. Any emissions from outside the computer or even outside the processor package are extremely unlikely to cause an error. Awareness of the problems caused by radioactive isotopes has led to them being almost completely eliminated from semiconductor packaging, but a more difficult problem is eliminating sunlight.

Sunlight contains high-energy neutrons, which can also cause soft errors. Because neutrons are uncharged, they pass easily through solid material. When passing through a semiconductor, they can create free charges in the same manner as alpha particles (Fig. 7-31). Trying to shield processors is simply not practical. A 1-ft thick concrete shield might only reduce the soft error rate (SER) from neutrons by 25 percent. Miles of atmosphere do block out many neutrons, reducing soft error rates at lower altitudes. Computers in Denver, which has an altitude of 5000 ft, have SER four times those found at sea level.[2] Notebook computers being used in airplanes would suffer even more soft errors.

A common way soft errors are reduced is by increasing the capacitance of sequential storage nodes. With each new fabrication generation, smaller dimensions tend to reduce the capacitance on all nodes. This allows higher

---

[2]Ziegler et al., "IBM Experiments in Soft Fails."

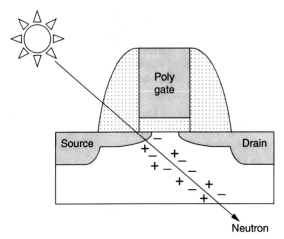

Figure 7-31 Neutron induced soft error.

frequency and lower power, but for sequential storage nodes the problem of soft errors creates the need for a minimum capacitance per node. The larger the capacitance of a storage node, the more electrons will be required to switch its voltage, making soft errors less likely. In addition, most large memory arrays on microprocessors include parity bits or error-correcting bits. These are redundant memory bits, which allow the processor to detect or even correct when one of the bits has the wrong value. The more redundant bits added, the less likely that two errors in the same data block will mask each other. Of course, these extra bits take up die area and add to the cost of the processor, so an important design decision is which memory arrays on the processor will have redundant bits and how many.

Any of the sources of noise lead to propagated noise. This is noise on the output of a gate due to noise on its input. Perhaps the noise on the input of a gate is not enough to flip the node entirely, but that noise may contribute to a failure on the gate's output. Designers must simulate the way noise propagates through circuit paths in the same way they simulate timing. Almost all circuits can trade off noise immunity and speed. By adjusting P-to-N ratios to make gates switch after very little movement of their inputs, the circuit path becomes faster for real transitions but more susceptible to false transitions. Noise and timing must be considered together to create a robust design.

Power

Most processors designed in the 1980s or earlier gave little or no thought to power. Even in the 1990s, typically only processors intended for portable applications where battery life was a concern would seriously design for power. Since 2000, power has become a major design concern for all processors.

Shrinking transistor sizes reduces the power required to switch a transistor, but increasing frequencies and numbers of transistors have lead to steadily more total switching power. In addition, reducing transistor size increases leakage currents, which flow even when the transistor is switched off. The combination of these two has lead to dramatic increases in processor power. Increasing transistor density exacerbates the problem by making processor power per area increase even more quickly than total processor power.

For portable applications, the impact on battery life has always made power a concern. For desktop applications, power is limited by the ability to deliver power into the processor and the need to remove the heat generated by that power. A high-performance processor might consume 100-W, the same a bright light bulb. However, a 100-W processor is far more difficult to support than a 100-W light bulb. Smaller transistors have lower maximum voltages; because the processor is operating at a much lower voltage, it might draw 100 times more current than the light bulb at the same power level. Building a computer power supply and motherboard that can deliver such large currents is expensive, and designing the wires of the package and processor die to tolerate such currents is a significant problem. A 100-W light bulb becomes too hot to touch moments after being turned on, but the surface area of the processor die might be 100 times less. The same power coming out of a smaller area leads to higher temperatures; without proper cooling and safeguards modern processors can literally destroy themselves with their own heat.

The total power of a processor (P_{total}) consists of the switching or active power (P_{active}), which is proportional to the processor frequency, and the leakage power ($P_{leakage}$), which is not.

$$P_{total} = P_{active} + P_{leakage}$$

The **active power** is determined by how often the wires of the processor change voltage, how large the voltage change is, and what their capacitance is.

$$P_{active} = \text{activity} \times C_{total} \times V_{dd}^2 \times \text{frequency}$$

In the equation for active power, the activity term measures the percent chance that an average node will switch in a particular cycle. This is a strong function of what application the processor is currently running.

An application that has very few dependencies and very few cache misses will be able to schedule many instructions in parallel and achieve high average activity and high power. An application that has long stalls for memory accesses or other dependencies will have much lower activity and power. Average node activity can be reduced by careful logic design including clock gating. Clock gating uses added logic to stop clock signals from switching when not needed. For example, when running an application that performs no floating-point operations, most of the clocks in the floating-point blocks of the processor can be switched off. Clock gating adds logic complexity but can significantly reduce active power.

The term C_{total} includes the capacitance of all the processor nodes that could switch. C_{total} tends to scale with the linear dimensions of the fabrication process, so compacting a processor design onto the new process generation and making no changes reduces this component of active power. However, adding more transistors for added functionality will increase the total switching capacitance. Circuit designers can reduce C_{total} by using smaller-width transistors. Unfortunately these smaller transistors will tend to switch more slowly. Eventually the maximum processor frequency will be affected. Careful circuit design can save power by reducing transistor widths only on circuit paths that are not limiting the processor frequency. Paths with maxdelay margin are wasting power by running faster than needed. Of course, it is impossible to make all the circuit paths the exact same speed, but neither should every path be sized for minimum delay.

The voltage and frequency of operation have a large impact on active power. The voltage of the processor impacts the speed that transistors switch, so it is not possible to reduce voltage without reducing frequency, but operating at reduced voltage and frequency is much more power efficient. The power savings of scaling voltage and frequency together are so large, that many processors are now designed to dynamically scale their voltage and frequency while running. This allows the processor to operate at maximum frequency and at maximum power when required by the workload, and to run at lower power when the processor is not being fully utilized. Transmeta was one of the first companies to implement this feature in their processors, calling it LongRun™ power management. In AMD processors, this ability is called Cool'n'Quiet®, and in Intel processors it is called SpeedStep® technology.

In addition to the active power caused by switching, the leakage currents of modern processors also contribute significantly to power. **Leakage power** ($P_{leakage}$) is determined by the leaking transistor width and the leakage current per width (I_{off}per$W$).

$$P_{leakage} = \text{total width} \times \text{stacking factor} \times I_{off}\text{per}W \times V_{dd}$$

As transistor dimensions are reduced, the total transistor width will be reduced, but adding new transistors will increase total width and leakage. Transistors that are switched off and in series with other off transistors will have dramatically reduced leakage. This reduction in leakage can be modeled by a stacking factor parameter of less than 1.

The largest factor in determining leakage is the leakage per transistor width of the fabrication technology. As transistors have been reduced in size, their leakage currents have increased dramatically. Scaling of processor supply voltages has required reductions in transistor threshold voltages in order to maintain performance, but this dramatically increases **subthreshold leakage**.

Figure 7-32 shows transistor current versus voltage on a logarithmic scale. This shows that the threshold voltage of the transistor is not really where the current reaches zero, but merely the point below which current begins to drop exponentially. The rate of this drop is called the *subthreshold slope* and a typical value would be 80-mV/decade. This means every 80-mV drop in voltage below the threshold voltage will reduce current by a factor of 10. This makes currents in the cutoff region so much smaller than the transistor currents in the linear or saturation regions, that when simulating small circuits the cutoff region is usually approximated as having zero current.

However, small currents in each of millions of transistors rapidly add up. The subthreshold slope also determines the impact of changing the threshold voltage from one process generation to the next. An 80-mV reduction in threshold voltage would mean that off transistors would leak 10 times as much current at the same voltage.

Subthreshold leakage current flows from source to drain in transistors that are off. There is also **gate leakage** through the gate oxide in transistors that are on. Electrons do not have precise locations; instead,

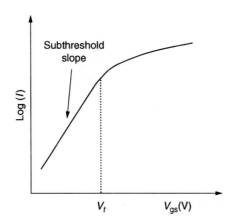

Figure 7-32 Subthreshold currents.

they merely have a certain probability of being within a particular volume of space. This volume is large enough that in transistors with gate oxides less than 5-nm thick, there is a significant probability that an electron that started on one side of the gate oxide might appear on the other side. If there is an electric field, the electron will be pulled away from the gate oxide and be unable to get back. The result is a flow of electrons, a current, through the gate oxide. This phenomenon is called *gate tunneling current* or just *gate leakage*.

The easiest way to control gate leakage is to use a thicker gate oxide, but this will reduce the on current of the transistor unless the electric permittivity of the gate oxide (ε_{ox}) can be increased. New high-*K* gate dielectric materials seek to increase gate permittivity to allow for more on current without increasing gate leakage. Tunneling currents set a lower limit on how thin an insulator can be and still effectively block current flow. Even if a 1-molecule-thick gate oxide layer could be manufactured, its high leakage would probably make it impractical for use.

In the same way a designer can trade off noise immunity for speed, most circuits can easily be made faster at the cost of higher power. Higher voltages, larger device widths, and lower threshold voltages, all enable higher frequencies but at some cost in power. Modern processor designs must carefully analyze timing, noise, and power and consider the interactions between them to perform successful circuit design.

Conclusion

The processor's circuit design gives a complete description of all the transistors to be used and their connections, but in order to be fabricated these transistors must be converted into layout. Creating layout is discussed in detail in the next chapter, but some aspects of circuit design depend upon layout. Sizing gates requires knowing what their output capacitance is, and this capacitance will come in part from wires. The length of the wires will be determined by the relative placement of gates and the overall size of the layout. The cross talk noise on wires is affected by the spacing to its neighboring wires and whether these wires can switch at the same time. This will also be determined by layout. The capacitance of the wires will have a large contribution to active power.

To perform circuit design, some assumptions must be made about the layout. This is called *prelayout circuit design*. The circuit designer and mask designer together plan out the overall topology of a circuit block, so that the various wire lengths and spacings can be estimated. Timing, noise, and power checks are then performed using these estimates. The initial circuit design is then created in layout.

Postlayout circuit design then begins. All the checks done prelayout must be performed again to make sure the initial assumptions about the layout match the result. Inevitably there are changes that must be made in the circuit design as a result of simulations using real layout. These circuit changes force changes in the layout. The process repeats until all the circuit checks are passed using real layout data.

Experienced circuit designers will be deliberately conservative in their prelayout assumptions in order to minimize the number of times the circuit and layout must be revised, but overly pessimistic initial assumptions will lead to over design. Layout and circuit design must work together to make a circuit with the necessary speed and robustness that does not waste die area or power.

Earlier steps in the design flow treat transistors as ideal switches that are either on or off. Circuit design simulates the electrical properties of transistors to predict their true behavior. ON transistors draw only a limited amount of current. OFF transistors do not draw zero current. Transistor size matters. It is possible for circuits to be both too fast and too slow. Voltages are sometimes between the maximum and minimum values. Taking into account all these nonideal behaviors allows circuit design to bridge the gap between the logical world of 1's and 0's and the real world of transistors.

Key Concepts and Terms

1T memory cell	Leakage power
Active power	Mindelay, maxdelay
Clock skew, clock jitter	MOSFET
Cross talk	NMOS, PMOS, CMOS
Cutoff, saturation, linear	P-to-N ratio
Fanout	Required window
Flip-flop	Subthreshold leakage
Gate leakage	Supply noise
Gate, source, and drain	Threshold voltage
Inverter, NAND, NOR	Transfer curve
Latch	Valid window

Review Questions

1. What has to be true for a MOSFET to be in the linear region?
2. What is the significance of a MOSFET's threshold voltage?
3. Draw the transistor diagram for a CMOS gate implementing the function NOT(*WX* + *YZ*).
4. Should a three-input NOR have a higher or lower P-to-N ratio than a two-input NOR? Why?

5. What is the difference between a speedpath and a race?
6. What is the difference between clock skew and clock jitter?
7. What is the difference between a valid window and a required window?
8. What are four sources of circuit noise?
9. What is the difference between active power and leakage power?
10. What is the difference between subthreshold leakage and gate leakage?
11. [Lab] Measure the current versus voltage of a MOSFET. What is the threshold voltage of the MOSFET?
12. [Bonus] Calculate the optimum fanout for a gate with $T_{fan} = 1$ and $T_{base} = 1$. What is the optimum if $T_{base} = 2$?

Bibliography

Bakoglu, H. *Circuits, Interconnections, and Packaging for VLSI*. Reading MA: Addison-Wesley, 1990.

Bernstein, Kerry et al. *High Speed CMOS Design Styles*. Norwell, MA: Kluwer Academic Publishers, 1998.

Glasser, Lance and Daniel Dobberpuhl. *The Design and Analysis of VLSI Circuits*. Reading MA: Addison-Wesley, 1985.

Harris, David. *Skew-Tolerant Circuit Design*. San Francisco, CA: Morgan Kaufmann Publishers, 2001. [A great book describing how clock skew affects different types of circuits and how its impact can be minimized.]

Mano, Morris. *Computer Engineering: Hardware Design*. Englewood Cliffs, NJ: Prentice-Hall, 1988.

Roy, Kaushik and Sharat Prasad. *Low Power CMOS VLSI: Circuit Design*. New York: Wiley-Interscience, 2000.

Sedra, Adel and Kenneth Smith. *Microelectronic Circuits*. 4th ed., New York: Oxford University Press, 1997.

Sutherland, Ivan, Robert Sproull, and David Harris. *Logical Effort: Designing Fast CMOS Circuits*. San Francisco, CA: Morgan Kaufmann, 1999. [The best book I've seen on fanout and sizing CMOS gates.]

Uyemura, John. *Introduction to VLSI Circuits and Systems*. New York: Wiley, 2001.

Weste, Neil and Kamran Eshraghian. *Principles of CMOS VLSI Design: A Systems Perspective*. Reading MA: Addison-Wesley, 1988. [One of the most commonly used circuit design textbooks, this book is an excellent introduction and includes information on layout missing from most circuit design books.]

Ziegler, James et al. "IBM Experiments in Soft Fails in Computer Electronics (1978–1994)." *IBM Journal of Research and Development*, vol. 40, no. 1, 1998, pp. 3–18. [IBM has probably done more study of soft errors than any other company, and this article is a detailed review of their findings.]

Chapter 8

Layout

Overview

This chapter shows how circuit design is converted into layout. The chapter describes the different layers of material that must be drawn to define transistors and wires, the typical design rules that must be followed, and some common pitfalls in creating layout.

Objectives

Upon completion of this chapter the reader will be able to:

1. Visualize the three-dimensional structure to be manufactured from a piece of layout.
2. Describe the different layout layers used to draw a transistor.
3. Draw the layout for a CMOS gate.
4. Understand how design rules influence the size of layout.
5. Review layout for some of the most common mistakes.
6. Determine drawn, schematic, and layout transistor counts.
7. Understand three different methods of shielding wires.
8. Describe electromigration.

Introduction

Layout is the task of drawing the physical dimensions and positions of all the layers of material that will ultimately need to be manufactured to produce an integrated circuit. Layout converts a circuit schematic to a form that is used to create the photolithography masks required to pattern each layer during the manufacturing process. As a result, layout specialists are often called mask designers. The area of the entire layout will determine the die area, which makes layout a critical step in determining the cost of manufacturing a given product.

Mask designers are often more artists than engineers. Layout is not something that can be analyzed easily with mathematical equations or numeric analysis. Mask designers must be able to look at two-dimensional drawings of layout and visualize the three-dimensional structures that they will create. Creating a single transistor requires multiple layers of material. Making the interconnections between the transistors requires multiple levels of wires. Mask designers must picture in their minds how the different physical pieces of a circuit will fit together.

The layout must faithfully reproduce the circuit desired and should use no more area than required. At each layer there are design rules that specify the minimum widths and spaces that can be reliably manufactured. As if working on a three-dimensional jigsaw puzzle, the mask designer must fit together the pieces needed by the circuit designer while following the design rules of the process, which determine how the pieces are allowed to go together. This jigsaw puzzle, however, has many different possible solutions. Even a very simple circuit can be drawn in layout in an enormous number of different ways. How the layout is drawn will determine not only the die area required by the circuit, but will also impact speed and reliability. In designing the circuit, assumptions were made about the maximum lengths of wires and amount of capacitance on each node. If the layout is poorly done, these assumptions may not be met, forcing the circuit to be redesigned. This can have a huge impact on the schedule of a project, and because it affects schedule, cost, and performance so strongly, creating layout is a critical step for any integrated circuit design.

Creating Layout

Creating layout begins with a circuit schematic. The circuit designer will have determined the number of NMOS and PMOS transistors required as well as the desired length and width of each transistor and all the needed connections between transistors. The mask designer constructs the layout for this circuit beginning with the lowest layers and working up. This is also the order the layers will be created in manufacturing with each new layer being added on top of the last. Creating the layout for transistors requires drawing the well, diffusion, and poly layers.

Figure 8-1 shows a cross section of the three-dimensional structure of an NMOS transistor and the corresponding layout, which shows an overhead view. Creating a MOSFET begins with defining a well region.

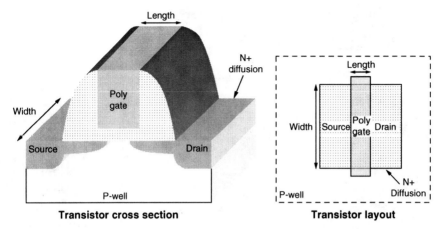

Figure 8-1 Transistor layout.

The well will determine the type of dopant in between the source and drain. For an NMOS device, the well will use P-type dopant, and for a PMOS, the well will be N-type dopant. The layout marks the boundaries of each well region so that during manufacturing each region will be implanted with the correct type of dopant. Inside the well, an area of diffusion meant for heavy doping is drawn. These regions are often labeled "N+" or "P+" to indicate heavier doping than the well. To make transistor source and drain regions, this diffusion is the opposite type as the well, N-type for NMOS devices and P-type for PMOS devices.

The poly gate layer completes the transistor. A poly wire is drawn across the diffusion region. The diffusion on one side of the poly wire will form the source of the transistor and the other side will be the drain. In manufacturing, the poly layer is deposited after the well implant but before the heavy diffusion implant that creates the source and drain regions. The poly gate blocks the diffusion implant from reaching the silicon directly below it, and this silicon stays the same type as the original well. This creates the transistor channel. The source and drain regions on either side of the poly gate are now separated by a narrow strip of silicon with the opposite type of dopant. The width of the poly gate line determines the separation between the source and the drain, which is the transistor length. The width of the diffusion region determines the width of the transistor. Increasing this width provides a wider path for current to flow from the source to the drain. However, this also consumes more area, power, and will slow signals driving the gate of this transistor. Anywhere a poly line crosses a diffusion region a transistor will be created. After the layout for the transistors is complete, the interconnecting wires are added.

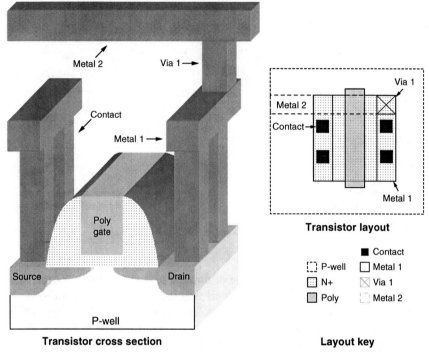

Figure 8-2 Transistor and interconnect layout.

Figure 8-2 shows the cross section and layout of a transistor with the first two metal interconnect layers added. The first layer created is the vertical **contacts** that will connect the diffusion regions or poly gate to the first layer of interconnect. These are columns of metal that form all the connections between wires and transistors. To aid in processing, all contacts are made of the same size. To contact wide transistors, multiple contacts are used. Above the contact layer is the first layer of wires, typically called *metal 1 (M1)*.

The metal 1 layer rests on top of a layer of insulation added on top of the transistors. This prevents it from touching the silicon surface or the poly layer. Metal 1 wires can create connections by crossing the horizontal distance between two contacts. The drain of one transistor can be connected to the gate of another by using a contact to go up from the drain to a metal 1 wire and then another contact to go down to the poly gate. However, metal 1 wires will always contact other metal 1 wires they cross. If the needed path is blocked by another metal 1 wire, the connection must be made using the next level of wiring.

Vias are vertical columns of metal between levels of wires. Just like contacts they form vertical connections, but between wires instead of between wires and transistors. The via1 layer creates connections between metal 1 and metal 2 wires above. A layer of insulation separates each wiring level so that connections are made between levels only where vias are drawn. This allows metal 2 wires to cross over metal 1 without making contact. Later, another via1 could take the electrical signal back down to metal 1 where a contact could then connect to the appropriate transistor.

Modern microprocessors might use seven or more layers of vias and metal wires to make all the needed connections between hundreds of millions of transistors. Each metal layer uses one type of via to connect to the layer below and another type to connect to the layer above. With the exception of metal 1, each metal layer typically has all the wires drawn in the same direction. For instance, metal layers 2, 4, and 6 might contain wires traveling east to west across the die whereas layers 3, 5, and 7 travel north to south. Alternating layers in this fashion allow all the wires on one level to be drawn in parallel, making the best use of area while still allowing connections to be made between any two points on the die by just using more than one layer. Adding more layers of metal interconnect may allow more efficient connections and a smaller die, but each new layer adds cost to the manufacturing process. Combining the layers required for transistors and those for interconnects allows any needed circuit to be drawn. Figure 8-3 shows the construction of the layout of an inverter.

At the top left of Fig. 8-3 is the circuit schematic of the inverter to be drawn in layout. The schematic specifies the dimensions of each transistor and shows all the needed connections. Below is shown the palate of seven materials that will be used to create the layout: P-type and N-type wells, P-type and N-type diffusion, polysilicon, contacts, and metal 1 wires. As drawn by the circuit designer, the circuit schematic is not meant to imply any particular orientation. The mask designer decides where each transistor and connection will be placed. Figure 8-3 breaks the construction of this layout into three steps with the top showing the circuit schematic oriented, as imagined by the mask designer, and the bottom showing the layout.

The first step is creating the transistors themselves. Step 1 shows the layout for a PMOS transistor and an NMOS transistor. Drawing a well with two rectangles of diffusion inside forms each transistor. One piece of diffusion forms the well tap, which provides an electrical connection to hold the well at a fixed voltage. The well tap is drawn as the same dopant type as the well. The other rectangle of diffusion is the opposite type of the well and forms the transistor source and drain regions. A piece of poly drawn across this diffusion forms the transistor gate. The width of the poly gate determines the length of the transistor and the width of diffusion it crosses determines the width of the transistor.

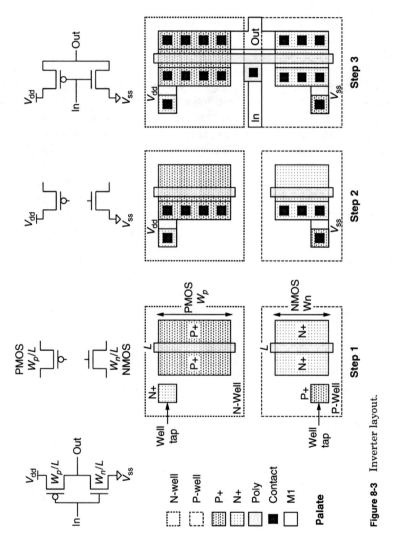

Figure 8-3 Inverter layout.

The PMOS device uses P-type source and drain in an N-well with an N-type well tap. The NMOS device is the exact opposite with N-type source and drain in a P-well with a P-type well tap.

In step 2, the connections to the supply voltage (V_{dd}) and the ground voltage (V_{ss}) are made. The mask designer chooses which side of the transistor to use as the source terminal and places as many contacts as there is room for along that side. Also, one contact is placed on each well tap. Two metal 1 wires are drawn to bring the needed voltage to the contacts. Presumably these metal 1 wires will be connected to other wires at a higher level when the inverter layout is used. Step 3 connects the inverter input and output. The input is drawn in metal 1 and goes through a contact to drive both poly gates. The output is also drawn in metal 1 and uses multiple contacts to connect to the drain diffusion of each transistor.

Figure 8-4 shows the same three steps creating the layout for a NAND gate. For simplicity, the wells and well connections are left out of this and most other figures in this chapter. It is important to remember that every transistor requires a well and every well requires a well tap, but many devices can share a single well. As a result, wells and **well taps** might be drawn only after the devices from multiple gates are placed together.

A NAND gate requires two PMOS devices and two NMOS devices. Each device could be drawn using separate areas of diffusion, but for transistors that will be connected at the source or drain, it is more efficient to draw a single piece of diffusion and cross it with multiple poly lines to create multiple transistors. Connecting both outside diffusion nodes to V_{dd} and the shared inner node to the output signal connects the PMOS devices of the NAND in parallel. The NMOS devices of the NAND are connected in series by connecting the outside side of one device to V_{ss} and the outside of the other device to the output signal. The intermediate node X of the NMOS stack requires no contacts. By drawing the two transistors from a single piece of diffusion the connection at node X is made through the diffusion itself without the need for a metal wire. This reduces the capacitance that must be switched and makes the gate faster.

The primary goal of layout design is to create the needed circuits in the minimum area possible. The larger the total die size, the more expensive the chip will be to fabricate. Also, smaller layout area allows for wires crossing these circuits to be shorter, having less resistance and capacitance and causing less delay. This creates a great deal of pressure to achieve the smallest layout possible, or put another way, the highest possible layout density.

Layout Density

Layout density is usually measured as the number of transistors per die area. Layout density targets allow early transistor budgets to be used

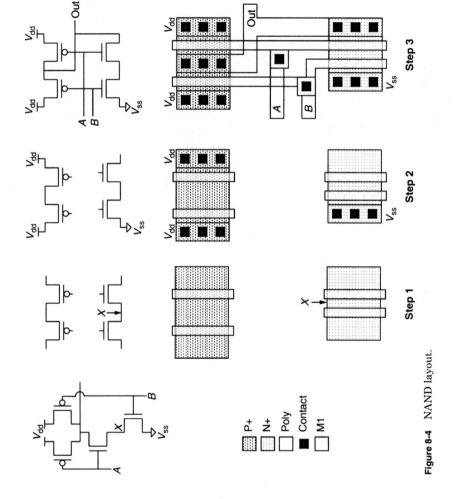

Figure 8-4 NAND layout.

in estimating the die area required by each portion of the design, which provides an estimate of what the total die area will ultimately be. Failure to meet these density targets can lead to products that are more costly or have less performance than expected. The number and size of transistors needed as well as the limits of the manufacturing process determine the layout area required by a circuit.

A wire drawn too thin may be manufactured with breaks in the wire, preventing a good electrical connection. A pair of wires drawn too close together may touch where they are not supposed to, creating an electrical short. To prevent these problems, process engineers create **layout design rules** that restrict the minimum widths and spaces of all the layers. Additional rules set minimum overlaps or spaces between different layers. The goal is to create a set of rules such that any layout that meets these guidelines can be manufactured with high yield. Table 8-1 shows an example set of design rules.

These example rules are written in terms of a generic feature size λ. A 90-nm generation process might use a λ value of 45-nm, which would make the minimum poly width and therefore minimum transistor length $2\lambda = 90$-nm. Of course, what two different manufactures call a "90-nm" process might use different values for λ and perhaps very different design rules. Some manufacturers focus more on reducing transistor size and others more on wire size. One process might use a short list of relatively simple rules but a large value of λ. Another process might allow a smaller λ but at the cost of a longer more complicated set of design rules. The example rules from Table 8-1 are applied to the layout of an inverter in Fig. 8-5.

TABLE 8-1 Example Layout Design Rules

Rule label	Dimension	Description
W1	4λ	Minimum well width
W2	2λ	Minimum well spacing
D1	4λ	Minimum diffusion width
D2	2λ	N+ to P+ spacing (same voltage)
D3	6λ	N+ to P+ spacing (different voltage)
D4	3λ	Diffusion to well boundary
P1	2λ	Minimum poly width
P2	2λ	Minimum poly space
P3	1λ	Minimum poly diffusion overlap
C1	$2\lambda \times 2\lambda$	Minimum contact area
C2	2λ	Minimum contact spacing
C3	2λ	Minimum contact to gate poly space
C4	1λ	Poly or diffusion overlap of contact
M1	3λ	Minimum M1 width
M2	3λ	Minimum M1 space
M3	1λ	M1 and contact minimum overlap

248 Chapter Eight

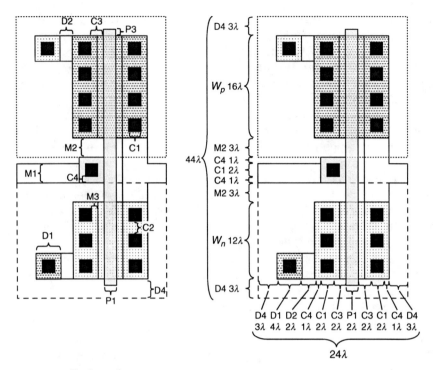

Figure 8-5 Design rules.

The left picture in Fig. 8-5 shows some of the widths and spaces that are limited by the design rules. The right picture shows the design rules and dimensions that ultimately determine the required height and width of this piece of layout. The height of 44λ is determined by the needed transistor widths and the width and space of a metal 1 connecting wire. The width of 24λ is determined by the size of the well taps, the poly gate, and contacts to the source and drain. Even small changes in design rules can have a very large impact on what layout density is achieved and how layout is most efficiently drawn.

In the example shown in Fig. 8-5, the height of the layout is determined in part by the width of the transistors. To create an inverter with wider transistors, the layout could simply be drawn taller and still fit in the same width. However, there is a limit to how long a poly gate should be drawn. Polysilicon wires have much higher resistance than metal wires. A wide transistor made from a single very long poly gate will switch slowly because of the resistance of the poly line. To avoid this, mask designers commonly limit the length of poly gates by drawing wide transistors as multiple smaller transistors connected in parallel. This technique is called *legging a transistor*. An example is shown in Fig. 8-6.

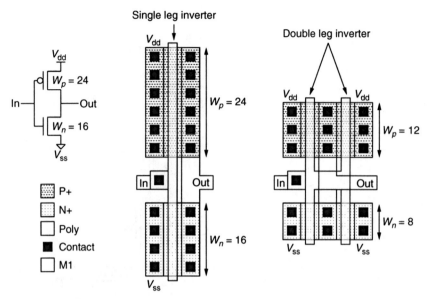

Figure 8-6 Legging of transistors.

In the example of Fig. 8-6, the circuit designer has asked for an inverter with a PMOS width of 24 and an NMOS width of 16. This could be drawn as a single leg inverter, but this would make the poly gates of the transistors relatively long. An alternative is to draw two devices (or legs) to implement each schematic transistor. Two PMOS devices of width 12 connected in parallel will draw the same current as a single PMOS device of width 24. This trick can be repeated indefinitely with gates requiring extremely large devices being drawn with many legs. The legging of schematic transistors often creates confusion when counting transistors in order to measure layout density. Transistors can actually be counted in three ways: drawn devices, schematic devices, and layout devices. These three counts are shown for the circuit and layout in Fig. 8-7.

The circuit of Fig. 8-7 is two separate inverters, one with input In(0) and output Out(0) and the other with input In(1) and output Out(1). It is very common for gates to be arrayed in this fashion when the same logic operation must be performed on every bit of a number. The **drawn device count** of this circuit is said to be 2, since the circuit consists of multiple copies of a single inverter with 2 transistors. Drawn device count is a good measure of circuit complexity. In implementing the two inverters, mask designers will not need to draw each separately. Instead, they will draw a single inverter and then place multiple copies of it. Clearly creating the layout for 50 identical inverters is far simpler than drawing layout for a

250 Chapter Eight

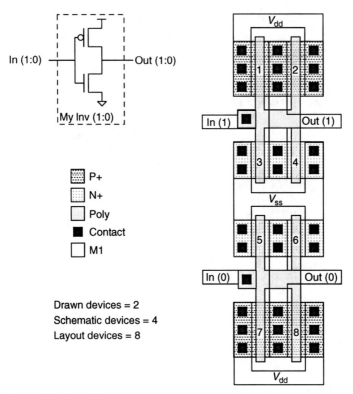

Figure 8-7 Transistor counts.

circuit with 100 uniquely sized and connected devices. A memory array design may contain a very large number of transistors, but most of them might come from a simple memory cell repeated a very large number of times. The better measure of complexity, both of the circuit and mask designs, is the number of unique drawn devices.

The **schematic device count** for two inverters is 4. Each inverter has two devices and two copies are needed, making four schematic transistors. However, because of **transistor legging**, the layout may have more than four devices. In Fig. 8-7, each transistor has been legged twice for a total of eight transistors, which have been numbered in the figure. The **layout transistor count** is the number of transistors that will actually be manufactured, which in this case is 8. When attempting to measure layout density, it is important to distinguish between schematic transistors and layout transistors. Because very wide schematic transistors may need to be implemented as many separate layout devices, layout density should be measured by counting layout rather than schematic devices per area.

Layout density is often used as a measure of how successful the mask designer has been at using the minimum area possible. However, this metric must be very carefully applied since the best possible density may be different for different types of circuits. Even when measured properly, layout density varies greatly from one circuit to another. Figure 8-8 shows the layout for two circuits each containing eight layout devices.

The layout on the left is smaller. It has a higher layout density, but it is not because it used different design rules or was drawn by a better mask designer. The circuit drawn on the left has fewer inputs. This means the total number of wires coming into the circuit is smaller, and primarily the devices needed determine the area required. This type of layout is said to be device limited. The circuit on the right requires more inputs, which means more wires. The area of the layout must be larger to accommodate the wires. This type of layout is said to be wire limited. For most layout, the number of devices and the number of interconnects together determine the minimum area required. Very rarely is layout

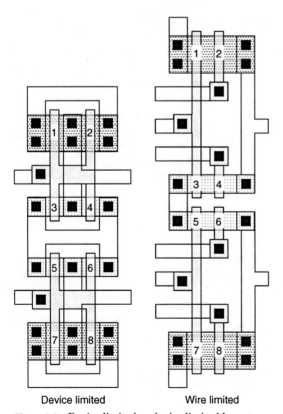

Device limited   Wire limited

Figure 8-8 Device limited and wire limited layout.

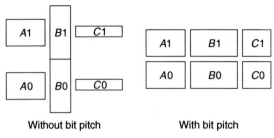

Figure 8-9 Bit pitch.

completely device or wire limited. Variations in layout density can be because of varying levels of skill and experience among different mask designers, but often differences in density are because of the different types of circuits being drawn.

To try and improve layout density and allow more reuse, most layout is drawn with a specific **bit pitch**. It is common that many different bits of data will have to go through similar logic. It is convenient if each row of layout that processes a single bit has a uniform height. Figure 8-9 shows how assembly is more efficient for cells of a uniform bit pitch.

If layout cells A, B, and C are each drawn with a different height, then a great deal of area is wasted when trying to assemble them into two identical rows. By creating each cell with the same height and only varying the width as needed, assembly is easier and more efficient. The bit pitch is typically chosen based on the number of wires expected to be routed horizontally within each row. The layout density of some cells will inevitably suffer as a result of being forced to maintain a uniform height. The highest density for each cell would be achieved by allowing each to be drawn only as high as was optimal for that cell alone, but after the cells are assembled the overall layout density will typically be far higher with a uniform bit pitch. The area saved in assembling large numbers of cells together can easily outweigh any area wasted inside the cells.

A common and painful mistake when estimating the space required for routing signals within each bit pitch is forgetting to allow tracks for a **power grid**. In addition to the signal lines routed to each logic gate, every gate must have access to supply (V_{dd}) and ground (V_{ss}) lines. All the power supplied to the die will come from the chip package to the top level of metal. The V_{dd} and V_{ss} voltages must be routed down through each level of metal wiring to reach the transistors themselves. All the lines running in parallel at each level are tied together by the perpendicular lines of the levels above and below, forming a continuous power grid.

This grid acts like the electrical power grid of a city, supplying power to each logic gate the way electricity is routed to each home. In cities

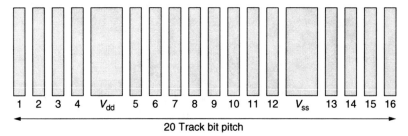

Figure 8-10 Bit pitch with power grid.

where the power grid is poorly designed or when demand for power is too great, brownouts can occur. Lights go dim and appliances may not work because the voltage being supplied by the grid has dropped below its intended value. The same can happen on a computer chip. If the power grid resistance is too high or many wide transistors in the same area draw current at the same time, the voltage in one region of the die can droop below its normal value. This will cause the logic gates in this region to switch more slowly and if severe enough can cause incorrect results when the slowest circuit paths fail to keep up with the fixed processor clock frequency.

To prevent this problem, the standard bit pitch may be designed with the V_{dd} and V_{ss} lines taking up 20 percent or more of all the available wiring tracks in every layer of metal wiring (Fig. 8-10). In a bit pitch technically wide enough to route 20 signals, there may be space only for 16 data signals once supply and ground lines are routed. The power lines are typically drawn wider than other wires to minimize their resistance. In the end, frequency is as greatly affected by layout as it is by circuit design.

Layout Quality

Transistor density alone is not a good measure of layout quality. In addition to optimizing area, mask designers must also consider speed and reliability. There are many ways layout can be drawn to reduce area, still meeting the design rules, but producing a circuit that will either be too slow or unreliable. Layout reviews where engineers and mask designers together review plots of layout for potential problems are still often the best way to assure layout quality. Figure 8-11 shows a circuit and corresponding layout with some of the comments that might come from a layout review.

The circuit shown in Fig. 8-11 is a NAND gate followed by an inverter. The layout has achieved good density. All the P-type and N-type devices

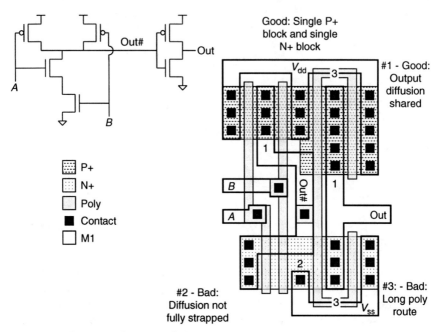

Figure 8-11 Layout review example #1.

needed are created from a single block of P+ diffusion and a single block of N+ diffusion. The output signals Out# and Out are driven by shared diffusion nodes. This is marked as note #1 in the figure. Using shared diffusion nodes will reduce the total amount of capacitance on these signals, allowing them to switch more quickly. However, in an attempt to achieve high density, this layout is left with a couple of potential problems. One of the N+ diffusion nodes is not fully strapped, meaning it is not as well connected as it could be (note #2). The N+ diffusion is wide enough for three contacts but one of the nodes connected to V_{ss} uses only a single contact. This is because another metal wire has been routed over the diffusion, using the space where more contacts would have been placed. The circuit will function correctly, but current from two N-transistors will have to travel through a single via and some distance through the diffusion alone. Diffusion has extremely high resistance, and without full strapping the transistors using this node will be slower than normal.

The inverter in this layout has been created with two N-type legs and two P-type legs to avoid single excessively wide transistors, but the poly gates for these devices have been drawn in series rather than in parallel (note #3). This makes a very long poly line just to reach the beginning of the poly gate of the second legs. The resistance of the poly will

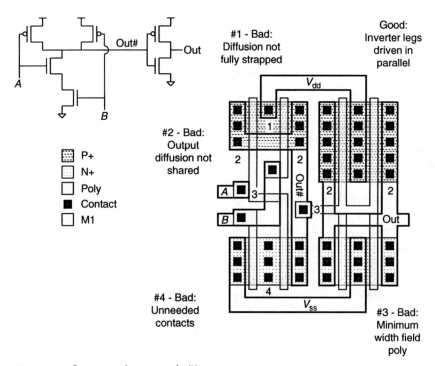

Figure 8-12 Layout review example #2.

make these second legs turn on somewhat after the first pair of legs and slow down the overall delay of the inverter. Figure 8-12 shows a second attempt at layout for the same example circuit.

The second example layout is slightly larger and has addressed some problems but created others. The main difference from the first example is that this layout uses separate blocks of diffusion to create the devices for the NAND gate and for the inverter. Doing this and moving the V_{ss} node for the NAND gate to the opposite side allows all the N+ diffusions nodes to be fully strapped. However, this layout now has one of the P+ diffusion nodes not fully strapped (note #1). The legged devices of the inverter are now driven in parallel, but the Out# and Out nodes are no longer driven by shared diffusion nodes (note #2). In the previous version of layout, Out# and Out each contacted just one N+ and one P+ diffusion node. In this layout, Out# contacts two P+ nodes and Out contacts two N+ and two P+ nodes. This will add capacitance and slow down these signals.

The legs of the inverter are driven in parallel by a length of field poly, meaning poly that is not over a diffusion region and therefore does not form a transistor gate. The required length of the transistor determines the width of gate poly, but field poly does not affect any transistor parameters. Because of the high resistance of poly, field poly should be drawn at greater than minimum width where possible. In this layout, the long lengths of minimum width field poly will add resistance and slow down the switching of the gates (note #3). Finally, the intermediate node between the two N-devices of the NAND gate is drawn with contacts when this node does not need to be connected to any metal wires (note #4). The space between the two N-devices could be reduced if these unnecessary contacts were removed. Figure 8-13 shows a third version of layout for the same circuit.

The third example corrects the problems of the previous two examples. All the diffusion nodes are now fully strapped with the maximum number of contacts allowed and all the contacts connected to metal 1. The signals Out# and Out are both driven by shared diffusion nodes keeping their capacitance to a minimum (note #1). The legs of the inverter are driven in parallel and the field poly connecting them is drawn wider than minimum where allowed (note #2). Although the very first example of this layout achieved the smallest area, this final third example will make the fastest and most robust circuit.

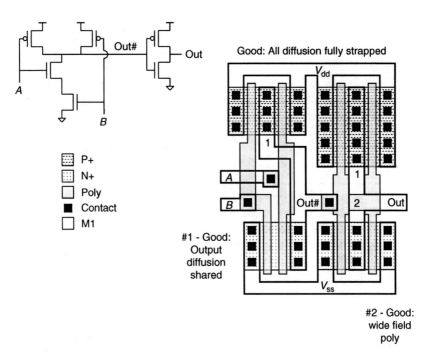

Figure 8-13 Layout review example #3.

Layout also has a large impact on circuit noise. As described in Chap. 7, one of the largest sources of electrical noise is cross talk between wires. The switching of one wire causes a neighboring wire to be pulled from its ideal voltage. Layout can alleviate this problem by paying special attention to which signals are routed next to one another. Routing a sensitive signal wire to minimize cross talk noise is called *shielding*.

Figure 8-14 shows three ways of reducing noise on the signal A (Sig$A$) wire by shielding. The simplest and most effective method is **physical shielding**. Physical shielding means routing a signal next to either a supply line or a ground line. Because the supply and ground lines never change in voltage, they will never cause cross talk noise. They also will reduce the noise induced by other sources by adding capacitance to a fixed voltage to the Sig$A$ line. Routing with a fixed voltage line on one side is called **half shielding** and routing with fixed voltage lines on both sides is called **full shielding**. Whether the supply or ground line is used does not matter. What is important is that the neighboring lines are carrying a constant voltage. Physical shielding is very effective, but also expensive in terms of wire tracks. A fully shielded signal line now occupies three wire tracks instead of just one. Trying to physically shield many signals will rapidly use up all the available wire tracks and require layout area

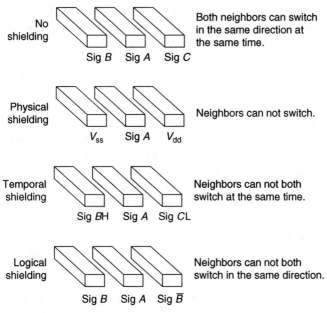

Figure 8-14 Shielding.

to be increased to allow more room for routing wires. Cross talk noise is reduced at less cost by using **temporal** or **logical shielding**.

The worst-case noise will be induced in a line when both of its neighbors switch in the same direction at the same time. Physical shielding prevents this by using power lines, which never switch. Temporal and logical shielding use signals that never switch in the same direction at the same time. For example, temporal shielding might route, as neighbors to a sensitive line, a signal that switches only during the high phase of the clock (SigBH) and a signal that switches only during the low phase of the clock (SigCL). Because the two neighbors never switch at the same time, the worst-case noise on Sig$A$ is reduced. Sig$A$ is effectively half shielded. Alternatively, the neighbors signal B (Sig$B$) and its logical opposite Sig $\overline{B}$ could be chosen. These signals may switch at the same time but always in opposite directions. The noise pulling in opposite directions will cancel each other out and Sig$A$ is effectively full shielded. Even if Sig$B$ and Sig $\overline{B}$ might switch at slightly different times because of different delay paths, the worst case of two neighbors switching in the same direction is avoided. By allowing signal lines to be routed next to Sig$A$, temporal and logical shielding reduce sensitivity to noise without the loss of as many wiring tracks. When drawing interconnects, in addition to noise, reliability must be considered.

When a thin wire carries large amounts of current, it will become hot. Too much heating causes wires to fail over time just like the filament in a light bulb. As the wire is repeatedly heated and cooled, it expands and contracts. Because not all the materials in an integrated circuit expand and contract with temperature at the same rate, stress is put on all circuit elements by this heating. Eventually breaks in the wire or between the wire and its contacts and vias are created, and the circuit will fail. Processes that use low-K insulation material between wiring levels may be more susceptible to these problems. Low-K insulation reduces the effective capacitance of the wires, making circuits faster, but these materials also tend to be soft. This makes them less able to hold wires firmly in place as they try to flex while being heated. To avoid the problem of overheating wires, the amount of current a signal wire will carry must be considered in sizing interconnects. An extremely wide transistor driving a long minimum width wire risks creating a failure after a very short time. A maximum current density for each wiring layer may be given as a guideline to mask designers in sizing interconnects.

The problem of high current densities is particularly bad in wires that always carry current in the same direction. The atoms that make up the wire tend to be pushed out of thin spots in the wire where current is the strongest, just like rocks being pushed along in a fast running stream. This phenomenon is called **electromigration** and causes thin spots in the wire with the highest current density to gradually become thinner. As the wire thins, current density is increased, and the process accelerates. Thin spots in the wire become even thinner until eventually a break is created. This is an additional reason to draw supply and ground lines wider than minimum. The same kinds of failure happen in the vias that connect different levels of wires. The more current carried, the more vias required for reliability.

The need for high speed, immunity to noise, and long-term reliability make quality layout much more than simply the layout with the smallest area.

Conclusion

Layout is one of the most time-consuming aspects of integrated circuit design and as a result has been the focus of many efforts at automation. Many of these efforts have met with limited success and much of the job of mask design is still very much a manual process. A consistent goal of layout automation has been to automate the process of compacting from one manufacturing generation to the next. Effort for compaction designs would be dramatically reduced if the layout of the previous design could be somehow automatically converted to the new process. If a new process were created using all the same design rules as the previous generation but with the basic feature size represented by λ simply made smaller, automatic conversion of layout would be trivial. All the dimensions of the layout could be shrunk by the same amount to create new layout for the new process. In reality, this is rarely the case.

Scaling all the design rules by the same amount allows the rule that could be reduced the least to determine the scaling for all the other rules. Inevitably, each new process generation finds some design rules more difficult to scale than others. Smaller metal 1 widths and spaces may be more easily achieved than smaller poly widths and spaces or vice versa. More area reduction is achieved by individually scaling each design rule as much as is practical, but this leads to new process design rules, which are not a simple constant scaling of the previous generation. If all the dimensions can not be scaled by the same amount, automatic layout compaction becomes much more difficult.

Making matters worse, new process generations often add new design rules that were not previously needed. Perhaps to achieve better control, a new process might specify that all poly gates must be routed in the same direction whereas an earlier generation allowed any orientation. Perhaps all metal 1 wires must be one of two specific widths whereas the earlier generation allowed any width above a minimum value. These new rules may allow the new process to achieve higher densities and higher yields, but they can easily make automated compaction impossible.

There has been more success in automating the placement and routing of custom-drawn layout cells. Synthesis and cell-based design flows allow layout from custom-drawn libraries to be placed and connected by automated tools. The library of cells itself must still be drawn by hand, but once created the effort of using this layout is greatly reduced. Still, the overall density achieved by using these tools is typically less than that could be achieved by full custom layout. The time and effort saved in layout can come easily at the cost of a larger die area. It seems likely that there will always be a strong need for experienced mask designers and hand-drawn custom layout.

Layout is the final design step before chips are manufactured. When the layout is complete, it is used to create the photolithography masks, and production of prototype chips begins. This date when the layout is complete is called **tapeout**, since in the past the layout of the chip would be copied on to magnetic tape to be sent to the fabrication facility. Tapeout is one of the most important milestones in any chip design, but it is by no means the last. The following chapters will describe the chip manufacturing, packaging, and testing and debugging, which must happen before shipping customers.

Key Concepts and Terms

Bit pitch	Power grid
Contact and via	Physical, temporal, and logical Shielding
Drawn, schematic, and layout transistor count	Tapeout
Electromigration	Transistor legging
Half and full shielding	Well taps
Layout design rules	

Review Questions

1. What is the difference between a contact and a via?
2. What does layout density measure? Why is layout density important?
3. Why should wide transistors be legged?
4. What are the uses of drawn transistor count and layout transistor count?
5. What is the difference between device-limited and wire-limited layout?
6. What is the difference between physical, logical, and temporal shielding?
7. [Bonus] Use colored pens or pencils to draw the layout for a two-input NOR gate. Use a different color for each different material.

Bibliography

Baker, Jacob. *CMOS Circuit Design, Layout, and Simulation.* 2d ed., New York: Wiley-IEEE Press, 2004.

Christou, Aris. *Electromigration and Electronic Device Degradation.* New York: Wiley-Interscience, 1993.

Clein, Dan. *CMOS IC Layout: Concepts, Methodologies, and Tools.* Boston, MA: Newnes, 1999.

Saint, Christopher and Judy Saint. *IC Layout Basics: A Practical Guide.* New York: McGraw-Hill, 2001. [A good plain language introduction to the basic concepts of layout.]

Saint, Christopher and Judy Saint. *IC Mask Design: Essential Layout Techniques.* New York: McGraw-Hill, 2002. [A follow-on to their layout basics book, this book gives specific do's and don'ts for anyone working with layout.]

Uyemura, John. *Introduction to VLSI Circuits and Systems.* New York: Wiley, 2001.

Weste, Neil and Kamran Eshraghian. *Principles of CMOS VLSI Design: A Systems Perspective.* 2d ed., Reading, MA: Addison-Wesley, 1994. [One of the reasons why this book is an industry and graduate school standard is because it covers layout along with circuit design.]

Chapter 9

Semiconductor Manufacturing

Overview

This chapter shows how layout is used to create integrated circuits. The chapter describes the manufacture of wafers and the processing steps of depositing, patterning, and etching different layers of material to create an integrated circuit. An example CMOS process flow is described showing chip cross sections and equivalent layout at each step.

Objectives

Upon completion of this chapter the reader will be able to:

1. Describe the different types of silicon wafers in common use.
2. Understand the advantages and disadvantages of different types of deposition.
3. Compare and contrast different types of etching.
4. Understand the creation of masks from layout and their use in photolithography.
5. Visualize the interaction of exposure wavelength, feature size, and masks.
6. Describe the use of ion implantation in integrated circuit fabrication.
7. Explain the steps of a generic CMOS manufacturing flow.
8. Understand how sidewalls and silicide are used in MOSFET formation.

Introduction

All the semiconductor design steps discussed in previous chapters are based upon simulations. This chapter shows how after all these steps are completed, actual integrated circuits are manufactured. To make a chip with millions of transistors and interconnections would not be practical if the devices had to be created in a serial fashion. The features to be produced are also far too small to be individually placed even by the most precise equipment. Semiconductor manufacturing achieves rapid creation of millions of microscopic features in parallel through the use of light sensitive chemicals called *photoresist*. The desired layout is used to create a mask through which light is shone onto the chip after it is coated in photoresist. The photoresist chemically reacts where exposed, allowing the needed pattern of material to be formed in the photoresist as if developing a photograph. This process is called **photolithography** (literally printing with light). Wires and devices are not individually placed in the correct positions. Instead, all that is needed is for a pattern of light representing all the needed wires to be projected onto the die. The use of photolithography is shown in Fig. 9-1.

Photolithography allows the pattern for a particular layer to be created in photoresist, but separate deposition and etching steps are needed to create that pattern in the required material. Various deposition steps create metal layers, layers of insulation, or other materials. Etching allows unwanted material to be removed, sometimes uniformly across the die and sometimes only in a specific pattern as determined by photolithography. Repeated deposition, lithography, and etching are the basis of semiconductor manufacturing. They give the ability to add a

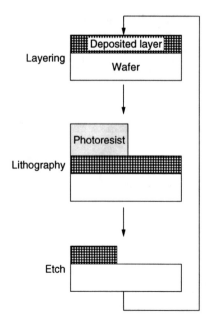

Figure 9-1 Deposition, lithography, and etch.

layer of the desired material, to create a pattern in photoresist, and then etch away unwanted material. Although semiconductor manufacturing creates all the wires or devices of a particular layer simultaneously, the individual layers are processed one at a time, gradually building the required three-dimensional structure.

As feature sizes have been scaled down, semiconductor manufacturing has grown steadily more sensitive to particles in the air far too small to be seen with the naked eye. A single particle 0.5 microns (μm) in diameter can easily cause shorts or breaks in a wire 1 μm wide. In a typical city, a single cubic foot of outdoor air can easily contain millions of particles of this size. To make feature sizes below 1 μm possible, all manufacturing must be carried out in specially constructed **cleanrooms**. These facilities constantly blow air down from the ceiling and through special filters in the floor before returning the cleaned air to the room. The level of cleanliness achieved is measured as the average number of particles above 0.5 μm in diameter per cubic foot of air. Modern fabrication can require a class 10 environment with only 10 particles per cubic foot or even a class 1 cleanroom with less than 1 particle per cubic foot.

The most difficult part of achieving this incredibly low level of particles is allowing fab workers any access to the fab itself. Workers enter through air locks to protect the cleanroom from outside air, but a person in normal clothes standing motionless can still emit 100,000 particles per minute.[1] Bits of hair, skin, or clothing, even tiny water droplets in a person's breath, can all ruin modern chips. To keep the air clean, workers must be covered from head to toe in specialized garments, called *bunny suits*, designed to trap particles. To further limit contamination, the latest fabrication equipment can load wafers from airtight containers, process them, and then return them to their containers without ever exposing them to the cleanroom air.

This chapter will discuss some of the details of the many different types of deposition, lithography, and etching used in semiconductor fabrication. Any semiconductor process requires these basic capabilities. The chapter ends with these steps being used in an example CMOS process. The starting point for the manufacture of any integrated circuit is the creation of semiconductor wafers.

Wafer Fabrication

Today, the manufacture of almost all semiconductor chips starts with silicon. There are a small number of other semiconductor materials used,

---

[1]Van Zant, *Microchip Fabrication*, 77.

but none is as inexpensive and easy to process as silicon. To provide uniform electrical properties throughout, the silicon must be almost totally pure and a perfect crystal. In a crystal, the arrangement and spacing of neighboring atoms around each atom are the same. This means that the energy levels available to electrons will be the same throughout the material. Defects in the crystal hinder current flow where it is needed and allow current to flow where it is not needed.

The most common method for making silicon wafers is called **Czochralski (CZ) crystal growth** (after its Polish inventor). A small silicon seed crystal is dipped into a heated crucible filled with molten silicon. As it is spun, the seed crystal is very slowly pulled from the melt. Liquid silicon clings to the crystal, and as it cools it freezes in place, causing the crystal to grow larger. The atoms naturally arrange themselves to follow the pattern of the original seed crystal. As the crystal is pulled from the melt, a cylindrical crystal silicon ingot is formed, which can be several feet long. By varying the pull rate of the crystal (typically only a few millimeters per hour), the diameter of the ingot is controlled. The largest ingots are up to 12 in (300 mm) in diameter. The ingot is shaped to the correct diameter and sliced into circular wafers about 1 mm thick. All the later processing will create circuits in just the top couple of microns of the wafer. The full thickness of the wafer is required only to allow it to be handled during processing without breaking.

The wafers must then be carefully polished to create an almost perfectly flat surface. A typical variation in height might be only a couple of microns across a 300-mm diameter wafer. This would be equivalent to a football field that varied in altitude by less than a hundredth of an inch. This smoothness is critical for later photolithography steps, which must be able to focus a clear image on the surface of the wafer. Wafers produced in this fashion are called **bulk wafers**. These wafers can be used directly for integrated circuit manufacturing or can go through more processing to create different types of wafers.

It is extremely difficult to make silicon that is neither N-type nor P-type. Almost inevitably, some impurities cause the silicon to have excess electrons or excess holes. Rather than leave this to chance, N-type or P-type dopants added to the melt ensure wafers with a specific doping level and electrical resistivity. A low doping level makes it easy for later processing steps to create regions of the opposite type from the bulk. If a P-type bulk wafer is being used, then any region that is to be made N-type must first add enough dopants to counteract the P-type dopants already in the wafer. A low background doping makes precise control of the net doping in each region after processing easier. However, CMOS circuits on lightly doped wafers can be susceptible to latchup.

Latchup occurs when electric current flows between wells of different types. This changes the well voltages and causes the heavy diffusion regions in the wells to emit charges. These charges diffuse from one well to the other and produce still more current between the wells. Once started, this positive feedback will continue driving large amounts of current beneath the surface of the chip. The only way to return the chip to normal operation is to switch its power off altogether and restart. A heavily doped bulk wafer helps prevent latchup by providing a low resistivity path between the wells beneath the surface. This prevents the voltage difference between the wells that triggers latchup. However, creating transistors may require very lightly doped regions, and this is difficult when the background doping is high. A common solution is epitaxial wafers.

To create an **epi-wafer**, a lightly doped layer of silicon is deposited on top of a heavily doped bulk wafer. Epitaxy is a special type of deposition process that forms a crystalline layer. The silicon atoms deposited on the wafer naturally line up with the already existing structure, maintaining the crystal pattern throughout the wafer. Epi-wafers provide a lightly doped surface layer on top of a heavily doped bulk. This enables optimal conditions for the creation of MOSFETs on the surface while helping to prevent latchup beneath the surface.

A third type of wafer sometimes used for CMOS circuits is a *silicon-on-insulator* (SOI) wafer. **SOI wafers** eliminate latchup entirely by building the transistors and wells in an extremely thin layer of crystalline silicon on top of a layer of insulator. Transistors made on SOI wafers have reduced diffusion capacitance, which can improve switching speed, but creating SOI wafers is more costly. Although epitaxy can add a crystalline layer to an already existing silicon crystal, silicon deposited on top of an insulator will have no crystal pattern to follow and will not form a good crystal structure. Instead, SOI wafers can be made by starting with two bulk wafers.

One wafer has a layer of insulation grown on its surface. It is then turned upside down and bonded to the surface of the other wafer. This creates a double-thick wafer with a layer of insulation in the center. The top wafer is then polished down to reduce its thickness until only a very thin layer of crystal silicon remains above the layer of insulation. Wafers created this way are called *bonded and etched-back SOI (BESOI) wafers*. Cross sections of **bulk, epi, and SOI wafers** are shown in Fig. 9-2.

Bulk wafers are the least expensive with epi-wafers and SOI wafers offering progressively more performance and immunity from latchup at progressively more cost. Manufacture of wafers is a specialized business with most companies that make integrated circuit products choosing to purchase their wafers from other businesses rather than making their own. Starting with a blank wafer, processing to create a specific integrated circuit product begins with depositing new materials onto the wafer.

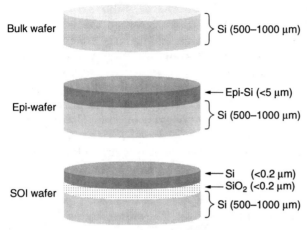

Figure 9-2 Wafer types. (Baker, *CMOS Circuit Design*, 100.)

Layering

Layering steps are those that add new material on top of or into the silicon wafer. Dopants must be added to the silicon to create transistors. Layers of insulating and conducting materials must be added to form wires. Other materials may be added to aid in processing, to provide better electrical contacts between layers, to block diffusion of material between layers, or to provide better adhesion. The electrical, mechanical, and chemical properties of each new layer must be carefully considered both alone and in interaction with all the other needed materials. One of the most important layering steps in creating transistors is doping.

Doping

Some of the earliest steps in a CMOS process involve doping the silicon to create the wells and diffusion regions that will form the transistors. Doping requires adding precise amounts of N-type or P-type dopants to specific regions of the wafer. The simplest way this is accomplished is through solid-state diffusion.

Diffusion in liquids or gases is something that we encounter all the time. Put a drop of ink into a glass of water, and it will slowly expand until eventually all the water is a uniform color. At all times, the molecules of the ink are moving in random directions. If there is a high concentration of them in one region, the chance of a molecule of ink leaving the region is higher than the chance of a molecule of ink entering the

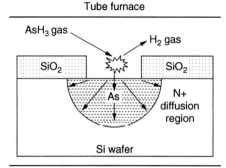

Figure 9-3 Solid state diffusion.

region. This is true only because there are more molecules inside the region than outside. This causes regions of high concentration to diffuse into regions of low concentration. Once the concentration is uniform everywhere, the chances of a molecule of ink leaving or entering any region is the same, and there is no longer any net diffusion. Diffusion will always occur when there is a nonuniform concentration of material and there is sufficient energy for it to move inside the medium. The concentration of ink in the glass could be fixed in some nonuniform distribution by quickly freezing the water before uniform concentration could be reached. Heating the water would cause the rate of diffusion to increase.

Heating solids allows diffusion to occur within them as well. At 800°C to 1200°C, many different dopant atoms will readily diffuse through silicon. This allows doped regions to be created within a silicon wafer by heating it and exposing it to a gas containing the desired dopant.

Figure 9-3 shows solid-state diffusion forming an N-type region in a silicon wafer. The wafer first has a layer of silicon dioxide (SiO_2) grown upon it and is then patterned so that holes exist where N-type regions are needed. The wafer is heated in a furnace where a gas breaks down at the surface of the wafer to release dopant atoms. These diffuse through the silicon, gradually moving from regions of high concentration to low concentration. The speed at which diffusion occurs depends upon the temperature and the type of dopant being used. Common N-type dopants include arsenic (As), phosphorous (P), and antimony (Sb). The only effective P-type dopant for silicon is boron (B).

Solid-state diffusion is simple but has some serious limitations on the types of dopant profiles that can be created. The highest concentration of dopants will always be at the surface of the wafer. Allowing diffusion to occur for a longer time or at a higher temperature will make the change in concentration with depth more gradual, but the peak concentration will always be at the surface. Figure 9-4 shows some dopant profiles possible through solid-state diffusion.

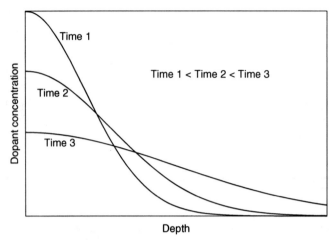

Figure 9-4 Diffusion dopant profiles.

The dopants will also diffuse in all directions. Any extra time or temperature used to allow the dopants to travel deeper into the silicon will also allow them to travel farther sideways from the original exposed area of wafer. The time and temperature of the diffusion step is chosen to try and give the desired profile. To gain more control over the three-dimensional distribution of dopant atoms, most doping today is performed by **ion implantation**.

One way to add a bit of lead to a brick wall would be to fire a machine gun at it. Bullets would become embedded beneath the surface at a distance that would depend on how fast they came out of the gun. More lead could be added just by firing more bullets. This method would also cause a fair bit of damage to the wall. Ion implantation adds dopant atoms to silicon in the same fashion.

First a gas containing the desired dopant is ionized by bombarding it with electrons (Fig. 9-5). This will produce charged atoms (ions) of the dopant. This process will also produce ions of other types, which must be screened out. A mass spectrometer uses a magnetic field to force ions to travel in an arc. The radius of the arc will depend upon the mass of the ion and the strength of the magnetic field. By carefully adjusting the field, the desired ions are given the correct arc to exit the mass spectrometer. Ions that are heavier will curve too wide and ions that are lighter will curve too sharply to escape. The dopant ions are then accelerated through a very strong electric field and fired at the wafer. A patterned layer of photoresist or other material is used to block the dopant atoms from areas where they are not needed.

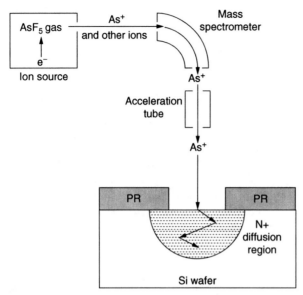

Figure 9-5 Ion implantation.

In areas of silicon that are exposed, the ions crash into the surface at high rates of speed. They will ricochet in random directions off the silicon atoms, eventually coming to rest somewhere beneath the surface of the wafer. Each ion will have a slightly different path of collisions resulting in a distribution of depths, but the depth of the center of the distribution will be determined by how much the ions were accelerated. The total dose of dopant depends upon how many ions are fired into the silicon.

Ion implantation damages the silicon crystal by knocking silicon atoms out of their proper positions in the crystal. Dopant atoms can also come to rest in gaps between silicon atoms rather than proper positions within the crystal lattice. To repair this damage, wafers must be annealed after implantation. Applying heat allows the dopant atoms and silicon atoms to diffuse to proper crystal positions. This will also spread the concentration of dopants left by implantation. Some possible ion implantation dopant profiles are shown in Fig. 9-6.

The profile for implant 2 in Fig. 9-6 could be created by using a higher acceleration voltage than implant 1 and heating the wafer longer after implantation to allow more diffusion. Ion implantation gives far more control over the distribution of dopant atoms. Doping can be performed deep beneath the surface without having to allow long diffusion times, which lead to a lot of sideways diffusion. Of course, deep is a relative thing, with a deep implant going only 1 to 2 μm beneath the surface of a 1-mm thick wafer.

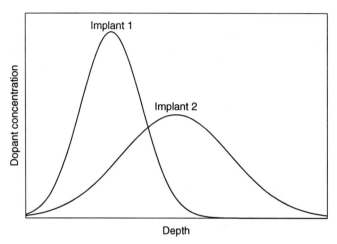

Figure 9-6 Ion implantation dopant profiles.

Ion implantation allows the maximum concentration of dopants to be beneath the surface. This is particularly useful in creating wells where a high concentration is desirable at the bottom of the well to prevent leakage currents, but a low concentration is needed at the surface to allow good transistor drive current. Wells created through implantation with concentration increasing with depth are sometimes called *retrograde wells* because the concentration changes with depth in the opposite manner as wells created through diffusion alone.

After doping, any step that heats the wafers will cause further diffusion. This must be taken into account when designing the dose and energy of the initial implant. The final distribution of dopants will be determined by the parameters of the implant itself and the amount and duration of all future heating of the wafer. In addition to adding dopants into the wafer, layering steps add new materials on top of the wafer.

Deposition

Deposition steps add a new continuous layer of material across the surface of the wafer. Later, patterning and etching steps may remove material in selected regions, but initially the layer will be deposited across the entire wafer. The four most common methods of deposition are spin-on, sputtering, chemical vapor deposition (CVD), and electroplating. See Fig. 9-7.

The easiest way to add a new material to a wafer is simply to pour it on. All that is required is that the material can be made liquid or dissolved in a liquid at a suitably low temperature and somehow later cured to be

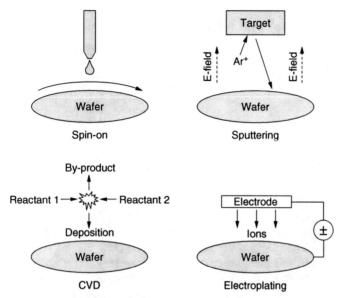

Figure 9-7 Deposition.

made solid. A thin layer of uniform thickness is achieved by spinning the wafer at high speeds (thousands of RPM). **Spin-on deposition** is commonly used to apply photoresist and sometimes layers of insulation. *Spin-on glass* (SOG) is a liquid solvent containing an organic compound of silicon and oxygen. When heated, the solvent evaporates and the compound breaks down leaving behind solid SiO_2, an excellent insulator and the main ingredient in window glass. Many other types of *spin-on dielectric* (SOD) are being investigated to provide lower permittivity than SiO_2 while still maintaining its high reliability. In general, a SOD is easier to apply than insulating layers created with CVD, but less reliable for extremely thin layers.

Spin-on deposition is fast, simple, and cheap, but most materials cannot be deposited this way. Copper becomes liquid only at temperatures above 1000°C, and to prevent it from freezing the moment it struck the wafer, the wafer would have to be heated to similar temperatures. In addition to causing any dopants to diffuse away from their intended locations, this would also of course melt any copper wires that had already been applied. Layers of conducting metal are instead typically deposited through sputtering.

In theory, **sputtering** can be used to apply any material to any other. The wafer to be deposited upon is placed in a vacuum chamber with a target made of the material to be deposited. A small amount of chemically neutral gas (like argon) is released into the chamber and ionized.

Figure 9-8 Keyhole void.

An electric field then accelerates the gas ions at the target. As they strike, these high-speed ions knock loose atoms of the target. These particles spread out through the chamber and gradually coat everything inside, including the wafer and the inside of chamber itself. If the target is a mixture of materials (like a metal alloy), the same mixture will be deposited. No high temperatures are required. Sputtering is commonly used for deposition of aluminum and copper.

Sputtering is convenient because it can deposit virtually any material or mixture of materials, but it does not provide good films of uniform thickness over vertical features on the wafer. In particular, vertical via holes are difficult to fill because sputtering tends to block the opening to the hole before the hole itself is fully filled. This results in what are called *keyhole gaps* or *voids* (shown in Fig. 9-8), which dramatically increase the resistance of vias. These problems are called *poor step coverage* and *gap fill*.

Some materials can be reflowed after sputtering to improve step coverage. A slight heating of the material may allow enough movement of atoms to smooth out the deposited layer and fill gaps. Another way of achieving good step coverage is to use **chemical vapor deposition (CVD)**.

In CVD, the wafer is placed in a chamber with one or more reactant gases. The reactants go through a chemical reaction, which produces the desired material and possibly some by-products. Some example reactions are given as follows:

Depositing Si	$SiCl_4 + 2H_2 \rightarrow Si + 4HCl$
Depositing SiO_2	$Si(C_2H_5O)_4 \rightarrow SiO_2 + 2H_2O + 4C_2H_4$
Depositing Si_3N_4	$3SiH_2Cl_2 + 4NH_3 \rightarrow Si_3N_4 + 6HCl + 6H_2$
Depositing W	$WF_6 + 3H_2 \rightarrow W + 6HF$

Silicon is deposited as part of creating an epitaxial wafer or to create polysilicon gates. CVD silicon dioxide is used as an insulator to separate interconnect layers. Silicon nitride is used as a diffusion barrier, for creating gate sidewall spacers, and as an etch stop layer. Tungsten can fill via holes or be deposited on top of silicon to create tungsten silicide (WSi_2), which greatly lowers the resistance of source and drain regions and poly gates.

The temperature and relative pressure of the reactant gases determine the rate of deposition. CVD can be carried out at normal atmospheric pressure (APCVD) and can achieve high deposition rates at relatively low temperatures. However, the flow of the reactant gases must be very carefully controlled. Because of the high pressure, the reactant chemicals diffuse more slowly through the chamber, and it is easy to have a higher concentration in one part of the chamber than the other. This will lead to different rates of deposition and different film thickness on different wafers.

Performing CVD at low pressures (LPCVD) produces higher-quality films and makes the flow of gases less of a concern, but higher temperatures must be used to achieve the same deposition rate. To achieve high deposition rates at low pressures and low temperatures, many chambers use plasma-enhanced CVD (PECVD). In this process, microwaves are used to strip electrons off the reactant gas molecules, turning the gas into plasma. In this state, reactions proceed much more quickly even at low temperatures. PECVD is used commonly enough that silicon dioxide deposited in this fashion is often referred to as *plasma oxide (PO)*.

The biggest benefit of CVD over sputtering is that it provides excellent step coverage and gap fill. Deposition will occur at the same rate on every surface regardless of its orientation, so filling deep via holes is not a problem. The biggest disadvantage is that it is not practical to deposit some materials by CVD. To use CVD, a suitable chemical reaction must be found, which produces the desired deposition material at a sufficiently low temperature, without any unsuitable by-products. Some reactants may be too expensive to manufacture. Some reactions may occur too slowly at temperatures low enough to not damage already deposited layers. Some by-products contaminate the wafers or are too dangerous to work with safely at a reasonable cost. By-products like gaseous HCl and HF are extremely corrosive and toxic, and H_2 is explosive. Handling these chemicals safely adds significantly to the cost of CVD.

A second disadvantage of CVD is difficulty in depositing mixtures of materials. Aluminum is commonly used as a conducting layer, but silicon tends to diffuse into aluminum wires, causing the aluminum to spike into the silicon surface and potentially short shallow diffusion layers to the well in which they sit. A common solution is to add 1 percent silicon to the aluminum material. Another problem with aluminum is poor electromigration. The aluminum atoms move easily out of regions of high current. Adding 0.5 percent copper can dramatically improve the reliability of aluminum wires. When sputtering, all that is needed is a target with the desired mix of materials. Depositing such a mixture with CVD would require multiple simultaneous reactions proceeding at precisely controlled rates. As a result, the preferred method for depositing alloys is sputtering.

TABLE 9-1 Deposition Methods

	Cost	Step coverage & gap fill	Common materials
Spin-on	Low	Good	Photoresist, SiO_2
Sputtering	Medium	Poor	Al, Cu, alloys
CVD	High	Good	Si, SiO_2, Si_3N_4, W
Electroplating	Low	Good	Cu, Pb, Sn

The last common method for deposition is electroplating. The wafer is submerged in a liquid solution containing ions of the metal to be deposited. An electrode of the same metal is placed in the solution and a voltage source connected between the electrode and a conducting layer on the surface of the wafer. Metal ions are attracted to the negatively charged wafer, and the positively charged electrode emits more ions as those in the solution are deposited. Electroplating is low cost and provides good step coverage and gap fill. Its biggest limitations are that only metals can be deposited and that the wafer must already have a conducting layer on its surface to be electroplated. A common solution is to first use sputtering to deposit a thin layer of metal, and then use electroplating to add metal to this initial layer. In the end, there is no one method of deposition that is best for all materials. Table 9-1 summarizes the different methods.

The trade-offs between different methods of deposition cause most process flows to use a mixture of all four, picking the most suitable method for each individual layer. A special form of CVD that deserves its own discussion is the creation of thermal oxide layers.

Thermal oxidation

Thermal oxidation is a method of creating a layer of SiO_2 on silicon. It is a form of CVD in that the wafer is exposed to gases that cause a chemical reaction, creating the desired material. What makes thermal oxidation special is that one of the reactants is the silicon wafer itself. Silicon has a number of properties that make it useful as a semiconductor; one of the most important is that SiO_2 is an extremely good insulator. Silicon dioxide tolerates extremely high electric fields, is chemically inert, is mechanically strong, and any silicon wafer will naturally form a high-quality layer of this insulator if simply exposed to oxygen or steam.

$$Si + O_2 \rightarrow SiO_2 \quad \text{Dry oxidation}$$
$$Si + 2H_2O \rightarrow SiO_2 + 2H_2 \quad \text{Wet oxidation}$$

Dry oxidation, which uses oxygen, gives the highest quality films. Wet oxidation, which uses steam, gives the fastest deposition rates. The

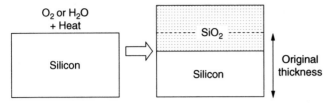

Figure 9-9 Thermal oxidation.

overall thickness of film created is determined by the temperature and length of time allowed for oxidation. The volume of silicon almost doubles upon oxidation, so the thermal oxide is "growing" the wafer. However, because the chemical reaction is consuming the wafer itself, about half the new thermal oxide layer will grow beneath the original surface of the silicon. See Fig. 9-9.

Any silicon surface (including polysilicon) will grow an oxide layer in the presence of the correct gases and at sufficient temperature. One way to create a thermal oxide layer, which covers only part of the die, is to grow a layer over the whole die first and then after patterning with photoresist, etch away the unwanted regions. If the oxide layer to be grown is relatively thick, it is difficult in later depositions to reliably cover the vertical step between the etched and unetched regions. Another commonly used process for growing thick oxide layers is called *local oxidation of silicon (LOCOS)*. In this process, silicon nitride is deposited and then removed from regions where thermal oxide is needed. Neither oxygen nor steam can diffuse through silicon nitride, so covered regions will not grow thermal oxide. See Fig. 9-10.

Oxide will diffuse underneath the edges of the nitride layer, and the growth of oxide there will tend to bend the nitride layer upward. The shape of the oxide projecting under the nitride layer is sometimes described as the "bird's beak." This provides a gentle slope for layers deposited on top of the oxide, but too long a bird's beak will require a large horizontal space between thick and thin oxide regions. Because it

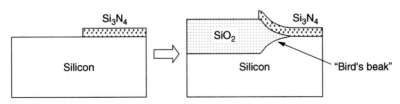

Figure 9-10 Local oxidation (LOCOS).

allows the creation of extremely thin, very reliable oxide layers, thermal oxide is the most common method for creating the gate oxide between the polysilicon gate and the silicon channel. The ability to easily create high-quality gate oxides is one of the most important factors in the overall success of silicon MOSFETs in the marketplace.

Planarization

A common step after the addition of any material layer is planarization to create a flat surface for the next processing step. This has become more critical as more layers of material are used. Starting from a nearly perfectly flat wafer, several layers of interconnect can make the surface a tiny mountain range. Photolithography must focus the image being patterned, but variation in the height of the wafer surface will blur this image. Abrupt variations in height are also more difficult to reliably cover in deposition steps. To prevent these problems from becoming worse after each new layer, most processes use **chemical-mechanical polishing (CMP)**. See Fig. 9-11.

The material to be polished is deposited in a layer somewhat thicker than needed to make up for the loss of material during polishing. An abrasive solution called a *slurry* is poured onto a polishing table while the wafer is pressed facedown against the table. The wafer is spun to grind the top layer flat, and the table is also spun to make sure the slurry is well distributed. Depending on the material being polished, different chemicals may be added to the slurry to aid removal through chemical etching.

This short description makes the process sound much cruder than it really is. Finding just the right spin rates, the right pressure to apply to the wafer, and the best composition of slurry is extremely difficult and far more an art than a science. In addition to planarizing insulation layers, CMP is commonly used to remove excess copper after deposition of interconnects. Before widespread use of CMP there was little use of copper interconnects and most chips were limited to two to three

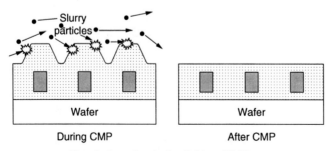

Figure 9-11 Chemical-mechanical polishing (CMP).

interconnect layers. Today copper is the interconnect material of choice and some chips have seven to eight interconnect layers. The development of CMP was a critical part of these process improvements.

Photolithography

Fabricating integrated circuits requires not only depositing multiple layers of material, but also patterning those layers. This is the job of photolithography. By exposing a light sensitive chemical (photoresist) through a mask, photolithography allows the desired pattern to be transferred onto the current top layer of material on a wafer. The steps required are shown in Fig. 9-12.

After a new layer of material is deposited, photolithography begins by spin-on deposition of photoresist dissolved in a liquid solvent. The wafer is then heated, in a step called *soft bake*, to evaporate the liquid solvent and form a solid layer of photoresist on the surface of the wafer.

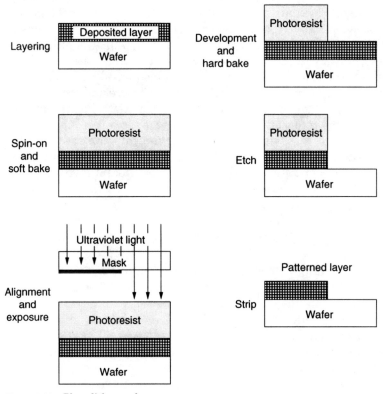

Figure 9-12 Photolithography.

The wafer must then be properly aligned before exposure. Each mask has special alignment marks (called fiducials) that will be patterned into each layer on the die. A laser is shone through the alignment marks on the mask and reflected off the surface of the wafer. By analyzing the reflection, the mask and wafer are precisely aligned. This is a process similar to how the laser in an audio CD player stays aligned with the proper track on the CD by monitoring its own reflected light.

Once aligned, shining light through the mask and focusing the image onto the surface of the wafer expose the photoresist. Most processing steps affect the entire wafer simultaneously, but it is not possible to create a focused image across the entire wafer. Instead, each exposure projects onto an area of only a few square centimeters. This could be enough to expose only a single die at a time or several, depending on the die area of the design. After each exposure, the wafer must be aligned again to expose the next region of the wafer. Because of this serial stepping from one area to the next to complete exposure of the entire wafer, the photolithography machines that perform this operation are commonly called *steppers*.

After the photoresist has been exposed, it is developed much like a photograph in a chemical bath. There are two basic classes of resists, positive resist and negative resist. Positive resists become more soluble when exposed to light, negative resists become less soluble. The example in Fig. 9-12 shows the use of positive resist, which is much more common. After development, the wafer is heated, in a step called hard bake, to cure the remaining resist to a hard lacquer-like surface.

The developed wafer is then etched. The photoresist prevents the etching of material it covers while material in exposed areas is removed. After etching, the remaining photoresist is stripped off and the original deposited layer is left behind, now patterned into the same shapes drawn on the mask. A modern fabrication process might employ as many as 20 to 30 separate masks for patterning different layers of material. The creation of these masks has become one of the most important steps in the manufacturing process.

Masks

Semiconductor manufacturing may start with making wafers, but blank wafers can be used to make any integrated circuit. The first step in making a particular design is making photolithography masks. Each layer of layout is made into a separate mask (also called a reticle) to be used to pattern that particular layer on the wafers. Creating masks uses the same manufacturing steps of deposition, patterning, and etching used to process the wafers. See Fig. 9-13.

The starting point for a mask is typically a 5-in^2 piece of quartz (crystalline SiO_2) a few millimeters thick. Sputtering is used to deposit a thin

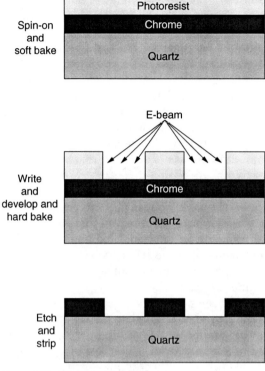

Figure 9-13 Mask making steps.

layer of chrome and then spin-on deposition adds a layer of photoresist. When processing a wafer, a mask would be used to expose the photoresist with the needed pattern, but of course when creating a mask there is no mask to be used yet. Instead, an electron beam or laser exposes the photoresist. First the layout database is broken down into individual polygons or pixels in a step known as *fracture*, and then the beam is swept back and forth across the face of the mask in a raster scan just like the electron beam used to draw the image on a traditional television screen. The main difference is one of scale. A high definition TV might paint an image of 10^6 pixels 60 times a second. Creating a mask requires features so small that an individual mask might have 10^{10} pixels. Painting an image of this resolution takes hours.

The time this takes shows why masks are so critical to mass-producing computer chips. E-beam lithography could be used directly onto wafers, eliminating the need for masks altogether, but it would be terribly slow. Using masks, a single flash of light can expose the resist for an entire die or even multiple small die all at once. With e-beam lithography each pixel must be individually exposed.

After the photoresist is "written," the desired image is developed and the newly exposed chrome etched away. Now the fractured layout image has been transferred to the chrome of the mask. Light projected through the mask will create a shadow of the desired pattern. An important final step is inspecting and repairing masks. It is expected when processing wafers that some percentage of the die created will have defects. Some of these defects will prevent the chips from working properly and these units will be discarded. Because the masks are used to expose the die, a defect on a mask will create a matching defect on every single die processed using that mask. It is critical that the mask is as near to perfect as possible, but the time and expense of making masks is strong incentive not to simply discard a mask with a defect.

Instead, automated machines inspect the masks pixel by pixel for defects. A defect is any place on the mask where chrome remains that should have been removed or where it has been removed that should have remained. If defects are found, some can be repaired. A high-energy e-beam is used to vaporize extra spots of chrome or to trigger CVD in spots that are missing chrome. Some very small defects may not be enough to affect processing. The stepper machines typically use lenses to focus the image of a 5-in^2 mask onto a 1-in^2 area of the wafer. This means all the dimensions on the mask will be 1/5 the size when created on the wafer, and very small defects will be made even smaller by this scaling.

Finally the mask is coated in a transparent protective layer called a *pellicle* and is ready for use. To create a full set of masks for a modern 90-nm process can easily cost millions of dollars. This all must be done before the very first chip is made.

Wavelength and lithography

As one of the most expensive and technically challenging steps in semiconductor manufacturing, lithography has often been seen as the greatest potential limiter to future process scaling and Moore's law. At the incredibly small dimensions of today's integrated circuits, the wavelength of light itself becomes a serious problem. To our perception light travels in perfectly straight and uniform lines, but below the micron scale the wavelength of light becomes a factor, and at these dimensions light bends and turns corners in ways very different from simple straight rays.

In the early 1980s, visible light was commonly used for photolithography. Conventional wisdom was that the smallest feature size that could be reliably patterned would be equal to the wavelength of light used. At this time, features of 3 µm were typical. Because violet light has a wavelength of only about 0.4 µm, the wave nature of light was not yet a problem. However, as features sizes steadily decreased year after year, it became clear that something would have to be done.

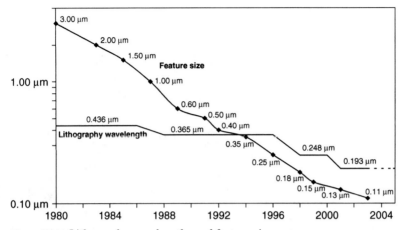

Figure 9-14 Lithography wavelengths and feature sizes.

Photolithography began using ultraviolet light with wavelengths too short for the human eye to see (<380 nm). Progressively shorter wavelengths provided better resolution, but a host of other problems as well. Creating ever-shorter wavelengths is difficult and expensive. Shorter wavelengths tend to be absorbed more when passing through lenses. Each time the light source is changed, new photoresist chemicals sensitive to the new wavelength have to be developed. If new light sources had to be chosen for every process generation, Moore's law would be forced to dramatically slow down.

To avoid these problems, modern lithography uses **optical proximity correction (OPC)** and **phase-shift masks (PSMs)** to take into account the wave nature of light and pattern features less than half the wavelength of the light being used. Lithography has continued to move to steadily shorter wavelengths of light but at a dramatically slower rate than the continued reduction of feature sizes. See Fig. 9-14.[2]

Figure 9-15 shows some of the features used in OPC and PSMs. At the left is shown a conventional mask where the image of a trench to be etched and later filled with metal is copied directly from the layout without alteration. When this mask is used in fabrication, the wire produced will not be exactly the same shape as on the mask. The ends of shapes and exterior corners will tend to be rounded with less material than the ideal shape. Interior corners will tend to accumulate extra material. Altering the mask to take the manufacturing process into account produces a wire closer to the ideal shape.

The center mask in Fig. 9-15 shows some of the common alterations performed by OPC. Hammerhead extensions on the ends of shapes and

---

[2]Pati, "Tutorial on Subwavelength Lithography."

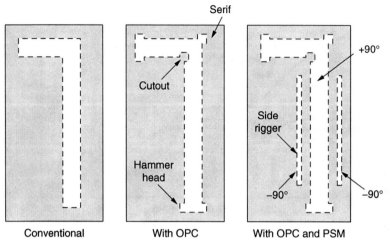

Figure 9-15 OPC and PSM.

serifs on exterior corners are added to compensate for rounding. Interior corners drawn with cutouts reduce the amount of extra material left behind. The right most mask in Fig. 9-15 shows PSM features as well. Normal masks project an image of bright and dark areas. For this reason, they are sometimes called *binary intensity masks (BIMs)*, since each region has two possible values, bright or dark. PSMs produce bright areas with the light at different points (or phases) in its wavelength. Bright areas with the same phase will reinforce each other through constructive interference, but bright areas of opposite phases will cancel each other out through destructive interference. By adding side riggers of the opposite phase, the width of a line is more precisely controlled. BIMs and PSMs are compared in Fig. 9-16.

On the left is shown a mask with no phase shifting. Ultraviolet light is projected through the mask and blocked by regions of chrome. However, the wavelength of light allows it to bend around the chrome barriers slightly. Light from both openings reaches the photoresist directly beneath the center piece of chrome and prevents the light intensity from being zero at that spot. The shadow projected by the mask is somewhat indistinct because of the diffraction of the light itself. On the right a PSM uses different phases of light to create a crisper shadow.

A PSM requires additional processing when manufactured to create regions of different phases. In this example, one of the regions free of chrome has been etched so that the mask is of a different thickness than at the other opening. Because the wavelength of light is different inside the quartz than in the air, light traveling through a different thickness of quartz will become out of phase. The light from PSM still diffracts around the central spot of chrome, but in this case the light

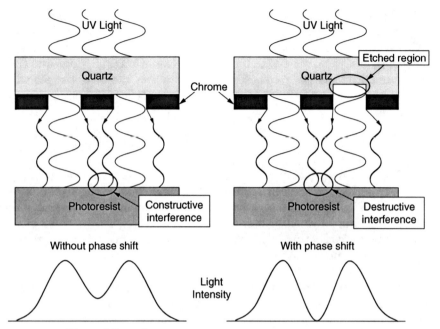

Figure 9-16 Phase-shift masks.

reaching the center from both sides is out of phase. Destructive interference cancels out the light from both sides, allowing the light intensity to reach zero at the center as desired.

By making the drop off of intensity more abrupt, the thickness of the lines patterned is more easily controlled. Small changes in the intensity of the light or the sensitivity of the photoresist will have less effect on the width of the line ultimately produced. Side riggers are made too narrow to expose the photoresist themselves, but they provide light of the opposite phase to help make the shadows of neighboring lines more distinct. The processing of adding OPC and PSM features is typically done during fracture when the layout is converted to the form to be used in making the masks. The result is masks that look very different from the layout from which they are derived and from the features they ultimately produce.

Although OPC and PSM have prevented the need for lithography wavelengths to match the pace of feature size reductions, concerns remain about the long-term future of lithography. Lithography at resolutions dramatically better than in use today has been demonstrated using electron beams, x-rays, or extreme ultraviolet (EUV), but it is feared that any of these alternatives could be vastly more expensive than current methods. Indeed, x-ray lithography has been proposed as the future of lithography since the late 1970s, but new ways of incrementally improving current methods have continually pushed back the need for more radical solutions.

The next likely incremental improvement is immersion lithography. Liquids have a higher index of refraction than air or vacuum, which allows light waves to be bent more sharply. Immersion lithography improves resolution by placing a drop of liquid between the lithography lens and the wafer enabling sharper focus.[3] This and other evolutionary changes will continue to be developed to try and prevent Moore's law from being slowed by lithography.

Etch

Etching, the removal of unwanted material layers, is the processing step typically performed after lithography. This removes the material uncovered by developing the photoresist while leaving behind the material still covered. Because deposition steps in general cover the entire wafer, etching is crucial to creating the desired pattern of material. Different etch processes are chosen based on etch rate, selectivity, and isotropy.

Etch rate is a measure of how quickly a given thickness of material can be removed. Time is money, and etch times of many hours would make high-volume manufacturing expensive. On the other hand, it wouldn't be possible to provide good control for an etch of only a few seconds. The etch process must be designed to give etch rates that are fast enough to not limit volume production while still being slow enough to give reproducible results.

Etch selectivity is the ratio of etch rates between the material to be removed and the material beneath it. Ideally, an etch process will be highly selective so that it will quickly etch through the unwanted material but only very slowly affect the layer underneath. Isotropy describes how directional an etch is.

At the left in Fig. 9-17 is a purely **isotropic** etch. This type of etch proceeds at the same rate in all directions. In the center of the figure is the opposite, a pure **anisotropic** etch. This type of etch cuts only vertically. At the right is shown a preferential etch that proceeds both horizontally and vertically but cuts more quickly in the vertical direction. Etch rate, selectivity, and isotropy must all be considered in developing a suitable etch for each layer. Etches are divided into two general categories: **wet etches** that use liquid solutions and **dry etches** that use gasses or plasmas.

Wet etches involve immersing the wafers in a liquid bath usually containing acids mixed with water. A wet etch removes material through a purely chemical process, so the key is finding the right chemical reaction. For example, hydrofluoric acid (HF) will readily dissolve SiO_2 but not silicon. This makes it possible to create a fast highly selective wet etch of

---

[3]Stix, "Shrinking Circuits with Water."

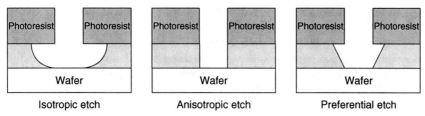

Figure 9-17 Etch isotropy.

SiO$_2$ on top of silicon. However, a chemical reaction will proceed on any exposed surface and makes no distinction between directions. Wet etches are isotropic, cutting horizontally and vertically at equal rates. Because of this, wet etches are not used for creating features sizes below about 2 µm. Wet etches are still useful since the upper levels of even modern processes may have features of this size or larger. Also, they are used to quickly remove an entire layer that was added to aid in processing. A wet etch of hydrogen peroxide (H$_2$O$_2$) mixed with sulfuric acid (H$_2$SO$_4$) is commonly used to strip photoresist after etching the layer beneath. Wet etches are also often used between processing steps to clean wafers of particles or chemical contaminants before the next processing step.

To etch feature sizes below 2 µm dry etches are required. Rather than using liquid solutions, dry etches occur in vacuum chambers under low pressure. The three basic types of dry etches are shown in Fig. 9-18.

Sputter etching (also called ion beam etching or ion milling) reverses the process of sputter deposition. When using sputtering to deposit a layer of material, argon ions are accelerated at a target, knocking atoms free to be deposited on the wafer. In sputter etching, the wafer is the target of the accelerated ions and material is gradually etched off its surface by the physical impact of the accelerated ions. This is a purely physical process, like atomic level sandblasting. This creates etches, which are extremely anisotropic, since the ions are accelerated vertically into the wafer, but not at all selective. The rain of ions will etch almost

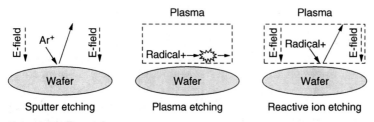

Figure 9-18 Dry etches.

any material at very similar rates. Selectivity can be achieved by using a chemical etching mechanism instead.

Plasma etching is a chemical process like wet etching, enabling it to be selective but it is also isotropic, cutting in all directions equally. Plasma etching reverses another deposition method: plasma-enhanced chemical vapor deposition (PECVD). In PECVD, plasma ions go through a chemical reaction at the surface of the wafer that results in deposition. Plasma etching uses other chemical reactions to remove material instead. Using plasma enables many chemical reactions that are not possible for wet etches. To create an etch process, which is both selective and anisotropic, sputter etching and plasma etching are combined in *reactive ion etching* (RIE).

RIE etches through both a physical and chemical process. A plasma of ions that will chemically react with the material to be removed is created and accelerated through an electric field at the wafer. The ions are accelerated at low enough energies that they typically will not remove material with which they cannot chemically react, but the chemical reaction will proceed much more quickly in the vertical direction with the ions all projected in that direction. Because it is both selective and anisotropic, RIE is the most common form of etching for layers with very small feature size.

The limits of etching must be taken into account in creating modern designs and fabrication processes. Imagine etching a layer of SiO_2 that has been deposited on top of a layer of wires and more SiO_2. It may be possible to find a chemical reaction to selectively etch SiO_2 and not the metal wires, but the chemical reaction cannot distinguish between the new layer of SiO_2 and the old layer of the same material beneath it. To help allow for selective etches, etch stop layers may be added to the process flow. These are layers deposited specifically to slow etch rates after the layer to be removed has been cut through. A layer of silicon nitride deposited before a new layer of SiO_2 is applied will make it far simpler to etch completely through one SiO_2 layer without cutting into the next. Etch stop layers do not perform an electrical or mechanical function, but they make the manufacturing process itself more reliable.

Another way of aiding manufacturing is through density targets. In any chemical etching process, the reactant chemical will tend to be consumed more quickly in regions of the die with a high density of the etched material to be removed. This causes the etch rate to be slow in regions of high density and high in regions of low density. To help keep the etch rate uniform across the die, density targets may be created for each layer. Extra unneeded poly lines can be added in regions of low density to aid in manufacturing. Density targets also make polishing steps more uniform by providing the same relative hardness across the die. As dimensions continue to push the limits of fabrication, the design and process flow will have to add more and more steps specifically to aid manufacturing.

Example CMOS Process Flow

This section shows how layering, photolithography, and etching steps are used to create CMOS devices. Figures 9-19 to 9-32 show the masks used and resulting cross sections of the die during each of the fabrication steps of a generic CMOS process flow. For brevity, some steps have been left out, but the major steps shown in this flow will be common to the fabrication of most modern CMOS integrated circuits. The major steps shown are:

1. Shallow trench isolation
2. N-well implantation
3. P-well implantation
4. Poly gate formation
5. N-source/drain extension
6. P-source/drain extension
7. Sidewall formation
8. N+ source/drain implant
9. P+ source/drain implant
10. Silicide formation
11. Contact formation
12. Metal 1 (M1) wire creation
13. Via1 hole creation
14. Via1 and metal 2 (M2) wire deposition

The figure for each step (starting with Fig. 9-19) in the upper left shows the mask (if any) used for patterning that step. For simplicity, any OPC or PSM processing of the masks has not been included. The lower left shows the processing in that step, specifically through a cross section labeled A. The upper right shows the layout of the layers processed so far, and below are shown the completed cross sections for that step along the A and B lines labeled in the layout. The process steps are additive, with each subsequent figure starting from the processed wafer of the previous figure.

The first step in creating transistors is to forming the insulation that will separate the regions that will later hold N-type and P-type transistors. The LOCOS process can be used to grow thermal field oxide, which will separate the active transistor regions; but to allow tighter spacing between the different types of transistors, trenches are etched into the silicon and filled with SiO_2 using CVD. This is known as *shallow trench isolation (STI)* to distinguish it from the deeper trenches that are used in the fabrication of DRAM. See Fig. 9-19.

Layers of silicon nitride and then photoresist are deposited and patterned. A dry etch cuts through the exposed silicon nitride and into the

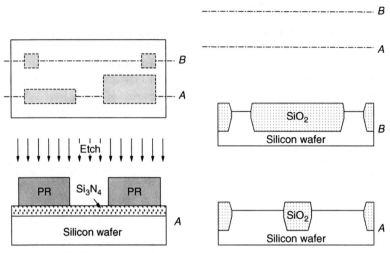

Figure 9-19 Shallow trench isolation.

silicon to form trenches. CVD then fills the trenches. CMP removes all the SiO₂ not in the trenches. The CMP step is designed to easily remove SiO_2 but not silicon nitride. After polishing, a wet etch removes the silicon nitride layer, leaving behind only the filled trenches.

The trenches of insulation help electrically isolate one well from another and prevent poly lines from creating unwanted transistors outside the active regions. Without thick oxide separating them from the silicon, poly wires would form transistors everywhere they were routed. The thickness of the STI and the doping beneath it make these unintentional transistors have such high threshold voltages that they will never turn on and therefore can be ignored.

After the isolation regions are formed, the silicon must be doped to prepare for different types of transistors. P-type transistors must be formed in a region of N-type silicon, and N-type transistors must be formed in a region of P-type silicon. These doped regions are the wells. Photoresist is deposited and then patterned before using ion implantation to create a retrograde well with the highest concentration of dopant beneath the surface of the silicon. See Fig. 9-20.

After the N-well has been formed new photoresist is deposited and patterned before implanting the P-well. It is possible to make use of the background doping of the wafer and only implant N-wells into a P-type wafer or only P-wells into an N-type wafer. However, the precise profile and concentration of dopants in each well can be optimized if both are implanted separately. Processes that implant both wells are called *twin tub processes*. See Fig. 9-21.

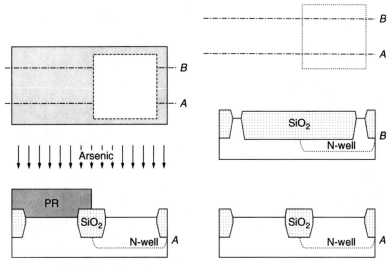

Figure 9-20 N-well implantation.

All the steps up until this point have prepared the die for the formation of transistors. In the next step, the first piece of the transistors, the gate, is formed. The die is heated to form an extremely thin layer of thermal oxide. This is the thin oxide or gate oxide that will separate the gate of the transistor from the channel. The thinner the gate oxide, the more current the transistor will provide, so great effort has been made to allow the gate

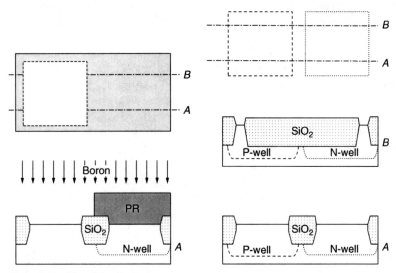

Figure 9-21 P-well implantation.

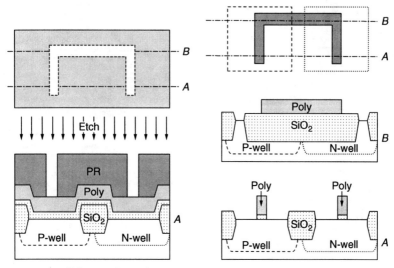

Figure 9-22 Poly gate formation.

oxide to be as thin as possible. In a 90-nm process, the gate oxide may be only 1.2 nm thick, making it less than 10 molecules across.[4] See Fig. 9-22.

CVD creates a layer of polysilicon on top of the oxide. Because the silicon is not deposited on top of crystalline silicon, it forms many small crystal grains rather than a single crystal structure. Photoresist is deposited and patterned and a dry etch cuts the poly layer into wires. The thickness of these wires is the most important dimension of the CMOS process since it will determine the channel length of the transistors. Where the polysilicon is on top of gate oxide, transistors will be formed, but where the polysilicon is on top of field oxide, it will be too far from the silicon to create a channel. These lengths of field poly are used to connect the lengths of gate poly that make up the transistors.

After the poly gates are created, the source and drain regions of the transistors are formed. See Fig. 9-23. The extensions of these diffusion regions that reach up to the channel should be very shallow to create transistors with low leakage. Photoresist is patterned to screen the N-well regions, and an N-type implant forms the extensions for the NMOS transistors. The poly gate itself acts as a screen for the implant into the P-well, blocking the dopants from reaching the silicon directly beneath the poly. This is a self-aligned process where the edges of the source and drain regions are determined by the patterning of the poly rather than a separate step.

---

[4]Thompson, "A 90nm Logic Technology."

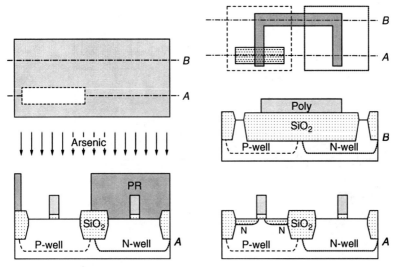

Figure 9-23 N-source/drain extension.

The creation of source and drain extensions is then repeated in the N-wells using a P-type implant to form extensions for the PMOS transistors. See Fig. 9-24.

Sidewalls are spacing structures created on either side of the gate, which allow further implants of the source and drain regions to be self-aligned to just outside the source/drain extensions. See Fig. 9-25. Silicon nitride is deposited in a layer that is created to trace the shape of the structures beneath it. This layer is the thickest on the vertical sides of the poly lines.

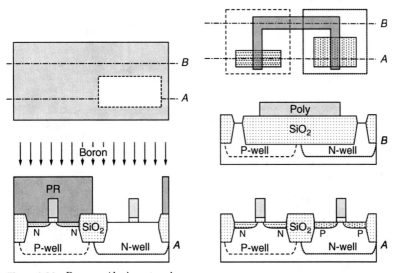

Figure 9-24 P-source/drain extension.

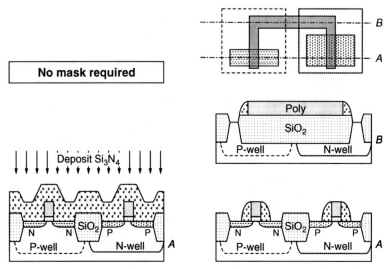

Figure 9-25 Sidewall formation.

A timed dry etch then leaves behind material only where the original layer was thickest.

After sidewall formation a heavier and deeper N-type implant into the P-wells completes the NMOS devices. The deeper more highly doped source/drain region will provide a low resistance path to the source/drain extension and a good location for metal contacts. In N-wells, the same N-type implant simultaneously forms N-well taps to provide a good electrical connection to the well. See Fig. 9-26.

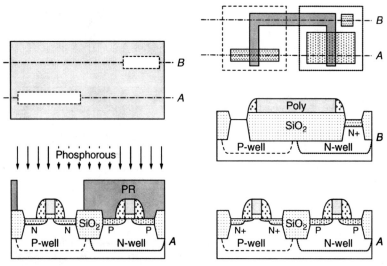

Figure 9-26 N+ source/drain implant.

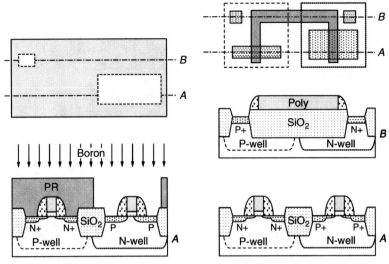

Figure 9-27 P+ source/drain implant.

Another P-type implant into the N-well region completes the PMOS devices and forms P-well taps in the P-wells. These deeper source/drain regions are also self-aligned using the sidewall spacers as an implant screen. A high-temperature anneal is required following implantation to repair damage to the silicon crystal. Polysilicon gates can withstand these temperatures that metal gates could not. See Fig. 9-27.

Even heavily doped polysilicon or source and drain regions still have much greater resistance than metal wires. To reduce the resistance of the gate wire and diffusion regions, a **silicide** layer is formed on top of them. A layer of refractory metal such as nickel, cobalt, or tungsten is sputtered onto the wafer. The wafer is then heated, causing a chemical reaction between exposed silicon or polysilicon and the metal. A wet etch then removes all the unreacted metal, leaving behind a layer of low resistance metal silicide. See Fig. 9-28. This material has still much higher resistance than pure metal but has far less resistance than doped silicon or polysilicon; it greatly improves the performance of the transistors.

After the formation of silicide, the creation of the transistors is complete. Now the creation of interconnects begins, starting with the contacts which will connect to the first level of metal. CVD is used to add a layer of SiO_2, which is then polished to form a flat surface. Photoresist is deposited and patterned, and a dry etch cuts holes through the oxide where contacts are needed. CVD of tungsten can then be used to fill in the contact holes. Tungsten has higher resistance than many metals but is conveniently deposited through CVD to fill high-aspect-ratio contact holes. Another polishing step removes any tungsten not in the contact holes. See Fig. 9-29.

296 Chapter Nine

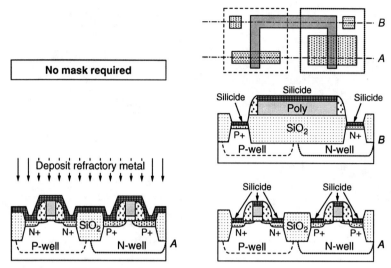

Figure 9-28 Silicide formation.

After contacts have been formed the first level of metal wires is created. Copper is the metal of choice for interconnects but has the disadvantage of diffusing very easily through SiO_2 and silicon. Copper atoms will contaminate the silicon and cause excessive leakage currents in the transistors unless carefully contained. To avoid this, copper wires are formed using a damascene process. Damascene refers to decorative designs (in the

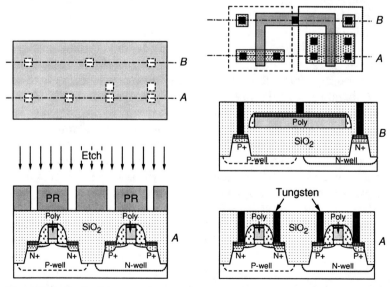

Figure 9-29 Contacts.

style of ancient Damascus) on metal or ceramics made by pouring metal into grooves carved in the surface. In semiconductor manufacturing, damascene processes contain copper by depositing the metal into grooves of insulation.

CVD adds another layer of SiO_2. Photoresist is applied and patterned, and a dry etch cuts trenches in the insulation where M1 wires are needed. A very thin layer of material is deposited by sputtering to line the trenches and act as a diffusion barrier to the copper. Copper is then electroplated onto the wafer filling the trenches. A polishing step removes the copper outside the trenches, leaving only the M1 wires. See Fig. 9-30.

Each layer of metal interconnect requires a new layer of vertical vias to connect to the layer above. This example process flow shows a **dual damascene** process where the metal for a layer of vias and interconnects is deposited simultaneously. On top of the M1 wires, layers of SiO_2, silicon nitride, and more SiO_2 are deposited. The silicon nitride will act as an etch stop for a later processing step. Photoresist is deposited and patterned and a dry etch cuts holes all the way through to the M1 layer. See Fig. 9-31.

Before the vial holes are filled with copper the dielectric is prepared for the next wiring level. Photoresist is applied and patterned. A dry etch cuts trenches where the M2 layer of wires is needed. The silicon nitride layer

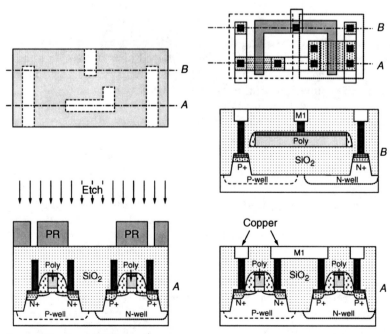

Figure 9-30 Create M1 wires.

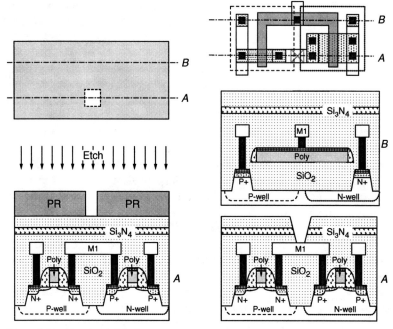

Figure 9-31 Create via1 holes.

acts as an etch stop, controlling the depth of the trench. A diffusion barrier is deposited and then copper is electroplated onto the wafer, simultaneously filling the via1 holes and the M2 trenches. A polishing step then removes all the copper not in the trenches. See Fig. 9-32.

This completes our example CMOS flow. A modern process would then repeat the last two steps, creating via holes and then filling in vias and a layer of wires as many times as needed. New wiring layers may be added indefinitely, although each added wiring level adds to the cost of production. As the number of transistors per die has increased, more interconnect levels have been added to allow for routing of the larger number of connections. A typical 500-nm process used four layers of wires whereas a 65-nm generation process might use as many as eight. Because the lowest levels are used to make the shortest connections, they are fabricated from thin layers that allow for very tight pitches but this makes their resistance high. Higher levels of wires are made of thicker metal layers to allow them to have lower resistance and make longer connections, but reliably fabricating the thicker wires requires a larger minimum width and space. Figure 9-33 shows a photo of a cross section of an eight metal layer process and one of its transistors.

Semiconductor Manufacturing 299

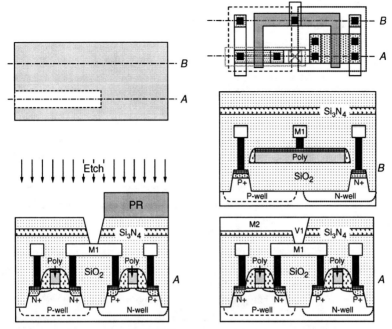

Figure 9-32 Create via1 and M2 wires.

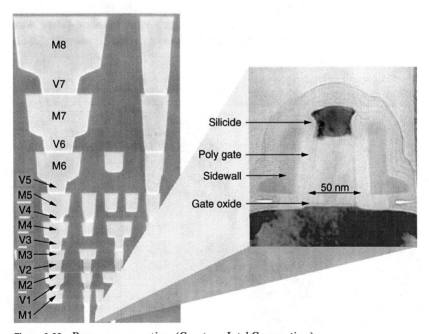

Figure 9-33 Process cross section. (*Courtesy: Intel Corporation.*)

Conclusion

The final product of this processing is a wafer of chips (Fig. 9-34). The chips on the wafer will be tested, cut into individual die, packaged, and then will go through further testing before they are shipped to customers. The following chapters describe the trade-offs in packaging and silicon test, which together with design and fabrication determine the product's ultimate performance and cost.

Design methods must improve and adapt to take advantage of the smaller feature sizes and larger number of transistors provided by the scaling of Moore's law, but advances in fabrication make this scaling possible. The industry has consistently moved in evolutionary rather than revolutionary steps, choosing incremental improvements to established processes rather than wholly new techniques wherever possible. These steady improvements have allowed many seemingly insurmountable technical hurdles to be overcome.

Future limits on improvements in fabrication may be as much financial as technical. Moore's law is driven as much by money as by technical innovation. Alternative lithography technologies offer the promise of vastly improved resolution, but questions remain as to their cost effectiveness. Larger wafers and smaller feature sizes can reduce the production cost per chip, but only if new fabs are fully utilized. The increased capacity of these fabs requires more total demand to keep them busy. Companies building new fabs are betting that the semiconductor market as a whole or their company's own market share will grow enough to make use of this added capacity. With fab construction

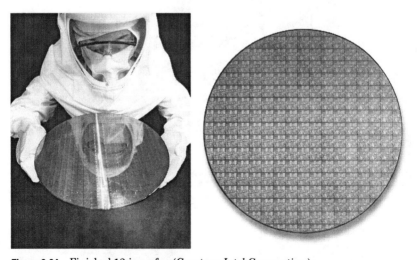

Figure 9-34 Finished 12-in wafer. (*Courtesy: Intel Corporation.*)

costs in the billions of dollars and still growing, there are increasingly few companies that can afford to take this type of risk.

Part of the risk is that semiconductor products are worthless without software. Processors are only useful for running programs. The future demand for processors and other semiconductor products depends greatly on what software is in demand. Processors are like cars that can go anywhere there are roads, but it is up to the software industry to build the roads to new destinations consumers actually want to visit. Continued investment by the semiconductor industry to drive Moore's law only makes financial sense if the software industry continues to find new ways of making use of new hardware capabilities. It is a huge leap of faith by the hardware industry to believe that providing more processing power, memory capacity, network bandwidth, and other hardware improvements will allow the creation of software with sufficient demand to pay for the costs of developing these capabilities (in addition to some profit).

Key Concepts and Terms

Bulk, epi, and SOI wafers	Isotropic and anisotropic
Chemical-mechanical polishing (CMP)	Optical proximity correction (OPC)
Chemical vapor deposition (CVD)	Phase-shift mask (PSM)
Cleanroom	Photolithography
Czochralski (CZ) crystal growth	Silicide
Diffusion and ion implantation	Spin-on deposition
Dry and wet etch	Sputtering
Dual damascene	Thermal oxidation

Review Questions

1. What are the advantages and disadvantages of different types of silicon wafers?
2. What materials are typically deposited with spin-on deposition, sputtering, and CVD? Why?
3. What is the difference between dry and wet etching?
4. What is the advantage of a retrograde well? How is it created?
5. Why is planarization important for integrated circuits with many levels of wiring?
6. How to OPC and PSM improve photolithography resolution?
7. How is local oxidation (LOCOS) different than shallow trench isolation?
8. How do silicide layers improve MOSFET performance?

9. What is meant by a damascene process? What is meant by dual damascene?
10. [Bonus] Make a list of as many elements as possible that are used in semiconductor manufacturing and what they are used for.
11. [Bonus] Draw a cross section of the layout in Fig. 9-32 for a cutline running through the center of the poly line forming the NMOS gate.
12. [Discussion] What are likely to be the most serious manufacturing obstacles to further process scaling?

Bibliography

Adtek Photomask. "Photomask Manufacturing: Concepts and Methodologies." 2002.
Allen, Phillip and Douglas Holberg. *CMOS Analog Circuit Design*. New York: Oxford University Press, 2002.
Baker, Jacob. *CMOS: Circuit Design, Layout, and Simulation*. 2d ed., New York: Wiley-Interscience, 2005.
Brand, Adam et al. "Intel's 0.25 Micron, 2.0 Volts Logic Process Technology." *Intel Technology Journal*, 1998, pp. 1–9.
Campbell, Stephen. *The Science and Engineering of Microelectronics Fabrication*. 2d ed., New York: Oxford University Press, 2000.
Carcia, Peter et al. "Thin Films for Phase-Shift Masks." *Vaccum and Thin Film*. Libertyville, IL: IHS Publishing Group, 1999.
Foll, Helmut. "Electronic Materials." 2005, http://www.tf.uni-kiel.de/matwis/amat/elmat_en. [Extremely accessible and extremely detailed, this is the best online semiconductor manufacturing reference I could find.]
Jones, Scotten. "Introduction to Integrated Circuit Technology." 3d ed., www.icknowledge.com.
Pati, Buno. "Tutorial on Subwavelength Lithography." *Design Automation Conference*, New Orleans, LA: 1999.
Plummer, James, Michael Deal, and Peter Griffin. *Silicon VLSI Technology Fundamentals: Practice and Modeling*. Englewood Cliffs, NJ: Prentice-Hall, 2000.
Roberts, Bruce et al. "Interconnect Metallization for Future Device Generations." *Solid State Technology*, 1995, pp. 69–78.
Singer, Pete. "The Interconnect Challenge: Filling Small, High Aspect Ratio Contact Holes." *Semiconductor International*, 1994, pp. 57–64.
Stix, Gary. "Shrinking Circuits with Water." *Scientific American*, July 2005.
Sze, Simon. *VLSI Technology*. 2d ed., New York: McGraw-Hill, 1988.
Thompson, Scott. et al. "A 90nm Logic Technology Featuring 50nm Strained Silicon Channel Transistors, 7 layers of Cu Interconnects, Low k ILD, and 1um^2 SRAM Cell." *IEDM*, 2002, pp. 61–64. [A detailed description of Intel's 90-nm fabrication process.]
U.S. Patent No. 6,329,118. "Method for Patterning Dual Damascene Interconnects Using a Sacrificial Light Absorbing Material." 2003.
Van Zant, Peter. *Microchip Fabrication: A Practical Guide to Semiconductor Processing*. 2d ed., New York: McGraw-Hill, 1990. [Great reference for details of the entire manufacturing flow.]

Chapter 10

Microprocessor Packaging

Overview

This chapter discusses how completed die are packaged for use. The chapter describes the trade-offs of different types of packages and presents an example package assembly flow.

Objectives

Upon completion of this chapter the reader will be able to:

1. Visualize how multiple levels of packaging are used in an electronic system.
2. Describe common package lead configurations.
3. Compare and contrast wire-bond and flip-chip packages.
4. Describe different types of package substrates.
5. Be aware of the trade-offs of important package design choices.
6. Understand thermal resistance and the factors that affect it.
7. Describe the trade-offs of using a multichip package.
8. Explain a typical package assembly process.

Introduction

The microscopic size of the features in integrated circuits makes a single chip the size of a thumbnail capable of enormous computational performance. This tiny size also means the same chip is damaged very easily. Most chips would be thoroughly ruined by simply picking them up. Any surface structure would be obliterated and even if coated in insulation, the salts common on everyone's skin could quickly contaminate the circuitry. This was true for even the earliest microprocessors, so all processors require some type of packaging.

The package first and foremost provides the chip protection from physical damage and chemical contamination. The package must do this while still providing electrical connections (leads) from the outside world to the chip. The interconnections on **printed circuit boards** (PCBs) are far larger than semiconductor chips, and the package must bridge this gap in scale. The package leads connect to the tight-pitched circuits of the processor die and spread apart to connect to the larger board traces.

For early packages, protection from the environment and simple connections between the die and the board were all that was required. Modern packages have grown far more complex as processor performance and power have increased. To allow for very high-frequency transmission over the data leads, their electrical characteristics (resistance, capacitance, and inductance) must be tightly controlled. Hundreds of leads may be required to deliver power to the processor, and all the power delivered will generate heat, which must be removed from the die quickly enough to prevent damage from extreme temperatures.

Because of its impact on data bandwidth, as well as on power in and power out limits, packaging has a large impact on the performance and reliability of a microprocessor. Package materials and assembly costs also contribute significantly to overall product cost. This means that there is no perfect package. Designers must balance the package cost and performance in choosing the best package for their product. The different needs of different processor products have led to a wide range of different packaging choices.

Package Hierarchy

A consumer cannot buy a bare silicon chip. Any piece of electronics inevitably has packaging to connect different components, and most products will have multiple levels of packaging for multiple chips. Packaging lets multiple die work together as part of a single product. Almost always a PCB makes the final level of connections. A personal computer would typically include a number of PCBs, the largest being called the *motherboard*. Architecture determines the processor's software compatibility, but the package dictates with what motherboards a processor may be used. Figure 10-1 shows a PCB and some ways of connecting components.

The most straightforward way of adding a silicon die to a circuit board is to simply solder it directly to the board. This is called **direct chip attach (DCA)** or just *chip on board (COB)*. This is the highest performance solution, adds the least possible weight, and takes up the

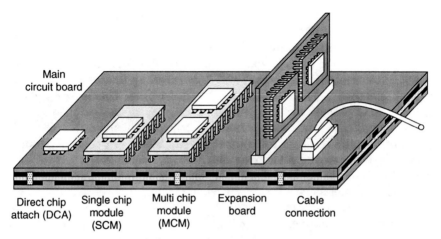

Figure 10-1 Packaging hierarchy.

least possible space. However, difficulties with this approach make it fairly uncommon. The most significant obstacle is fully testing unpackaged chips. This is referred to as the **known good die (KGD)** problem. Without soldering the die to a board it is difficult to test its full functionality, but removing a bad die after it has been soldered is awkward and can damage the board. In addition, the pitch of connections on the chip may be far tighter than on most PCBs. Manufacturing the board to match this pitch will add to its cost. Finally it is easy for bare die to be damaged or contaminated during assembly. For these reasons, most die are packaged separately before being attached to the board.

Packaging a single die or a small number of die in a separate module allows functionality to be fully tested before shipping and makes assembly of the board simpler. The package spreads the spacing of the die connections to leads whose pitch is matched more easily by the traces on the board. If the number of die required would make the packaging costs of a multichip module (MCM) too expensive, packaged components can be first connected on a separate PCB board (usually called an expansion card or daughterboard), which in turn connects to the motherboard.

Cable connectors are plugged into the motherboard to provide connections to components off the board. These could be internal drives inside the computer case, external drives or devices outside the case, or even a network connection to other computers. In the end, all of these cables will eventually connect to another PCB.

PCBs are most commonly made of a fiberglass material called **FR4** (flame retardant material type 4). Glass fibers composed of SiO_2 (the same insulator commonly used on die) are woven into mats, coated with an epoxy resin, and pressed together into solid sheets. Heating cures the epoxy and makes a stiff substrate to form the basis of a PCB. Sheets that have been already been impregnated with epoxy but not yet fully cured are called *prepreg*. To allow at least two levels of wiring in the board, the FR4 substrate usually has sheets of copper foil pressed on both sides. Photoresist is deposited and developed on each side, and the copper is etched to leave behind the desired traces. This is exactly the same type of process used to pattern wiring levels on die, but to keep costs down, PCBs rely on less advanced lithography and only wet etching. This means the tightest pitches on the board are typically much larger than the wiring pitches on die.

Drilling holes in the board and plating them with metal creates vias. Vias make connections between the wiring layers, and these same holes are used to make connections to pinned packages. Circuit boards with a large number of components or high-power components may require more than two levels of traces. In a multilayer board, typically some of the interconnect layers are dedicated to the high-voltage supply and some to the low-voltage ground. These layers are often continuous sheets of metal except where vias have been drilled through. Other layers are dedicated to signal traces, which make input and output connections between the components. Having sheets of metal at constant voltage above or below the signal traces helps reduce electrical noise and allows for higher bandwidth switching. The power and ground planes also help provide a low-resistance path for delivering power to all the components. Multiple two-sided boards can be combined to create multilevel boards.

Figure 10-2 shows the steps to create a six-layer board, with six layers of metal traces. The starting point for a multilayer board is multiple copper plated FR4 sheets (step #1). Each two-layer board has its traces etched separately and vias drilled if needed (step #2). Layers of prepreg are added to separate the traces of the different cores (step #3) and heat and pressure fuses the boards together (step #4). The outermost layers are patterned and the board may be drilled again to create vias, also called plated through holes (PTH), all the way through the multilayer board (step #5). Vias that start from an outer layer but do not go all the way through the final board are called **blind vias**. Vias where both ends connect to inner layers are called **buried vias**. Supporting these different types of vias allows for more efficient wiring of the board but adds cost by adding more drilling steps.

The last step before adding components is to apply a layer solder mask to the outside of the board. This is a material to which solder will not stick. It is patterned to leave exposed only those points where components will later be soldered to the board. The most common solder mask is green in color, giving most PCBs their distinctive color. The silicon dioxide cores are actually translucent. Finally components are mounted on the board (step #6). Sometimes the completed board with components attached is referred to as a *printed circuit assembly (PCA)*.

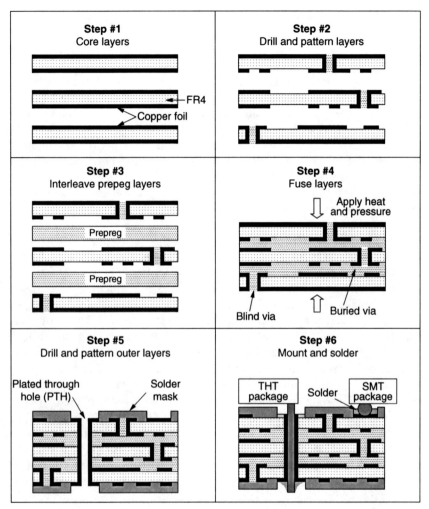

Figure 10-2 Multilevel PCB.

Components are added to the board in three basic ways, **through hole technology (THT)**, **surface mount technology (SMT)**, or a **socket**. The component on the left in step #6 of Fig. 10-2 has been mounted using THT. Pins on the package are inserted through the metal-plated holes in the board. The pins are made long enough to protrude from the bottom side of the board. The connections are then soldered by a process called *wave soldering*. By vibrating a pool of liquid at the right frequency a standing wave is created. A conveyor belt carries the board through the wave bringing just the bottom side of the board in contact. First the board is passed through a wave of flux, a chemical that will remove any oxide on the metal contacts and allow for a good electrical connection. Then the board is passed through a wave of molten solder, which only clings to the board where there is no solder mask. The result is all the pins being connected by solder to their vias.

The component on the right in step #6 has been mounted using SMT. SMT uses a stencil to apply a paste mix of solder and flux to contact points on the surface of the board. Components are then placed with their contacts touching the paste. The board is heated to allow the solder to form electrical connections between the component and the board. The biggest advantage of SMT is that it does not require drilling holes through the board. This reduces the cost of producing the board and allows components to be mounted directly on opposite sides of the board. This makes SMT particularly attractive for mobile or handheld products where packing components onto the smallest board possible is critically important.

The biggest advantage of THT is its mechanical stability. A component with pins inserted through the thickness of the board and soldered into place will not come loose easily. SMT components can tear free of the board when heated. The materials of the package and PCB all expand when heated but some expand more than others. The rate of this swelling is measured as the **coefficient of thermal expansion (CTE)**. If the CTE of the board and an SMT package are very different, heating the component will cause stress to build in the soldered connections to the board. With enough stress, the component will break free of the board.

The most expensive but most flexible choice is to provide a socket for the package. This is a connector designed to allow packaged components to be added without requiring solder. This allows expensive components to be replaced or upgraded easily. Unused sockets allow functionality to be added to a PCB after sale. To make sure removing a component will not pull the socket itself loose, sockets are typically connected to the board using THT. Sockets also allow cables to be connected to the board when the end product is assembled. Ideally, any components to be added or connections to be made after the board is manufactured will be provided sockets to allow solderless assembly of the final product.

Whether a die is attached directly to the main circuit board, packaged and then attached, or packaged and attached to a daughterboard, some level of packaging is essential to being able to make use of a semiconductor chip. For microprocessors, which often operate at high frequencies and high-power levels, packaging design choices have as much impact on their use as silicon design choices.

Package Design Choices

The processor designer must make many choices when selecting a package. These choices will determine limits on not only the processor die but also the motherboards the processor can be used with. The most important package design choices are:

1. Number and configuration of leads
2. Lead type
3. Substrate type
4. Die attach method
5. Use of decoupling capacitors
6. Use of heat spreader

Many of these choices are independent of one another, and over time almost every possible combination has been tried. The result is what seems like an overwhelming number of different types of packages (each with its own acronym). The zoo of modern packages is understood more easily by looking at each of these design choices separately. The most obvious characteristic of any package is the number and configuration of leads.

Number and configuration of leads

In 1965, Gordon Moore first proposed his law predicting the growth over time of the maximum number of components on an integrated circuit. Five years earlier, E.F. Rent of IBM was investigating the number of interconnections required between discrete circuits. He found that the number of interconnections required by a circuit is a function of the total number of components. This relationship came to be known as *Rent's Rule*.[1]

Rent's rule:
$$\text{Num leads} \propto (\text{num components})^R$$

As logic gates are added to a processor to increase performance or add new functionality, the processor consumes data more quickly. Additional leads are required to prevent the processor's performance from being unduly limited by its ability to communicate. The interaction of Moore's law and Rent's rule has been the biggest driver in the evolution of processor packages. If the number of components on a microprocessor is increasing exponentially, we should also expect an exponential increase in the number of leads. The value of the Rent's exponent R varies for different types of integrated circuits but is usually estimated for microprocessors as about 0.5.[2] If the number of logic transistors on a microprocessor doubles about every 3 years, we should expect the number of leads to

---

[1]Landman and Russo, "Pin versus Block Relationship."
[2]Bakoglu, *Circuits, Interconnections, and Packaging*, 418.

double every 6 years. The processor industry has seen exactly this kind of increase with the first microprocessor having 16 leads in 1971 and high performance processors in 2005 having lead counts between 500 and 1000.

Some of these new leads are data leads, which carry information into or out of the processor. A wider data bus allows higher data bandwidths but at the cost of adding more leads to the package. Many more of the additional leads are power leads, which supply current to power the processor. In modern processors, it is common for more than half of the leads to be used simply to power the processor. To draw large amounts of power at low voltages requires very large currents, which would melt a single lead. If the maximum safe current per lead were half an ampere, then a processor able to draw 100 A would require 200 leads for current flowing into the processor and another 200 for current flowing out of the processor. Rapid increases in power have therefore contributed to the rapid increase in the number of leads required.

Leads are typically placed on the package in three basic configurations, dual in-line, quad, and grid array packages (Fig. 10-3 and Table 10-1).

For low lead counts, the leads can be placed along only two edges of the package, so called a *dual in-line package* (**DIP**). If more leads are needed, they can be placed on four sides, making a **quad** package. For very high lead counts, **grid array packages** use multiple rows of leads on all four sides. The spacing between leads (lead pitch) is commonly between 1 and 0.4 mm. For a given lead count, smaller pitches allow a smaller package but require a more expensive PCB with traces that match the tighter pitch.

The number and configuration of leads for a processor package must be chosen very carefully. These choices will determine the number of data leads through which the processor communicates, the number of power leads through which it draws power, and which motherboards will be able to use the processor. Compatibility with existing motherboards is a huge advantage for new processors that are able to use the same package as

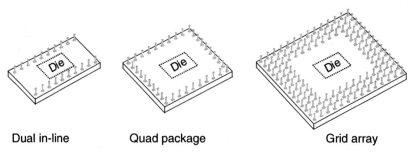

Dual in-line   Quad package   Grid array

Figure 10-3 Lead configurations.

Microprocessor Packaging

TABLE 10-1 Lead Configurations

Lead configuration	Typical number of leads	Example packages
Dual in-line	<50	Plastic dual in-line package (PDIP)
		Ceramic dual in-line package (CERDIP)
Quad package	50–250	Plastic leaded chip carrier (PLCC)
		Plastic quad flatpack (PQFP)
Grid array	100–1000	Pin grid array (PGA)
		Ball grid array (BGA)

a previous processor design, but new designs often require more leads than the current package uses. In order to use a single package design for multiple processor generations, the package may be designed with more leads than are currently needed to provide some room to grow, but extra leads add to the package cost. As always, there is no perfect choice. Once the number and configuration of leads are chosen, the next important package choice is the type of lead to use.

Lead types

Leads must provide low-resistance electrical connections between the metal traces in the package and the metal traces on the circuit board. The package traces will ultimately connect to the die. The board traces will connect the processor to other components on the board. The type of lead also affects how the package is mounted on the board, either THT, SMT, or with a socket. Some common lead types are shown in Fig. 10-4.

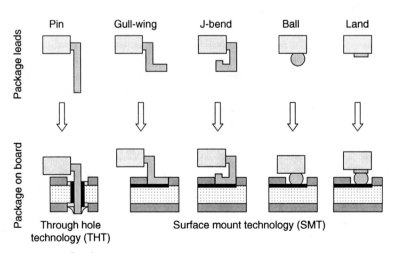

Figure 10-4 Lead types.

Straight pins used to be the most common type of lead. They are required for THT mounting and give the package a solid physical as well as electrical connection to the board. To allow packages to be mounted on the surface of the board, the pin can be bent out horizontally once it clears the bottom of the package. This is a gull-wing pin and is the simplest pin to use with SMT. The flat portion of the pin gives a large area to make a good contact to the board. The disadvantage of gull-wing pins is that these outward pointing pins cause the package to take a large amount of board area. Bending the pins inward instead of outward saves area on the board. These are J-bend pins, named since their shape resembles the letter "J." A disadvantage of J-bend pins is that the soldered connections to the board are now under the package, making them far more difficult to inspect. Likewise soldering by hand or removing J-bend components is far more difficult than a component with gull-wing pins.

The smallest and most direct connection possible is made using either a ball lead or a land pad. Components using either of these types of connections are sometimes called *leadless packages* because they lack pins, although technically a lead is any electrical connector between the package and the board. Ball lead packages have solder balls attached to package. After the component is placed with these balls resting on metal pads on the board, heat is applied to reflow the solder and make the connections.

Land leads are simple circular pads of metal. The board has solder paste applied before the component is placed to form the needed solder ball connections between the package land and the board. Although the final connection is similar for ball lead and land lead packages, land lead packages are less costly to produce but make assembly more difficult.

In theory, any of these types of leads could be used with a socket, but in practice gull-wing and ball packages are rarely used that way. Sockets must apply significant pressure in order to make a good electrical connection without solder. Gull-wing pins bend easily and soft solder balls are damaged easily. Straight pins are ideal for sockets. A clamping mechanism is used to grab the sides of the pin creating the electrical connection and holding the package in place. Sockets for J-bend packages are made that press in on the vertical sides of the pins. Placing pins on the board instead of on the package creates a land lead socket. The pins point up nearly vertical from the board and bend parallel to the board at their tips. The package rests on top of the pins with its pads touching the flat tips. A mechanical clamp forces the package down onto the pins and holds it in place.

Sockets add cost but allow the end user of the board to replace bad components or make upgrades easily. It is a huge advantage for a processor to be able to use the same socket as a previous generation. Sockets also help with rapid prototyping. If many different chips are to be tested with the same board, designing with a socket may be more of a convenience for the board manufacturer than the customer.

Substrate type

The types of interconnects used between the package leads and the die are determined in part by the choice of substrate. The most common package substrates are molded plastic, ceramic laminates, and organic laminates. Plastic packages are the least expensive but can be used only with low-power, low-performance chips. Ceramic laminates have the best performance but are the most expensive. Organic laminates offer intermediate performance and cost.

Plastic packages are formed by attaching the die to a leadframe and then encasing the die and part of the leadframe in protective case of hard plastic. The basic steps to forming a plastic package are shown in Fig. 10-5.

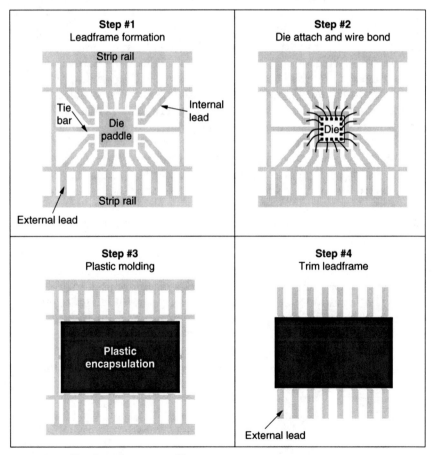

Figure 10-5 Plastic package assembly.

The assembly process begins with creation of the leadframe (step #1). The leadframe is punched out of metal sheets, usually an alloy of copper or nickel and iron. Alloy 42 is the most common material used. It is an alloy of 42 percent nickel and 58 percent iron. Nickel-iron alloys are stronger than copper and have a CTE that more closely matches silicon. However, copper has lower electrical and thermal resistance.

The leadframe consists of a die paddle (also called a flag), which will support the die during assembly, and internal and external leads. Internal leads will ultimately be encased inside the plastic package. External leads will remain outside and form the pins of the package. Leadframes for multiple packages are connected into strips by strip rails. Tie bars, which also connect it to the strip rails, support the die paddle.

Die attach and **wire bonding** (step #2) make the electrical connections between the leadframe and the die. The die is glued face up to the die paddle, and bonding pads on the die are connected to the internal leads. The wire-bond connections are made with gold or aluminum wires. A machine touches the wiring to a bonding pad on the die and uses heat and ultrasonic vibration to attach the wire. The tool then draws out a length of wire to the corresponding internal lead and bonds the wire again. The wire is then cut and the process repeated for the next bonding pad.

During plastic molding (step #3), the leadframe strip is placed inside a mold with cavities around the die and internal leads of each individual leadframe. Liquid plastic is forced into the molds and allowed to set. The strip rails and external leads are left protruding from the plastic. Finally the strip rails and excess metal are trimmed away (step #4), and the individual packages are separated from the strip rails. The pins are then bent to form either straight vertical pins for THT or gull-wing or J-bend pins for SMT.

Plastic packages are the least expensive but have a number of disadvantages. The plastic material cannot tolerate very high temperatures and does not pass heat readily. The only easy path for heat to escape the die is through the package leads themselves. This is fairly inefficient. To prevent overheating, die in plastic packages are limited to low power levels. The leadframe also offers only a single level of interconnects, which must be fairly large to have sufficient mechanical strength to be reliably punched out of sheet metal. Better thermal properties and interconnects are possible with ceramic laminate packages.

Ceramic substrates are made of multiple layers of ceramic tape printed with traces and vias and fused together. The steps for creating a ceramic package are shown in Fig. 10-6.

Constructing a ceramic substrate begins with an uncured (or green) flexible tape composed of ceramic powder, usually **alumina** (Al_2O_3), mixed with a binding chemical. At this point, the tape has the feel of a piece of leather. Blanking (step #1) is where the strip of tape is cut into

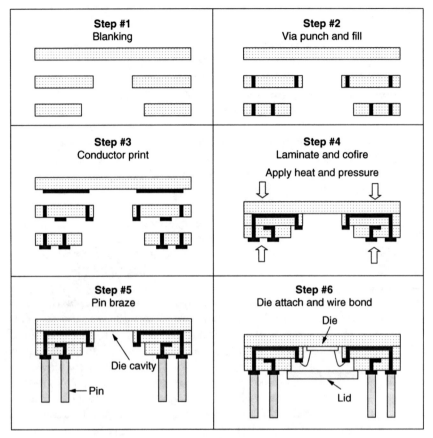

Figure 10-6 Ceramic package formation and assembly.

sections of the proper size to form the substrate. Some pieces may have a hole cut from the center that will later form a cavity in the package to hold the die. Via punch and fill (step #2) cuts holes in the layers using either a mechanical punch or a laser and fills them with metal. These will form electrical connections between the different layers of traces. Conductor printing (step #3) applies the traces through screen printing. A porous screen covered with a pattern of nonporous material is held over the pieces of tape. A squeegee forces a viscous mix of metal powder and binder chemicals through the open portions of the screen, leaving behind metal "ink" in the shape of the desired interconnects.

The different layers are then pressed together and heated to very high temperatures (step #4). This cofiring fuses together the ceramic powder of the green tape layers into a single rigid piece and fuses the metal powder into continuous wires. This process of fusing powdered material together without melting is called *sintering*. The very high temperatures required (as much as 1500°C) would melt most metals, so metals with high melting points like tungsten or molybdenum must be used for the traces. These metals have unfortunately much higher resistance than the copper or nickel-iron interconnects used in leadframe packages.

Pin brazing (step #5) attaches the pins to the substrate using a hard high temperature solder. The die is then glued into the cavity of the package and wire bonded to the package traces (step #6). A protective lid is fixed in place over the die cavity once wire bonding is complete. The package shown in Fig. 10-6 is a "cavity down" package where the die cavity faces the package pins. This has the advantage of placing the backside of the die in direct contact with the top of the package. A heat sink placed on top of the package can then readily remove heat generated by the die. A "cavity up" package places the die cavity on the top of the package. This allows pins to be placed across the entire bottom face of the package but means there will be an air gap between the die and any heat sink placed on the top of the package.

Ceramic packages transmit heat much more readily than plastic packages and tolerate much higher temperatures. In addition, the ceramic CTE is very close to that of silicon so that high temperatures will cause little stress between the die and the package. Multiple levels of traces allow high current levels to be delivered to the die with little electrical noise. All of these reasons make ceramic packages (especially cavity down packages) ideal for high-power chips. However, there are disadvantages. Ceramic packages require the use of high-resistance metal interconnects and have high dielectric permittivity leading to high capacitance. Most importantly ceramic packages are much more costly to make than plastic packages. Organic laminate packages provide better thermal and electrical performance than plastic packages at lower cost than ceramic packages. The steps for creating an organic package are shown in Fig. 10-7.

Organic laminate packages begin with the same epoxy fiberglass sheets as PCBs (step #1). These are typically about 1 mm in thickness and plated with copper. The sheets are individually drilled and the holes plated to form vias. The copper sheets are then etched to leave copper only where interconnect traces are needed (step #2). Sheets of prepreg are then interleaved with these package core layers and laminated together. Afterward, there may be another drilling step to create vias through all the layers (step #3).

For some packages, all the metal and dielectric layers are formed this way just as they are for a PCB. However, to achieve tighter wiring pitches some packages add buildup layers. The use of these thin layers and tight pitches is known as *high-density interconnect (HDI) technology*. Spin-on dielectrics and sputtering of metals may be used, or *resin coated*

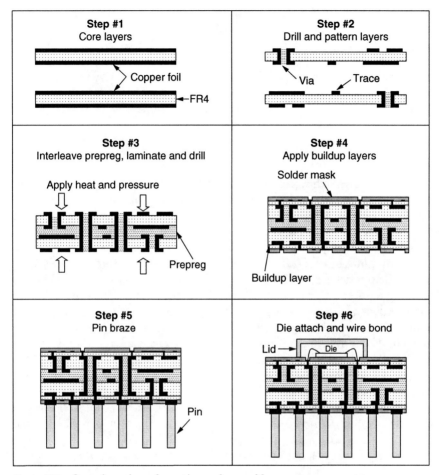

Figure 10-7 Organic package formation and assembly.

copper (RCC) foil may be applied to add these layers. These layers may be only 50 to 100 μm thick and are far too thin to be handled separately and mechanically drilled as the core layers are. Instead, buildup layers must be added on top of already processed core layers. Lasers are used to cut holes in buildup layers, and after being filled with metal these vertical connections are called **microvias** to distinguish them from the much larger drilled vias. On the outside layers, a solder mask is applied as in PCBs (step #4). It is important to realize that Fig. 10-7 is not drawn to scale because at true scale the core layers would be 10 to 20 times thicker than the buildup layers.

Once the substrate layers are complete, pins are brazed (step #5) and the die attached by wire bonding beneath a protective lid (step #6). The package shown in Fig. 10-7 could be described as a six-layer package because of the six separate layers of interconnect. More specifically, it might be called a 1-4-1 layer package because the package has 1 buildup layer on top, 4 core interconnect layers in the center, and 1 buildup layer on the bottom. The number of layers needed is dictated by the total lead count of the package as well as the package size. More layers allow for denser interconnect and a smaller package but increase the cost of packaging. Once again there is no perfect package.

Die attach

Die attach refers to how the die will be physically and electrical connected to the package substrate. The previous section has described the die attach method of wire bond. Wire bond is the most commonly used and the cheapest method of die attach, but it has a number of disadvantages. In wire bond, the placement of the bonding pads on the die and on the substrate must be carefully chosen to avoid wires to different pads touching. Multiple tiers of bonding pads in the substrate can help, but the die bonding pads are still restricted to only the perimeter of the chip. For die with a large number of leads but a small number of transistors, this can lead to the die size being limited not by the transistors needed but by the number of bonding pads required. For these reasons, wire bond is typically not used with die requiring more than a few hundred leads. The length of the wires used to connect die to substrate is also a concern. These wires will add capacitance and inductance to the leads and limit switching speed. These problems are fixed in another method of die attach called *controlled collapse chip connection (C4)* or **flip-chip**. Wire bond and flip-chip packages are compared in Fig. 10-8.

Flip-chip packages get their name because the chip is flipped over to attach the die with the transistor side facing the substrate. Rows of

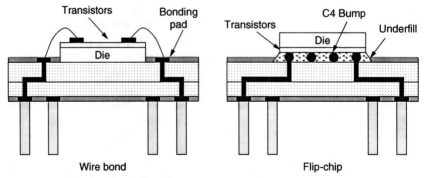

Figure 10-8 Wire bond and flip-chip.

solder bumps (also called **C4 bumps**) are created across the face of the die, and these bumps match pads connecting to the substrate interconnect. Whereas wire-bond connections must be made one at a time, flip-chip connections are made all at once as the die is pressed into the package and heated to allow the solder bumps to reflow. Because the solder bumps are placed across the entire face of the die, thousands of connections can be made. The inductance and capacitance of these connections is minimized since there are no flying lead wires connecting the die to the substrate.

Although the electric connections made by flip-chip are superior, the actual physical connection to the substrate is more of a problem. In a wire-bond package, the backside of the die is glued to the substrate before any lead connections are made. The die rests directly on the substrate allowing for a good bond. In flip-chip packages, the die rests on top of the solder bumps, which rest on the package substrate. If the CTE of the die and the package are not well matched, then the solder bumps alone will not be sufficient to hold the die on the package. As the die heats up, stress could cause the solder bumps and die to tear free of the package. To prevent this, the die must be glued to the package, but coating the package with glue before attaching the die will prevent good electrical connections by the solder bumps. The common solution is to use *capillary underfill* (CUF) materials.

After the die is placed facedown on the substrate and the solder bumps are reflowed, a CUF epoxy is spread at the sides of the die in the gap between the die face and the package. Through capillary action, the **underfill** material naturally flows between the C4 bumps and fills the gap between the die and the package. After the epoxy has cured, it will hold the die in place and protect the solder bumps from external contaminants.

Decoupling capacitors

Many packages include some passive components in addition to active chips, and one commonly added component is decoupling capacitors, also called *decaps*. As processor clock frequency and power have increased, providing a steady supply voltage to the processor has become increasingly difficult. The amount of electrical current drawn by the processor can change dramatically in just a few clock cycles as different parts of the processor begin or finish calculations. Power-efficient designs work hard to switch only those gates that are being used at that particular moment. This reduces the average power of the processor but will increase the maximum change in current that is possible from one cycle to the next.

Sudden increases in current cause the supply voltage to droop, which slows the speed of the logic gates. Too large a droop and some circuit paths on the die may not be able to operate properly at the processor

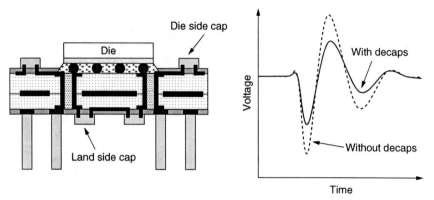

Figure 10-9 Decoupling capacitors.

frequency. The processor will produce incorrect results because these results were not yet complete when the processor clock arrived to capture their values. If a sudden decrease in current causes the supply voltage to spike too high, the gate oxide of transistors may break down and the processor will be permanently damaged.

Adding decoupling capacitors to the package, as shown in Fig. 10-9, reduces the noise on the supply voltage. The capacitors are connected with one terminal to the high voltage and one to the ground and work by storing charge. When the supply voltage droops, the capacitors will supply extra current and reduce the voltage droop. When the supply voltage spikes, the capacitors will absorb charge and reduce the peak voltage. *Die side capacitors* (DSC) are placed on the same side of the package as the die. *Land side capacitors* (LSC) are placed on the opposite side of the package. LSC placed directly beneath the die are more effective than DSC, but they also take up room that could otherwise be used for more leads. DSC must be used by SMT packages where there is not enough space between the bottom of the package and the PCB for decaps.

Although they add cost, by reducing supply noise decoupling capacitors improve processor frequency and reliability. Passive elements like decaps can be built into the processor die and may even be more effective when made this way. However, they take up large amounts of die area. Overall, adding components to the package that allow the smallest possible die size may reduce product costs.

Thermal resistance

In addition to supplying clean power into the processor, the package is designed to help efficiently move heat out of the processor. As processors have increased in frequency, their power has roughly doubled every

3 years.[3] As transistors have scaled to smaller dimensions, that power is concentrated in a smaller area. High-performance processors may produce as much as 100 W/cm^2. Without proper cooling, many processors could quickly reach temperatures that would cause permanent damage.

Heat is never destroyed; it can only be moved. A refrigerator does not create cold, it merely moves heat from its interior to the exterior. Leave the refrigerator door open and in the end the room will get warmer from the heat of the refrigerator's motor. The ability of the package and system to cool the processor is therefore determined by how easily heat can move. This is typically measured as thermal resistance and has an important impact on the performance of the processor.

As the power of the processor increases, its temperature will increase. The difference between the processor temperature and the ambient temperature of its surroundings will increase linearly as more power is applied. This rate of temperature increase (usually written in degrees Celsius per watt) is the thermal resistance of the processor. Relatively low-power processors sometimes rely on heat being transmitted solely through the leads of the package into the PCB. Higher-power processors must provide a mechanism for efficiently dissipating heat into the surrounding air. Figure 10-10 shows the parts of a high-power processor package that determine thermal resistance.

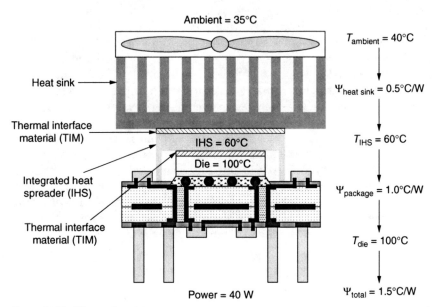

Figure 10-10 Thermal resistance.

---

[3]Mahajan, "Emerging Directions for Packaging."

Transistors on the die surface release heat. In a flip-chip package, the transistors will be on the side of the die facing the package substrate. Some heat will flow out through the solder bumps and package leads. The rest will diffuse through the silicon chip to the backside of the die. Packages for high-power processors may include an **integrated heat spreader (IHS)**. This is a metal plate that touches the backside of the die. It helps cool the die by allowing the heat of the die to rapidly diffuse over a much larger area.

To make sure there are no gaps between the die and the heat spreader, a polymer paste is spread across the backside of the die. This type of layer, added specifically to allow heat to carry between separate pieces of a package, is called *thermal interface material (TIM)*. Heat from the die passes through the TIM into the IHS. The heat then moves through the metal of the IHS before reaching a second layer of TIM. This layer fills the space between the IHS and the heat sink.

Heat sinks are fabricated of aluminum or copper fins to have as much surface area as possible. They allow heat to be dissipated into the air around the processor. Many heat sinks use a built-in fan to blow air through the fins and remove heat more rapidly. Although a processor in a package might occupy less than 1 cm^3 and weigh only a few grams, the heat sink required to use the processor might be 150 cm^3 and weigh 500 g. Some heat sinks are attached to the processor package, but many are so large and heavy that they must be attached instead to the PCB or directly to the external case. Being bolted directly into the product case gives a heat sink more support but prevents it from being attached until the final assembly of the product. The motherboard and processor must be installed in the case before the heat sink, and if the processor is to be replaced, the heat sink must be unbolted from the case first. The size, weight, and noise generated by a computer system are all significantly impacted by the processor heat sink.

The thermal resistance measures the effectiveness of the package and heat sink in cooling the die. It is the ratio temperature rise versus power. In the example in Fig. 10-10, the ambient air around the heat sink inside the product case is assumed to be 40°C. When the processor is using 40 W of power the hottest spot on the IHS is measured as being 60°C. This means that the thermal resistance of the heat sink and its TIM is 0.5°C/W, allowing a 20°C rise in temperature when 40 W is applied.

$$\Psi_{\text{heat sink}} = \frac{(T_{\text{IHS}} - T_{\text{ambient}})}{\text{power}} = \frac{(60°C - 40°C)}{40 \text{ W}} = 0.5°C/W$$

At the same time, the die temperature is measured as 100°C (near its reliability limit). The thermal resistance of the package is therefore

1°C/W, allowing a 40°C rise when 40 W is applied. The overall product thermal resistance is the sum of the thermal resistance of all the layers giving a total of 1.5°C/W, allowing a 60°C rise on the processor die from the ambient air around the heat sink at 40 W of power.

Thermal resistance could be improved by thinning the die, but if it is too thin, it may crack during the assembly process. Applying the IHS with more pressure thins the first TIM layer, but this again risks cracking the die. Applying the heat sink with more pressure thins the second TIM layer, but this can crack the PCB itself. An important function of the IHS is to protect the die from the mechanical stress of attaching the heat sink. A larger heat sink reduces thermal resistance but adds volume and area. Running the heat sink fan at higher speeds also helps but produces more noise. External airflow over the board components also affects die temperature. Adding more case fans cools the ambient air in the case, but this again adds noise. Some heat sinks use liquid forced between the fins instead of air, but liquid cooling solutions are more expensive and difficult to make sufficiently reliable.

Compounding the problem of cooling individual components is the trend of smaller packages that allow more components to be placed closer together on a single PCB. The overall power density of the board is made higher, and a large heat sink that cools one component may actually impede airflow to other components on the board. For systems where space constraints require the use of very small or no heat sinks, thermal resistance will be dramatically higher. This can easily limit the processor performance that is possible without the die reaching temperatures beyond its reliability limit. Processors can operate with on-die temperatures above 100°C, but at higher temperatures the transistors, wires, and solder bumps degrade over time until eventually the processor fails. As future processors use transistors with smaller dimensions and increased power density, improvements in thermal resistance will be required to maintain the same power levels and temperatures.

Multichip modules

There is a constant drive to add more functionality to each die. Performance is improved and the required board space reduced by combining multiple die into a single larger die that implements the same functions. However, yield loss for very large chips limits how large a die can be manufactured at a reasonable cost. An intermediate solution is to combine multiple separate die into a single package, a **multichip module (MCM)**.

The performance of an MCM is improved over that of separate packaged die by having short high-quality interconnects between the die built into the MCM. Although the performance is not as good as a single large die, silicon cost is reduced. A single defect can cause the loss of a die, but if a processor is composed of multiple separate die, the impact of each defect is less. Whether overall costs are reduced depends upon the defect density, total die area required, and the relative costs of single die and MCM packages. Figure 10-11 compares the cost of a two-die MCM versus a single-die solution.

Splitting functionality into multiple die always reduces silicon costs, but the impact is much greater if the total silicon area is large or the defect density is high. Very large die are especially costly to produce because they are more likely to have defects. Splitting a design into multiple die prevents a single defect from causing the loss of all the silicon area, and the total silicon cost grows much more slowly as the total area needed increases.

The package and assembly costs are not impacted by the silicon area but are much higher for an MCM compared to a single-die package. Whether costs are reduced by using an MCM depends on whether the increase in package cost is larger than the savings in silicon cost. Because silicon costs are small for small die, below a certain total silicon area a single-die solution will be cheaper. However, if a single die would be very large, dividing the design into multiple die and creating an MCM product can ultimately be less expensive to manufacture.

The additional cost of an MCM package depends upon the type of substrate used. MCMs do not use molded plastic substrates, making the

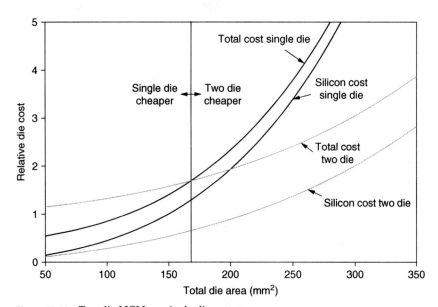

Figure 10-11 Two-die MCM vs. single-die costs.

least expensive solution a laminate substrate (MCM-L). These use FR4 layers and are manufactured just like a standard PCB. They may use bare die applied to the board or already packaged die. As with single-die packages, a ceramic substrate (MCM-C) provides better thermal resistance at higher costs. The most costly and highest performance solution is to use thin film deposition and patterning techniques identical to those used on die to build a deposited substrate (MCM-D). Many MCM packages will combine these techniques by depositing thin buildup layers on top of thicker core layers of FR4 or ceramic.

The scaling of Moore's law means that any product that is less expensive when produced as an MCM will in a short number of generations be more cheaply produced as a single die. In the 0.6- and 0.35-μm technology generations, processor frequencies reached high enough values that two levels of cache memory were needed for good performance. However, transistor dimensions were still large enough to make it impractical to build a single die that contained the processor and all the cache memory. This led to the use of MCMs, which allowed designs that would be too expensive to fabricate as a single chip to be divided among several chips while keeping the performance loss to a minimum. In later generations, smaller device dimensions have allowed the level 2 caches and processor to be built on a single die. New MCM solutions have followed where two complete processor die are combined in one package. These are called *multichip processors* (MCP). This is one way to implement a "dual-core" processor, by packaging two-processor die together. As scaling continues, it will become more cost effective to manufacture a single "dual-core" die, but undoubtedly there will always be new functionality and performance, which could be added by making use of MCM packages.

Example Assembly Flow

The final steps to create a working microprocessor combine the silicon die with its package. The previous chapter described the steps to process a CMOS wafer. We are now ready to complete our example processor product by combining the die with the processor package. The common steps for assembling a flip-chip package are:

1. Bumping
2. Die separation
3. Die attach
4. Underfill
5. Lid attach

The first step in preparing die for a flip-chip package is bumping, the creation of solder balls that will provide electrical connections between the die and the package. Figure 10-12 shows the steps of a common method of bumping, electroplating.

The starting point is a processed wafer with all the transistors and metal interconnect layers already finished. The drawings in Fig. 10-12 show only the top few microns of the wafer and are not drawn to scale. Compared to solder bumps that might be 100 μm in diameter, even the largest interconnects on the die are tiny. The process of bumping begins with depositing and developing photoresist to allow etching to cut through

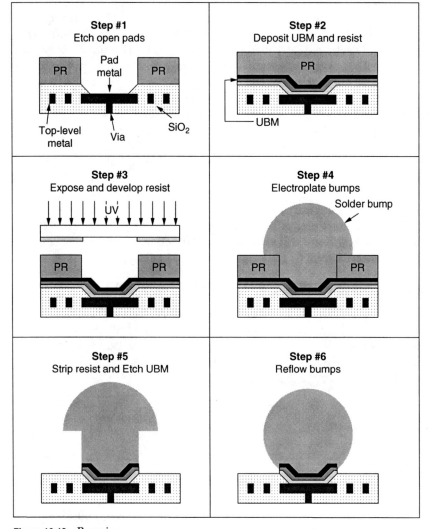

Figure 10-12 Bumping.

the insulation to the metal pads, which will ultimately connect the bumps to the circuitry of the die (step #1). Sputtering is then used to coat the die in layers of metal called the **under-bump metallization (UBM)** and a new layer of photoresist is applied (step #2).

The lead and tin of the solder bumps are not very compatible with the metal layers and insulation of the die. The material of the solder bumps does not cling well to the die and diffuses into the interconnects and insulation. The UBM is deposited to provide a better interface between the die metallization and the solder bumps. There is not a single good choice of material that clings well to the die metal pad, blocks diffusion of the solder, and clings well to the solder. As a result, UBM usually comprises three or more layers of metal. An adhesion layer provides a good mechanical connection to the die. A diffusion barrier prevents contamination of the die, and a solder "wettable" layer provides a surface to which molten ("wet") solder will cling. Common materials for these layers are alloys of titanium, chromium, and nickel.

The new layer of resist is exposed and developed to open holes over the pads for formation of the bumps (step #3). One method of depositing the material of the solder bumps is electroplating. The wafer is placed in a solution containing a probe coated in the material to be deposited. A voltage is attached across the probe and the UBM coating the wafer. Ions of metal from the probe are electrically attracted to the exposed regions of UBM. As they attach to wafer, they become electrically connected to the UBM and attract still more ions from the solution. The bump material is rapidly built up past the level of the photoresist layer (step #4).

After electroplating the resist is stripped away and a wet etch removes the UBM wherever it is not covered by solder. The result is a mushroom-shaped structure of solder material left behind (step #5). Applying heat allows the solder material to flow into its natural spherical shape (step #6). The die are now ready to be packaged, but first they must be separated from one another.

Die separation is the process of cutting the wafer into individual die. Before separation, each die on the wafer goes through preliminary testing. Contact probes touch the C4 bumps and allow at least basic functionality to be tested. Defective die will be discarded and not packaged. For die separation, the wafer is mounted on a circle of adhesive plastic that will hold the individual die in place as the wafer is cut. A protective coating is sprayed over the die, and a saw cuts through the wafer but not through the entire thickness of plastic. After repeated cuts have divided the entire wafer, the cutting debris and protective coating are washed off. Die that tested as functional are picked off the plastic sheet to be packaged and defective die are left behind. The remaining steps in packaging the processor are shown in Fig. 10-13.

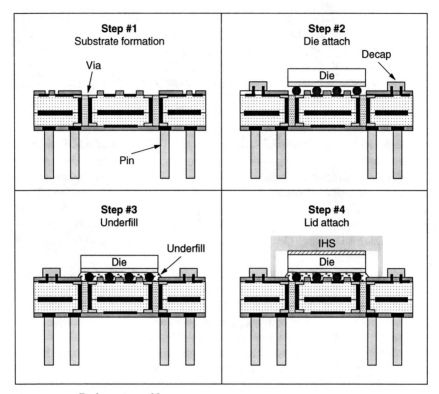

Figure 10-13 Package assembly.

The package substrate has been manufactured in parallel with the processing of the wafer (step #1). After die separation, flux and solder are applied to the package substrate through a stencil, and the die and any decoupling capacitors or other passive components are attached (step #2). The flux cleans oxide off the metal contacts, and heating allows the solder to flow and make good electrical contacts.

Underfill epoxy is then spread on the substrate at the edges of the die and flows underneath the die, filling the spaces between the C4 bumps (step #3). The package is heated again to cure the underfill and solidly glue the die to the substrate. Finally, a thermal interface material is spread on the backside of the die and the IHS is attached (step #4). Sealant fixes the IHS in place and the packaged processor is now complete.

Conclusion

The importance of the physical, electrical, and thermal characteristics of the package makes it possible to differentiate microprocessor products based on the package alone. It is common for identical die to be placed into multiple different packages and sold as different products. For a desktop package, balancing cost and performance might be the highest priority. For a server product that will be sold at a very high price, a more expensive package allowing for more performance may be justified. For mobile products, the size of the package and enabling of dense placement of components on the PCB is critical.

Recent packaging trends have focused on reducing the overall size of the package to allow for smaller, more portable electronics. Any package with an area less than 1.5 times that of the die is called a *chip scale package (CSP)*. These packages are created by the same methods as larger packages but use smaller lead pitches and smaller wire traces to minimize their size. One possible approach for reducing package size still further is embedding the die inside the package substrate. A bumpless buildup layer (BBUL) package replaces a FR4 core layer with the die itself. The die does not need C4 bumps to contact the package substrate because buildup layers of interconnect are formed around the die. This approach has the disadvantage that the substrate can not be manufactured in parallel with the die, but brings packaging size down to a bare minimum.

As processors continue to increase in performance and power while being used in an increasing variety of products, packaging will continue to be a key part of microprocessor design.

Key Concepts and Terms

Blind and buried vias
C4 bump
Coefficient of thermal expansion (CTE)
DIP, quad package, grid array package
Direct chip attach (DCA)
FR4 and alumina
Integrated heat spreader (IHS)
Known good die (KGD)
Microvia

Multichip module (MCM)
Printed circuit board (PCB)
Socket
Surface mount technology (SMT)
Through hole technology (THT)
Under-bump metallization (UBM)
Underfill
Wire bond and flip-chip

Review Questions

1. What are the main design choices when choosing a package?
2. Compare modeled plastic, ceramic laminate, and organic laminate packages.
3. Why are the CTE of package materials important?
4. What does Rent's rule predict for the scaling of microprocessor lead counts?

5. What are the advantages of flip-chip packages compared to wire-bond packages?
6. What are the three most common lead configurations?
7. How can using an MCM package reduce costs?
8. What factors affect the thermal resistance of a processor?
9. [Lab] Looking at a PC motherboard identify as many packaging types as possible.
10. [Bonus] Assuming an ambient temperature of 40°C and a maximum allowed on die temperature of 100°C, what is the thermal power limit for a processor in a packaging giving a total thermal resistance of 0.5°C/W? What is this limit for a thermal resistance of 3°C/W?

Bibliography

Bakoglu, H. B. *Circuits, Interconnections, and Packaging for VLSI.* Reading MA: Addison-Wesley, 1990.

Corbin, John. et al. "Land Grid Array Sockets for Server Applications." *IBM Journal of Research and Development*, vol. 46, 2002, pp. 763–778.

Davis, E. et al. "Solid Logic Technology: Versatile, High Performance Microelectronics." *IBM Journal of Research and Development*, vol. 8, April 1964, pp. 102–114.

Hannemann, Robert et al. *Semiconductor Packaging: A Multidisciplinary Approach.* New York: Wiley-Interscience, 1994.

Hasan, Altaf et al. "High Performance Package Designs for a 1 GHz Microprocessor." *IEEE Transactions on Advanced Packaging*, vol. 24, November 2001, pp. 470–476. [Description of the slot 2 and socket 370 packages used for the Intel Pentium III.]

Klink, Erich et al. "Evolution of Organic Chip Packaging Technology for High Speed Applications." *IEEE Transactions on Advanced Packaging*, vol. 27, Feburary 2004, pp. 4–9.

Landman, Bernard and Roy Russo. "On a Pin versus Block Relationship for Partitions of Logic Graphs." *IEEE Transactions on Computers*, vol. C-20, December 1971, pp. 1469–1479. [The original paper citing Rent's rule.]

Larson, Shawn. "HDI-Buildup Technology and Microvias: No Longer a Mystery." *EDN*, March 6, 2003, pp. 111–114.

Lau, John and Shi-Wei Lee. *Chip Scale Package: Design, Materials, Process, Reliability, and Applications.* New York: McGraw-Hill, 1999.

Lau, John. *Flip Chip Technologies.* New York: McGraw-Hill, 1996.

Lau, John. *Low Cost Flip Chip Technologies for DCA, WLCSP, and PBGA Assemblies.* New York: McGraw-Hill, 2000.

Lii, Mirng-Ji et al. "Flip-Chip Technology on Organic Pin Grid Array Packages." *Intel Technology Journal*, vol. 4, August 2000, pp. 1–9.

Mahajan, Ravi et al. "Emerging Directions for Packaging Technologies." *Intel Technology Journal*, vol. 3, May 2002, pp. 62–75.

Chapter 11

Silicon Debug and Test

Overview

This chapter discusses how completed die are tested. The chapter describes the validation of the design, the root causes and the fixing of design bugs, and the testing flow to identify manufacturing defects.

Objectives

Upon completion of this chapter the reader will be able to:

1. Understand the importance of design for test (DFT) circuits.
2. Understand how scan sequentials are used.
3. Describe the different types of silicon bugs.
4. Understand how a shmoo diagram is used in debug.
5. Describe some of the tools used in silicon debug.
6. Describe the different ways of fixing silicon bugs.
7. Understand the silicon test flow used to identify manufacturing defects.

Introduction

The most exciting single day in the life of any semiconductor product is **first silicon**, the day when the first chips complete fabrication. This day comes after what may have been years of simulation and is a suspenseful test of whether anything crucial has been overlooked in the design or fabrication of the product. Some of these first chips are immediately packaged and plugged into circuit boards for testing. While engineers cross their fingers and hold their breath, the product is powered up for the first time. In the best scenario, the chips may behave exactly as simulations predicted they would. In the worst case, the only result of applying power may be smoke from a suddenly melting circuit board. Most first silicon designs fall somewhere between these extremes, being somewhat functional but not behaving exactly as expected. Silicon debug and test is the story of what happens next.

Even design teams lucky enough to have their first chips perform perfectly are left with the nagging question, "Are we ready to ship to customers?" If the product successfully runs 10 programs, does that prove the design is correct? Is that sufficient to begin sales? If not, is testing a hundred programs sufficient, or a thousand? This is the job of post-silicon validation, testing of finished parts to find design bugs. A silicon design bug is any unexpected behavior from the part, and first silicon chips inevitably have some bugs.

Pre-silicon validation attempts to find and correct as many bugs as possible before first silicon is created, but the slow speed of pre-silicon simulations requires years to test only minutes of actual operation. In hardware, millions of transistors are switching in parallel to determine the behavior of the processor. In simulation, the behavior of one imaginary transistor must first be determined before simulating the next and then the next. This serialization of transistor behavior dooms simulations to being millions of times slower than hardware operation. Because of this difference in speed, the shortest time to market is usually not to try and produce a perfect first silicon design. Instead, the first silicon chips will be produced at the point when it is believed that post-silicon validation can find the remaining bugs and bring the product to market more quickly than continued pre-silicon simulation.

When bugs are detected, their cause must be found and fixed. The process of finding the root cause and eliminating design flaws in silicon is called *silicon debug*. The time spent in silicon debug has a direct impact on how quickly a product ships to customers; for a new processor design, debug may last from a few months to a couple of years.[1,2] For products that have life cycles of a few years at best, reaching the market a few months early because of effective silicon debug has a huge impact on profits.

Even after design flaws have been fixed, there will always be some die that do not function properly because of manufacturing defects. If all manufacturing defects could be eliminated, the process dimensions would be reduced to cut costs and increase performance until some defects began to appear. As a result, there will always be some nonfunctional die, and there must be a method for quickly and reliably identifying these die.

---

[1]Carbine and Derek, "Pentium® Pro Processor Design."

[2]Bentley, "Validating the Intel® Pentium® 4."

This is the process of silicon test. Only an extremely small fraction of die will ever go through the added steps of silicon debug to try and characterize design flaws. Every die manufactured must go through silicon test before being shipped. It is critical that this process is quick to keep costs down and is thorough to avoid shipping defective die.

To aid in post-silicon validation, silicon debug, and silicon test, all modern processors add some circuits used specifically for testing. These are called **design for test (DFT)** circuits.

Design for Test Circuits

DFT circuits are special in that the customer never uses them. Every processor die uses some amount of area and some amount of power for circuits that are never used during the normal operation of the chip. However, without these circuits, validation and test would be almost impossible.

The problem is the decreasing observability and controllability of processors. Without DFT circuits, the only signals that can be directly observed or controlled are the processor pins. The number of processor pins is increasing but far more slowly than the number of total transistors. This makes it more difficult with each processor generation to directly read or write a specific node within the design by the use of the pins alone. DFT circuits help by expanding observability and controllability. DFT increases the cost of the design by taking up die area and consuming some leakage power. DFT circuits also impact processor frequency, and their complexity can increase the design time to tapeout. However, the benefits of DFT circuits during post-silicon validation are so great that all modern designs use at least some DFT circuits. By aiding silicon debug, DFT reduces the number of design steppings required to fix bugs. This reduces the time from tapeout to shipping and saves costs by reducing the number of mask sets created. Also, shipping costs are reduced by shorter test time, allowing for less test equipment and fewer defects shipped.

DFT circuit trade-offs are as follows:

Disadvantages	Advantages
Increased design time to tapeout	Reduced time from tapeout to shipping
Added die area	Fewer post-silicon design revisions
Added leakage power	Shorter test time to identify defects
May reduce frequency	Fewer defects shipped

Early processor designs used primarily ad hoc methods to support testing. Major functional units might have switches added to allow inputs to come from their usual sources or special test inputs. Additional pins

334 Chapter Eleven

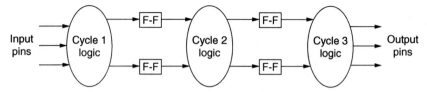

Figure 11-1 No scan.

might be added to provide extra inputs at test. Unfortunately, the design and usefulness of these methods varied a great deal between different microarchitectures. Today, the most commonly used and the most systematic DFT circuit has scannable sequential elements, also called **scan**.

Figure 11-1 represents a processor without scan. The cycles of logic gates are separated by flip-flop sequentials, which capture their inputs and drive new outputs at the start of each clock cycle. Test values are driven at the input pins and the results monitored at the output pins. However, if an output pin produces an unexpected value, it is not clear where in the processor pipeline the error occurred. This is a way of saying there is very little observability. Tests must be very long to allow for internal errors to propagate to the pins where they are detected. In addition, it is difficult to find the right combination of input signals to test for a particular error. This is a circuit with very little controllability. Tests become even longer when complicated initializations are needed to produce the desired values at internal nodes.

Figure 11-2 shows a processor with a simple implementation of scan. Each sequential element has a mux circuit added. A mux selects one of two or more inputs to route to its output. The muxes added to scan sequentials allow the input for each sequential to come from the previous cycle of logic during normal operation or a separate input during test. Each sequential output drives not only the logic required for standard operation but also the test input of another scan sequential. This connects all the sequentials into a serial scan chain. When testing, a single input pin provides a vector

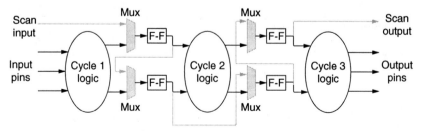

Figure 11-2 Destructive scan.

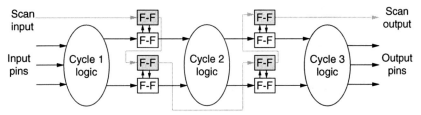

Figure 11-3 Nondestructive scan.

of values that are serially shifted into all the scan sequentials. A single clock cycle allows the values to be driven into the combinational logic, and the results are then serially shifted to an output pin to be checked. Both observability and controllability are vastly improved. A mismatch in captured scan values from what is expected will tell exactly which pipestage has the problem. One disadvantage of this implementation of scan is that it is destructive. This means that as scan values are serially shifted through the sequentials, the original values are lost. This makes it difficult to restart the processor after it has been stopped and the scan values read out.

Different implementations of scan may add complexity and die area in order to allow improved testing or have less impact on normal operation. Figure 11-3 shows the scan chain implemented as a separate set of sequentials. The circuits are designed so a special capture signal causes the scan sequentials to store the value of their associated normal sequentials. A load signal causes the normal sequentials to change to the value of the scan sequentials. This mechanism and a separate clock for the scan chain allows for the scan chain to capture and read out values without disturbing normal operation, a nondestructive scan implementation. Nondestructive scan enables rapid capture of many different points in time during a single test.

The best coverage is provided by making every sequential scannable. This is called *full scan*. However, the area penalty for this may be excessive. Many designs choose *partial scan* where some fraction of the sequentials are scannable. Faults may have to propagate through multiple cycles of logic before reaching a scannable element, which makes tests longer to get the same coverage and makes debug more difficult, but it is still vastly simpler than with no scan.

To keep test time to a minimum, it is useful to calculate the minimum number of test vectors needed to check for possible faults. Using CAD tools to generate these test vectors is called **automatic test pattern generation (ATPG)**. ATPG must assume a fault model, which describes how a design flaw or manufacturing defect may appear in functional tests. The simplest and most commonly used fault model is the **stuck-at fault** model. This assumes that defects can make any node appear to be permanently stuck at a low or a high voltage.

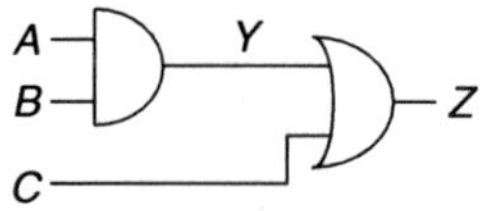

```
         A ─┐‾‾‾\
            |    )─ Y
         B ─┘___/    \___
                      )─── Z
         C ──────────/
```

Y Stuck at 0   A = 1, B = 1, C = 0   Z = 1
Y Stuck at 1   A = 0, B = x, C = 0   Z = 0

Figure 11-4 Stimulating stuck-at faults.

Figure 11-4 shows the vectors needed to test for stuck-at faults at node Y in an example circuit. Assume that node Y can neither be directly observed nor controlled, but inputs A, B, and C are controllable and node Z is observable. First the fault model assumes that node Y is stuck at a low voltage, a logical 0. To test for this, nodes A and B must be set to 1 to try and force node Y to a 1. In addition, node C must be set to a 0 to make the value of node Z be determined by the value of Y. If the circuit is working properly, Z will be a 1 for this set of inputs. The fault model then assumes that node Y is stuck at a logical 1. Setting A to 0 and B to any value will try to drive Y to 0. The input C is still driven to 0 to make the result of Y visible at node Z. If the circuit is working properly, Z will be a 0 for this set of inputs. By assuming stuck-at-one and stuck-at-zero faults at each node, an ATPG tool creates a list of test vectors to check for each possible fault.

The same type of testing is applied at the board level using **boundary scan**.[3] This is a special implementation of scan that connects the I/O pins of a die into a scan chain. By connecting the boundary scan chains of different die on a PCB in series, a board level scan chain including all the device pins is created. This scan chain is used to quickly check board level design and debug failures that occur only because of interactions between different components.

Scan is most effective when the processor and scan clocks are precisely controlled, allowing scan capture to happen at a specific point in a test and normal operation to resume for a controlled number of cycles before the next capture. To help with this, most designs allow the on-die clock generator to be bypassed during test and the processor clock signals to be driven directly from external sources. For very high-frequency designs, it may not be possible to transmit a full-frequency clock through the product pins. In these cases, on-die clock generation circuits are designed to be programmable, allowing the on-die generated clock to be started or stopped after a predetermined number of cycles.[4]

---

[3]IEEE 1149.1.

[4]Josephson, "Design Methodology for the McKinley Processor."

Breakpoint mechanisms give even better control of clocking during tests. Breakpoints monitor the processor state for specific events. These events could be things like a particular type of instruction, a certain number of branch mispredicts or cache misses, or a particular exception. The signaling of a breakpoint then causes the processor clock to stop automatically so that the scan state can be captured and read. This allows for the scan capture to be tied to a particular processor behavior rather than a particular clock cycle. Nested trigger events allow capture to occur after a specific sequence of events that indicates a particular complex failure.

Scan is appropriate for areas of the die that are a mix of logic and sequentials. Scan does not work well for memory arrays, which are almost entirely made up of sequentials. For large arrays, the area penalty of making all the array elements scannable would be excessive. However, it is very difficult to write functional tests that will stimulate all the memory cells of all the on-die arrays. A common compromise is adding array freeze and dump circuits. These allow the array state to be frozen by a breakpoint event. This prevents any further writes to the array. Then the array dump allows all the bits of the array to be read directly at the pins. These values are compared to expected values to check the array for defects.

For very large arrays, dumping the entire contents of the array may take excessive test time. Instead, the circuitry to test the array is built on die as part of the array itself. This is called **built-in self-test (BIST)**. BIST requires a stimulus generator, which writes values into the array, and a response analyzer, which checks for the correct read values. BIST adds significant area and complexity, but for the largest arrays on die the area of a BIST controller may be tiny compared to that of the array itself.

Some DFT circuits allow defects not only to be detected but also to be bypassed in order to allow the part to be shipped. The simplest method for doing this is to disable the part of the die that has the defect. A single processor die might be sold as two different server products, one with multiprocessor support and one without. If a die has a defect that affects only multiprocessor functionality, it may still be sold but only as the product that does not support this feature. The same die might also be sold as a desktop product with half the on-die cache disabled. Designing the cache to allow either half of the full array to be disabled allows any die with a single cache defect to still be sold by disabling the half with the defect.

The full cache size can be supported despite defects by adding redundant rows. These are extra rows of memory cells that are enabled to replace defective cells. When a defect is found, fuses are set on the die that cause cache accesses to the defective row to be directed to the redundant row instead. For processors where on-die cache makes up more than half of the total die area, the ability to tolerate a small number of defects in the cache has a huge impact on die yield and therefore on die cost.

Modern processors could not be reliably tested without DFT circuits. As the complexity of processors continues to grow steadily, more sophisticated DFT circuits will be required. DFT is crucial not only for detecting manufacturing defects but also for identifying design bugs during post-silicon validation.

Post-Silicon Validation

The most important task as soon as first silicon chips are available is looking for design flaws. Manufacturing defects will always affect some percentage of the die, but until the design is correct, 100 percent of the chips fabricated will be defective. Post-silicon validation tests real chips for design bugs. A silicon bug is any unexpected (and usually undesired) behavior detected in manufactured chips. There are five basic categories of silicon bugs:

1. Logic bug
2. Performance bug
3. Speedpath
4. Power bug
5. Circuit marginality

The most serious bugs are logic bugs: cases where the new design produces an incorrect result. The Pentium FDIV bug is probably the most famous example of a silicon logic bug. In one version of the Intel Pentium for some combinations of inputs, the result of a divide instruction was incorrect. Looking for logic bugs is the first and primary focus of post-silicon validation.

Designs with performance bugs produce the correct result but take more time than expected to reach it. Imagine a design that, because of a bug, predicts every branch as taken no matter what is its real behavior. This prediction will be wrong a large percentage of the time, causing programs to execute very slowly, but they will eventually produce the correct result. Validation must not only check the results of applications but also that the number of cycles to generate the result matches expectations.

Speedpaths are circuit paths that limit the frequency of a processor. They prevent the processor from producing the correct result at the target clock frequency. Of course, some circuit paths on the die will always be the slowest, but designers use pre-silicon timing simulations to attempt to make the worst circuit paths in all parts of the die meet the same target frequency. If a small number of paths are limited to a much lower frequency than all the others, this is considered a speedpath bug. These paths may have to be improved before the product can ship at a reasonable frequency.

The notion of a power bug is relatively new, but new high-power processors or processors meant for portable applications may struggle to fit within the power limits of their platform. The correct result and good performance and frequency are not sufficient. The processor must be able to achieve these at a certain target power level. If the power required to run an application is much more than expected, this may have to be corrected in order to ship. For example, circuits with unintended contention cause power bugs. A design flaw might cause two gates to simultaneously drive the same wire toward opposite voltages. Depending on the relative sizes of the gates, the wire may end up at the logically correct voltage, but meanwhile the circuit is consuming dramatically more power than it should.

Circuit marginality checks look for any of the other types of bugs that appear only at extremes of voltage, temperature, or process. The die may function perfectly below a certain temperature, but begin to show logical failures above this temperature. The design is fundamentally correct but not robust. To be reliable, the processor must be able to operate successfully throughout a range of process and environmental conditions.

Logic bugs and performance bugs are caused by unexpected behavior from the RTL model or from silicon that does not match the RTL. The design flaw may have existed in RTL for years without being noticed because the particular sequence of instructions needed to trigger it was never attempted. If the RTL does not show the bug, it may be that the circuit design has not faithfully reproduced the behavior of the RTL. Speedpaths, power bugs, and circuit marginality problems arise because circuit behavior does not match simulation. There are pre-silicon checks for all of these problems, but these simulations are not foolproof. Flawed inputs to the simulator or software bugs in the simulator itself can prevent design bugs from being detected pre-silicon. Even when used perfectly, these simulations are only approximations of reality. In order to keep run time within reason, the simulators must make simplifications that are true in the vast majority of cases but not always. The ultimate test of a design's behavior is to test the chips in a real platform.

Validation platforms and tests

To find silicon bugs, tests are run on validation platforms. These are systems with much or all of the functionality of the real systems that will ultimately use the processors. They often also include extra equipment specifically to help with post-silicon validation. A logic analyzer may be used to directly monitor and stimulate the processor bus. The system can also be designed to give access to DFT circuits on the die. Special voltage supplies and cooling equipment are used to control the voltage and temperature of the processor independently of the other system components. The motherboard will include socket for the processor so that parts are swapped easily without soldering. Sockets may also be included for peripheral chips such as the chipset if the processor must be validated with more than one type.

Tests are run by executing a series of instructions on the processor and comparing the final processor and memory state with the expected result. Directed tests are handwritten by the design team to create unusual conditions and to try out the operation of specific processor functions. Random tests, which are random series of valid processor instructions, are generated by CAD tools. These tests force all the different functional areas of the processor to work together and can find bugs caused by unusual interactions between areas. New processors with the same architecture as a previous generation may inherit ancestral tests that were created to test an older design. If the new processor is compatible, it should produce the same result.

The quality of tests created for post-silicon validation is critical since even in hardware it is only possible to test a tiny fraction of all the possible processor operations and operands. For example, a simple integer add might take two 32-bit inputs. This means there are 2^{64} possible combinations of inputs, which on a 1-GHz processor would require more than 500 years to execute.[5] Exhaustive testing is simply impossible, so skills must be applied in creating tests that are likely to find design flaws but still have a reasonable run time. Luckily this effort is done in parallel with the design of the processor itself. The post-silicon validation team should have a complete set of tests written and a plan for what order they will be attempted before the first chips ever arrive.

Any test requires not only an instruction sequence but also the final processor and memory state that indicates the test has run correctly. If an already validated architecturally compatible processor is available, then by definition the logically correct behavior for the new processor is to match the results of the previous generation. The older design is used to determine the correct result of each test. For new architectures or architectural extensions, the test results must be simulated. RTL simulations show the expected behavior cycle by cycle, but these simulations run so slowly that they can generate results only for very short tests. Faster architectural simulators may be used to generate the correct final processor state, at the expense of not being cycle accurate. FPGA emulators use programmable hardware to vastly increase simulation speed and allow the creation of much longer tests.

---

[5]Josephson and Bob, "The Crazy World of Silicon Debug."

A bug's life

Bugs can be created at any point during the design. Imagine you are a logic designer on a processor design project. Tapeout is still 2 years away but the RTL model must be completed long before then. After a late night of coding and somewhere between the first and second coffee of the day, your fingers fumble as you type the name of a variable in your RTL code. By chance, the misspelled name is a valid variable, although a different one than you intended. The RTL compiles successfully and a bug is created. This may seem like carelessness or merely bad luck, but when writing thousands of lines of code these types of mistakes inevitably happen. Any project that plans assuming its engineers will never make a mistake is surely doomed. Some common sources of logic bugs are:[6]

Goofs. RTL typos and errors introduced when copying code from among similar functional areas.

Microarchitectural flaws. RTL faithfully reproduces the specified microarchitectural behavior, but this behavior itself does not produce the correct architectural result.

Late changes. RTL code was correct but was changed late in design, creating a bug. The reason for a late change is often in order to fix a different bug.

Corner cases. Common cases are implemented correctly but a combination of different uncommon circumstances produces an incorrect result.

Initialization. Most RTL simulations always start in the same state. Real silicon will have random values when powered up, which may cause bugs if not properly initialized.

Pre-silicon validation attempts to detect and correct these bugs as soon as they are created, but the pre-silicon validation process is no more perfect than the engineers. Some bugs will lay undetected within the design until silicon testing. From the point of view of post-silicon validation, a bug's life begins the day it is detected and proceeds through the following steps:

1. Detection
2. Confirmation
3. Work-around
4. Root cause and fix
5. Check for similar bugs

---

[6]Bentley, "Validating the Intel® Pentium® 4."

The detection of a post-silicon bug often begins with a validation system showing the *blue screen of death* (BSOD). When the Windows® operating system detects an error from which it cannot recover, it displays a warning message on a solid blue screen before restarting. Any processor with a bug triggered by the operating system may fail in a similar way. The same type of failure is also produced by an application hitting a bug that causes corruption of the operating system's instructions or data.

Another common symptom of a bug is a hung system. If functional unit A of the processor is waiting for a result from functional unit B, and B is at the same time waiting for A, the processor can stop execution altogether. This condition is called **deadlock**. Similarly unit A may produce a result that causes unit B to reproduce its own result. If this causes unit A to also reproduce its result, the processor again becomes stuck. In this case, the processor appears busy with lots of instructions being executed, but no real progress is being made. This condition is called **livelock**. Correct designs will have mechanisms for avoiding or getting out of deadlock or livelock conditions, but these fail-safes may not work for the unexpected circumstances created by design flaws. As a result, silicon bugs often appear as a hung processor.

A bug may appear as a test that completes successfully but does not produce the expected results. Detecting these bugs requires that the correct results for the test have been created using a compatible processor, hardware emulation, or software simulation. Performance or power bugs may appear as a test that produces the correct results but uses much more time or power than expected.

Any of these symptoms must be treated as a sign of a possible silicon bug. Of course, every time Windows crashes or hangs, this does not necessarily mean that the processor design is faulty. Software bugs in the operating system or even the test itself often cause these types of symptoms. The processor may be correctly executing a test that because of a software bug does not actually do what was intended. The test may violate the architecture of the processor in some way. For example, a flawed random test generator might produce an invalid instruction. The processor attempting to execute this test may produce a different result than the previous generation processor, but the architecture may not guarantee any particular behavior when executing nonsense code.

In addition to software problems, hardware design flaws outside the processor may also cause bugs. A chipset or motherboard flaw never triggered by the previous generation processor might appear as a processor bug. The most difficult bugs to analyze are those caused by interactions between multiple software and hardware problems. If the processor will be used with multiple operating systems or hardware configurations, then each of these possible systems must be tested. Bugs caused by circuit marginalities will often appear sporadically. To search for these bugs, the frequency, voltage, and temperature of the processor must be swept through the full range allowed. Extremes of the manufacturing process are tested by creating "skew" wafer lots where chips are intentionally fabricated at the edges of the normal range of variation.[7]

For any bug, the first step after detection is to confirm the symptom as being caused by a real bug by trying to reproduce the failure. Early in post-silicon validation, silicon test may not be able to fully screen out manufacturing defects. This means a simple broken wire, rather than a design flaw, may be causing a failure symptom. Reproducing the same symptom with other parts rules this out. If the bug can be reproduced in the RTL model, this vastly simplifies analysis. This requires picking out a small enough segment of the failing test that can be simulated in RTL in a reasonable time. Once reproduced in simulation, the values of all the RTL nodes are available, giving far more information than can be measured from the silicon. Unfortunately many bugs cannot be reproduced in simulation because they would require too long a test sequence or they are the result of circuit problems. In any case, if a bug is confirmed on multiple parts, the next task is finding a work-around.

It is important that the root cause of each bug is ultimately determined, but in the short term what is more important is that the search for more bugs continues. It's extremely unlikely that any particular bug will be the last one found, and a bug that prevents tests from running successfully may be hiding many more bugs. This is especially true for bugs that prevent the processor from completing reset or booting operating systems. These bugs may prevent any post-silicon validation at all being performed until a work-around is identified.

A common work-around is turning off some of the processor functionality. If branch prediction or the top-level cache are faulty and are preventing the processor from completing the reset sequence, these features might be disabled temporarily. The processor will execute exceedingly slowly but validation work can continue. Temporary updates to the BIOS or operating system may allow a bug to be avoided. Failing these, further testing may be limited to only certain platforms or certain voltage and temperature conditions. In the worst case, some tests may have to be avoided until a work-around is found or the bug is fixed.

Finding bugs is in many ways easier after first silicon. Many more test cycles can be run than were possible in simulation. However, finding the root cause of bugs is far more difficult. Pre-silicon tools typically look for one type of problem at a time. Separate simulations are used to check for logic bugs, speedpaths, and circuit marginalities. In the real world, these may all interact to produce sometimes baffling behavior. Locating the specific design

---

[7] Josephson, "Design Methodology for the McKinley Processor."

flaw producing a bug requires special equipment and methodologies. Silicon debug is the specific task of identifying the cause of silicon bugs and determining how they can be fixed. This is discussed in the next section.

After silicon debug has identified a fix, it must be determined how each silicon bug was missed by pre-silicon validation. This is not for the purpose of assigning blame, but because if a flaw in the pre-silicon tools or methodology allowed one bug to escape into silicon, it is extremely likely that there are other bugs in silicon that were missed for the same reason. Ideally, the pre-silicon tools will be fixed and rerun on the design to look for more bugs that may have not yet been found in silicon. Just as important, continually improving overall pre-silicon validation will reduce the number of silicon bugs on future projects. As processor designs continue to grow in complexity, validation before and after first silicon will remain a critical part of the design flow.

Silicon Debug

Once post-silicon validation has detected a reproducible bug, even as workarounds are investigated, silicon debug begins at once. The silicon debug team must discover not only the cause of the bug, but also a way to fix the problem. This task can feel like a frantic search for a needle in a haystack. It may be that a single transistor out of 100 million is connected or sized incorrectly. This one design flaw may be all that is preventing the product from shipping.

The primary tools of the silicon debug team are specialized machines designed specifically to test integrated circuits. These testers are called **automatic test equipment (ATE)** and can cost more than a million dollars each.[8] The same type of testers will be used to find manufacturing defects during silicon test. Unlike a validation platform, which operates in much the same way as the end-user system, ATEs are stored-response testers. For each test, the ATE stores a cycle-by-cycle pattern of what values should be driven on each of the processor input pins and what values are expected on each of the output pins. The ATE does not respond to the processor as a real system would. If an output pin is driven differently than expected, the ATE does not react in any way other than to record the mismatch. A test of even a few seconds may require loading a pattern of billions of cycles, so ATEs are not practical for running large numbers of very long tests. As a result, most post-silicon validation cycles will be run on validation platforms rather than testers. However, ATEs have several significant advantages compared to a standard platform.

---

[8]Nelson, "Affordable ATE: From Lab to Fab."

The tester can alter the voltage or timing of inputs on a per pin basis. It also monitors the voltage and timing of outputs on a per pin basis. Because there are no system interrupts or other competing applications, a tester pattern is expected to run in exactly the same way every time, while tests of platforms supporting peripherals and an operating system will always have some variation. Most importantly, ATEs have full access to all the die's DFT circuits where a validation platform may not. By being precisely repeatable and eliminating possible bugs due to the operating system or other hardware components, ATEs provide the ultimate test of whether a bug is due to a processor design flaw or not.

The first task of silicon debug is reproducing bugs found in validation platforms on the tester. This means finding the shortest pattern possible, which will still trigger the bug. What parts of the tester pattern can be removed without eliminating the failure gives clues about the possible causes of the bug. DFT circuits also provide a great deal of information. Debug engineers will look at values from the scan chain, at the results of array dumps, and may program BIST circuits to perform further tests. These results are often enough to determine the root cause of a bug, but for more difficult bugs more information may be needed.

One of the most common ways an ATE is used in silicon debug is to create a **shmoo** plot. This is a plot of the result of running a single test at many different combinations of two different parameters, usually voltage and frequency. The ATE drives a pattern of inputs to the processor and monitors the output pins for the expected values. If the processor behaves as expected, the test passes and that point on the plot is marked accordingly. The test is then repeated at a different combination of parameters until the entire range of operation is seen on the plot. The name shmoo comes from a pear-shaped character in the comic strip Li'l Abner, which ended syndication in 1977. Makers of core memories in the 1960s thought their shmoo plots resembled the character, and the name has been used ever since.

Figure 11-5 shows an example of three different shmoo plots. In this example, passing points are marked with an "X" and failing points are left

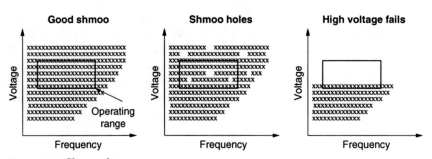

Figure 11-5 Shmoo plots.

blank. The leftmost plot is an example of a plot for a healthy chip. As with any processor, the part fails above a certain frequency, but this maximum frequency increases with voltage. This would be considered a healthy shmoo because the test passes at all points within the expected operating range. The middle plot shows shmoo holes. The part passes at most points within the operating range, but there are some specific combinations of voltage and frequency that fail. The rightmost plot shows a voltage ceiling. Above a certain voltage, the part fails at any frequency. The shape of shmoo plots helps debug engineers determine the cause of a bug as well as determine what conditions must be used to reliably reproduce the bug.

Multiple shmoos can be created for a single part by varying other parameters. In fact, the plots shown in Fig. 11-5 could have been measured from the same processor by varying temperature. The leftmost plot might have been taken at low temperature. As temperature is increased, the middle plot shows the beginning of a problem. At an even higher temperature, the rightmost plot clearly shows the problem. Because the failure is not frequency dependent, a speedpath bug is ruled out. Also, because the processor performs correctly at low temperatures, the problem is not a simple logic bug either. This is a circuit marginality bug caused by a race condition. Two signals are triggered by a single clock edge and one must arrive before the other for the processor to function properly. The two circuit paths are racing against each other. Higher temperature changes wire resistance enough to decide which path wins the race and whether the test passes or fails.

A shmoo can identify the general bug type and the conditions that trigger the bug. One tool for finding the location of a design flaw on the die is using an **infrared emissions microscope (IREM)**.[9] Silicon is transparent to infrared light and any transistor drawing current will emit infrared. An IREM image will show a picture of which transistors on the die were active when the image was taken and give a relative sense of how much current was being drawn by different regions of the die. Bugs are found by comparing IREM images of a passing part and a failing part or even a single part that passes a test at some conditions and fails at others. Switching back and forth between two IREM images, the debug engineer may notice a flashing spot. This is caused by a bright point of emissions on the image of the failing test run, which does not show on the passing test run. It is likely that the bug is affecting the devices at this point on the die.

Further information can be acquired by probing the die. Scan cannot show the values at every node and also cannot show how the voltages on different wires change within a cycle. Most wires on a processor are far too small to physically probe, but the fact that silicon is transparent to

---

[9]Bailon, "Application of Breakthrough Failure Analysis."

infrared is used to monitor individual transistors as well as taking photos of the whole die. An infrared laser focused on one transistor will be partially reflected and the amount it is reflected depends upon the voltage at the transistor. This allows the monitoring of the voltage within each cycle as the transistor switches. This process is called **laser voltage probing (LVP)**.[10]

Shmoo plots, IREM images, and LVP waveforms all provide additional information along with the output of DFT circuits, but in the end it is the ingenuity and reasoning of the debug engineers, which create a theory as to the cause of a bug. More measurements may be taken to test this theory, and ultimately a fix is proposed. Possible solutions to a silicon bug are listed as follows:

- Full-layer stepping
- Metal-only stepping
- Change fuse settings
- Microcode patch
- BIOS update
- OS update
- Document as errata

Many bugs can be fixed only by changing the processor design. Each revision of the design is called a *stepping*, and very few processors are ever sold on their first stepping. Typically about half of the mask layers required for fabrication are used to create the transistors and about half are used to create the interconnect. If the transistors and interconnect must be changed, this is a full-layer stepping. Money and time are saved if a design change affects only the interconnects and allows the same transistor level masks to be used. This is called a *metal-only stepping*. In either case, updated circuit schematics must be created and new layout drawn. The new layout is converted to masks, and the new design is fabricated. This whole process can take weeks or months, making it extremely important that the change actually fixes the bug and does not create new ones. One way to make sure of this is through **focused ion beam (FIB) edits**.[11]

A proposed design change may be implemented in hardware by physically altering the die. A FIB machine can cut through wires, disable transistors, or even trigger the deposition of new wires. FIB edits cannot create transistors, but in anticipation of this problem some designs will

---

[10]Yee, "Laser Voltage Probe (LVP)."

[11]Livengood and Donna, "Design For (Physical) Debug."

include a small number of unconnected logic gates scattered about the die. By changing interconnects, a FIB edit can make use of these spare gates. Compared to the original fabricated structure, FIB edits are extremely crude. A die with FIB edits would not be reliable in the long term and often would not be able to function properly at full frequency. Editing a single die takes hours or days. For these reasons, FIB editing could never be used for volume manufacturing, and altered die are never sold. However, days performing a FIB edit can demonstrate whether a proposed design change will actually eliminate a bug without creating unintentional side effects. This may save weeks that would be required by an extra design stepping.

Making changes to the silicon design is slow and expensive and should be avoided whenever possible. Most processors will include fuses that are set to enable or disable features. These fuses set the processor clock frequency and voltage, as well as turning on and off features. Some fuses may be included specifically to allow new architectural or microarchitectural features with high risk of bugs to be switched off. If features can be disabled with some performance loss but make the die functional, this allows the product to ship with the right fuse setting while debug continues. Experienced design teams will intentionally add these work-arounds where possible.

Another way some silicon bugs are fixed without changing the silicon design is through a microcode patch.[12] Microcode instructions are low-level steps executed by the processor when performing complex instructions, dealing with unusual events, or to initialize the processor during reset. Because microcode flows perform some of the most complex operations, they are likely to contain bugs. The microcode instructions are hardwired into a *read-only memory* (ROM) on the processor die. However, some processors are designed with the ability to "patch" the microcode. Every time the processor turns on, the reset microcode flow checks to see if a microcode patch has been loaded into flash memory on the motherboard. If so, the processor loads this patch into a RAM array that sits next to the microcode ROM on the processor die. Each time the microcode ROM is accessed, the patch RAM is checked to see if a new version of the microcode flow is available for that particular function. Running the patch can correct bugs caused by errors in the microcode ROM or allow new microcode flows to work around other bugs in the design.

Updates to the motherboard BIOS are also used to avoid some bugs. The BIOS is a ROM array that contains initialization flows and functions for dealing with special operations on the motherboard the same way the microcode ROM stores similar routines for the microprocessor. Storing the

---

[12]Sherwood and Brad, "Patchable Instruction ROM Architecture."

BIOS on flash memory chips allows it to be retained when powered down but also to be updated if needed. The BIOS controls the reset flow for the entire system and sets values in the control registers on the processor that turn on or off different options. If some specific optional modes of operation are known to have bugs, these can be disabled by the BIOS on system start-up.

An update to the operating system can also work around some problems. OS updates are commonly available to fix software bugs in the OS itself, but they are also used to avoid some hardware flaws. Updates to the microcode, BIOS, or OS all have the advantage of not requiring changes to the silicon, and these fixes are even possible after the processor has already begun shipping. Unfortunately, there are many bugs that cannot be fixed this way. Some of these bugs are never fixed but are documented as unintended behaviors called **errata**.

Real processors are never perfect. The actual implementation inevitably behaves in some ways differently than the original planned design, and as a result all processors have errata.[13,14] It is surprising to learn that processors are sold with known design bugs, but in reality this is true for all very complex manufactured products. If an automobile manufacturer discovered that in its latest model car the headlight high beams did not work when the car was in reverse, the manufacturer would probably not bother with a recall. There would be no safety issue and in normal driving using the high beams while in reverse is not necessary. Even if the design flaw caused the car to stall if the high beams were turned on when in reverse, this might be corrected on a later model but the original designs would probably still be sold. The manufacturer would simply make a note in the owner's manual that this particular model had this unintended behavior. Just like with processor errata, most customers would never encounter the bug (and most would never read the owner's manual).

Processor bugs that are good candidates for being treated as errata are those that either have insignificant consequences or are extremely unlikely to ever be encountered. Imagine a processor that when trying to execute a particular invalid instruction gives a divide by zero error instead the expected undefined instruction error. This is a design flaw. The processor behavior does not match the original specification. However, this particular flaw will have little to no impact on real applications. Properly compiled programs should not contain undefined instructions in any case. If a program with software bugs does try to execute an invalid instruction, an error will be reported although not the one that would normally be expected.

---

[13]"Intel® Pentium® Extreme Edition and Pentium® D Specification Update."

[14]"Revision Guide for AMD Athlon™ 64 and Opteron™."

Imagine tests of the same processor in a four-processor system find that when divide by zero errors are reported on all four processors in the same clock cycle, the system may crash. This is also clearly a design flaw, and the symptom is much more severe. However, this circumstance would probably never be encountered in a real system. Divide by zero errors are very rare and for four to occur simultaneously is astronomically unlikely. If system crashes caused by software errors are orders of magnitude more likely than encountering a hardware bug, fixing the hardware bug will provide no real gain in system reliability. For this reason, processor designs will always have errata.

Once silicon debug has found and corrected all the design bugs, which are severe enough to prevent shipping, the last step before being sold is developing a silicon test program, which will separate out manufacturing defects.

Silicon Test

Only a very small percentage of parts ever go through the detailed analysis of silicon debug, but every part sold must go through silicon test. Manufacturing defects prevent some die from functioning correctly, and there will always be manufacturing defects. If the process could be developed to the point where there were no more defects, the feature sizes would be shrunk to reduce cost and increase performance until manufacturing defects began to appear again. The most profitable point is always pushing the process to where some percentage of die fail, but this requires a reliable way of separating good die from bad. In addition to causing some outright defects, manufacturing variation will cause some die to run at lower frequencies or higher power than others. These parameters must be measured as well to make sure the processor will not only be logically correct but will also meet its power and performance specifications. Silicon test is the process of developing and applying a test program to guarantee the quality of the processors shipped. An example silicon test flow is shown in Fig. 11-6.

Silicon test begins even before the processors have completed fabrication. Because it can take weeks to process a wafer, it is important to be able to check the quality of the wafer before it is finished. **Electrical test (E-test)** measures test structures on a partially processed wafer. The processors themselves cannot be tested because they are incomplete, but test structures allow parameters such as transistor on-current, off-current, and wire resistance to be measured. These test structures are built into the die or made in the scribe lines in between the die, which will later be destroyed when the wafer is cut into individual parts.

The results of E-test are used as feedback to the fab. If measured E-test parameters begin to fall outside the expected range, fab work may be

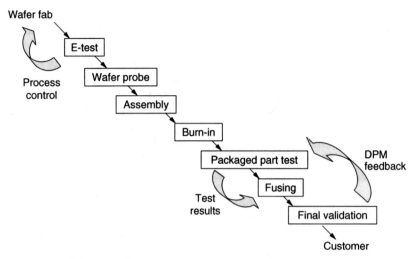

Figure 11-6 Silicon test flow.

halted until the problem is found and corrected. Without E-test, the fab might process thousands of wafers incorrectly before the finished processors were found to all be defective. With E-test, problems are found and corrected more quickly, leading to a higher percentage of good wafers.

After wafer processing is finished, the completed processors are tested while still on the wafer. This is the **wafer probe test** step. Electrical probes are touched to the top-level metal of the processor and simple functional tests are run. The processor's power and probable speed may also be measured. Before packaging it is usually not possible to completely test the processor or operate it at its full frequency. However, identifying and discarding obviously defective die before packaging reduces costs.

Die that passed wafer probe testing are assembled in packages and are sent to burn-in. This step runs the processors at elevated voltages and temperatures as a stress test. Some partial defects will allow a die to function correctly in initial testing but will cause a failure after a short time of operation. Burn-in attempts to trigger these latent defects. Operating at higher voltages and temperatures than normal simulates months of use in a few hours. Parts that still operate correctly after this stress are likely to function correctly for the full expected lifetime of the processor. The conditions and time of burn-in must be carefully chosen to weed out most of the die that will fail early in their lifetime while not significantly reducing the total expected lifetime of good die.

The most thorough testing comes after burn-in. At the **packaged part test** step, ATE machines use stored-response tests to find defective parts. These are the same types of tests used during post-silicon validation and silicon debug. However, where validation might run thousands of programs over hours of test time, silicon test must spend only a few seconds per part. Long test times can easily cause the total number of parts shipped to be limited not by the fab's production capacity, but by the number of ATE machines and the time to test each part. The most important function of DFT circuits is to provide good coverage with short test times. Silicon test uses pieces of the most important functional tests used in validation. Post-silicon validation may have found that die, which failed one particular test always failed three other specific tests as well. This means that at silicon test there is no need to include all these tests. The goal is to find the minimum set of tests that will guarantee sufficient quality for the parts shipped.

For parts that fail, tests may be repeated at lower frequencies. Most processor products include multiple frequency bins. Silicon test determines for each die the highest frequency bin at which it operates correctly. In performing speed binning, there is a great deal of pressure to neither over-test nor under-test. It would be easy to make all the testing extremely conservative. Worst case voltage, temperature, and system timings could all be assumed, and any part that passed tests under these conditions would be certain to work in a more typical environment, but overly pessimistic testing leads to lost profits. Die that could have functioned perfectly well at higher frequency bins are sold at lower speeds for less money. Optimistic test conditions will lead to more die passing tests at higher frequencies, but some of these die may fail after being sold if their actual use conditions are worse than what has been assumed at test. Silicon test must find a balance and test parts with enough conservatism to avoid shipping parts at frequencies that will cause too many failures and at the same time passing each part at the highest frequency it can realistically operate.

The results of packaged part testing and customer demand then determine how the part will be fused. Fusing will set the frequency bin for the processor and may enable or disable different features. If a part only passed testing at the minimum frequency bin with all optional features disabled, then it must be fused and sold with these settings. If a part passed at the maximum frequency bin with all optional features enabled, it can be sold with these settings for the highest profit. However, demand for different frequency bins and features never exactly matches production capacity. Some parts may be "down binned" and sold at lower frequencies if there is not sufficient demand to sell all parts at their maximum frequency. Of course, in the long term, prices and production capacity will be adjusted to try and bring demand and supply closer together.

After fusing, processors may go through a final validation test before shipping. This could involve running an example application in a real system, providing a measure of how many defective parts are not being found by the silicon test flow. A perfect flow would detect all defects, but this is not possible with reasonable test costs. In reality, all manufacturers ship some defective die. Customers that buy large numbers of parts are well aware of this. A typical target for defects shipped might be 500 **defects per million (DPM)** or 0.05 percent.[15] Increasing test time by a factor of 10 might reduce defects shipped to 100 DPM, but no amount of test time will ever bring the defects shipped to zero. The optimum solution for customers is to keep DPM low enough that dealing with defective parts is not overly expensive while not trying to force DPM so low that test costs drive up the processor's price unreasonably. By performing final testing of at least some parts in a real environment, the processor manufacturer can estimate the DPM being shipped. If too many defects pass the silicon test flow, new test vectors need to be added to catch the failures being found at final validation.

After silicon test is complete, the processor is shipped to the customer. The microprocessor design flow, which began with an idea, ends with the finished product being put to use. Semiconductor design is special because it starts with ideas and very inexpensive materials, including the main component of sand, and with a lot of hard work and effort (and the occasional $2 billion fab) creates something valuable and useful. Perhaps most importantly, processor design creates something for which people are finding new uses all the time.

Conclusion

What makes pre-silicon and post-silicon validation important is that silicon bugs are so expensive. A bug detected the same day a piece of faulty RTL code is turned in costs very little to fix. An e-mail from a validation engineer to the logic designer points out a failing test and perhaps with just an hour of effort the logic designer spots his mistake and turns in new code. As time goes on, bugs become steadily more expensive. The same bug discovered after circuit and layout work is complete will require a circuit designer and mask designer to rework the affected area. Circuit checks for logical equivalence, timing, power, and noise must be redone. Layout checks of circuit equivalence, design rules, and reliability must be redone. Changes in circuit timing or layout area may impact other areas, which will also have to be redone. If the bug has been found close to completion of the design, this rework can delay tapeout and first silicon. To avoid this, some designs may tapeout with known bugs planning to fix them in a later stepping along with any bugs found post-silicon.

The cost of bugs found post-silicon is dramatically more than those found pre-silicon. Bugs found through tests of silicon are often more difficult to

---

[15]Carbine and Derek, "Pentium® Pro Processor Design."

debug, and bugs found at this point will impact the time from first silicon to shipping the product. The expensive and specialized equipment of silicon debug is worth it because it reduces this time. Of course, the potentially most expensive bugs are those found after shipping has already begun. In the worst case, these lead to lost sales and disastrous recalls. The intent of all the effort of pre-silicon and post-silicon validation is precisely to avoid this type of bug. Missed bugs are often those created by extremely unlikely situations and therefore suitable to be ignored as errata. Others are fixed in the field by updates to microcode, BIOS, or the operating system, but every bug found after shipping has the potential for disaster. A bug with serious effects that is likely to be triggered by commercial software can turn a profitable product into an enormous loss. The exact same crisis is created with a bug-free design that has not been sufficiently tested for manufacturing defects. In the end, the consumer cannot tell the difference between a flawed design and an individual flawed die. All they see is a product that doesn't work.

The huge potential impact of bugs will continue to make rigorous pre-silicon and post-silicon validation, as well as a thorough silicon test program, key parts of any successful processor design. The work of silicon debug and test is most critical in the days after first silicon and before shipping has begun, but it never really stops. Even after a stepping has been created, which is suitable for sale, there are always improvements that can be made. Changes to the design can improve frequency, reduce power, or increase manufacturing yield. These changes are far more effective when directed by silicon debug rather than simulations. With each new stepping, the post-silicon validation must be repeated and any new problems debugged. New steppings may also require updates to the silicon test program. Throughout the product lifetime, the tasks of microprocessor design never really stop, and the promise of Moore's law is that there are always new design challenges and new possibilities in the future.

Key Concepts and Terms

Automatic test equipment (ATE)	First silicon
Automatic test pattern generation (ATPG)	Focused ion beam (FIB) edit
	Infrared emissions microscope (IREM)
Built-in self-test (BIST)	Laser voltage probing (LVP)
Deadlock, Livelock	Packaged part test
Defects per million (DPM)	Scan, boundary scan
Design for test (DFT)	Shmoo
Electrical test (E-test)	Stuck-at fault
Errata	Wafer probe test

Review Questions

1. Why do all designs find bugs post-silicon?
2. What are some common causes of logic bugs?
3. What is the difference between deadlock and livelock?
4. How does a stored-response tester work?
5. Describe five ways in which a design bug could be corrected.
6. Describe three types of post-silicon validation tests.
7. How are FIB edits used in silicon debug?
8. What types of silicon bugs are suitable for errata?
9. What is the purpose of burn-in testing?
10. What is the advantage of performing E-test and wafer probe testing?
11. [Discussion] What are the possible impacts of creating a processor design without DFT circuits? Would it be possible to validate a design using DFT circuits and then manufacture a version without them? What would be the advantages to and problems with this approach?
12. [Discussion] This text states that all microprocessors have errata. Is this true for other complex products? In what markets besides computers and under what circumstances are errata considered acceptable? When are they not acceptable?
13. [Discussion] Of all the steps in microprocessor design which is most impacted by the scaling of Moore's law?

Bibliography

Bailon, Michelle et al. "Application of Breakthrough Failure Analysis Techniques on 90nm Devices with an EOS Fail." *Proceedings of 12th IPFA*, Singapore: 2005. [Describes an example of silicon debug using IREM.]

Balachandran, Hari et al. "Facilitating Rapid First Silicon Debug." *International Test Conference*, Winchester, England: 2002.

Bentley, Bob. "High Level Validation of Next-Generation Microprocessors." *7th IEE International High-Level Design Validation and Test Workshop*, Cannes, France: October 2002.

Bentley, Bob. "Validating the Intel® Pentium® 4 Microprocessor." *Design Automation Conference*, Las Vegas, NV: 2001.

Carbine, Adrian and Derek Feltham. "Pentium® Pro Processor Design for Test and Debug." *International Test Conference*, Washington, DC: 1997.

Chandrakasan, Anantha et al. *Design of High-Performance Microprocessor Circuits*. New York: Wiley-IEEE Press, 2001.

Chang, Jonathan et al. "A 130-nm Triple-Vt 9-MB Third-Level On-Die Cache for the 1.7-GHz Itanium® 2 Processor." *IEEE Journal of Solid State Circuits*, vol. 40, 2005, pp. 195–203. [Describes the redundancy used to improve yields with an extremely large cache.]

Dervisoglu, Bulent. "Design for Testability: It Is Time to Deliver It for Time-to-Market." *International Test Conference*, Atlantic City, NJ: 1999.

"Intel® Pentium® Processor Extreme Edition and Intel® Pentium® D Processor Specification Update." Version 6, Intel Document Number 306832-006, August 2005.

IEEE 1149.1. "IEEE Standard Test Access Port and Boundary Scan Architecture." Institute of Electrical and Electronics Engineers, 2001.

Josephson, Don Douglas et al. "Design Methodology for the McKinley Processor." *International Test Conference*, Baltimore, MD: 2001. [A very readable and detailed description of the DFT circuits and methodology used in debugging the first Itanium 2. Includes good descriptions of example bugs, their discovery and resolution.]

Josephson, Doug and Bob Gottlieb. "The Crazy Mixed up World of Silicon Debug." *IEEE Custom Integrated Circuits Conference*, Orlando, FL: 2004. [A great overview of the challenges, methods, and "craziness" of silicon debug.]

Lee, Leonard et al. "On Silicon-Based Speed Path Identification", *Proceedings of the 23rd IEEE VLSI Test Symposium*, Palm Springs, CA: 2005.

Livengood, Richard and Donna Medeiros. "Design for (Physical) Debug for Silicon Microsurgery and Probing of Flip-Chip Packaged Integrated Circuits." *International Test Conference*, Atlantic City, NJ: 1999. [Describes using FIB editing for both probing and circuit edits.]

McCluskey, Edward. *Logic Design Principles with Emphasis on Testable Semicustom Circuits*. Englewood Cliffs, NJ: Prentice-Hall, 1986.

Nagle, Troy et al. "Design for Testability and Built-In Self Test: A Review." *IEEE Transactions on Industrial Electronics*, vol. 36, May 1989, pp. 129–140.

Nelson, Rick. "Affordable ATE: From Lab to Fab." *Test & Measurement World*, November 1, 2002.

Rabaey, Jan et al. *Digital Integrated Circuits: A Design Perspective*. 2d ed., Englewood Cliffs, NJ: Prentice-Hall, 2003.

Rajsuman, Rochit. *System-on-a-Chip Design and Test*. London: Artech House, 2000.

"Revision Guide for AMD Athlon™ 64 and AMD Opteron™ Processors." Revision 3.57, AMD Publication #25759, August 2005.

Rootselaar, Gert Jan van and Bart Vermeulen. "Silicon Debug: Scan Chains Alone Are Not Enough." *International Test Conference*, Atlantic City, NJ: 1999.

Sherwood, Timothy and Brad Calder. "Patchable Instruction ROM Architecture." *International Conference on Compiler, Architecture, and Synthesis for Embedded Systems*, Atlanta, GA: November 2001, pp.24–33.

Tam, Simon et al. "Clock Generation and Distribution for the First IA-64 Microprocessor." *IEEE Journal of Solid State Circuits*, vol. 35, 2000, pp. 1545–1552. [Describes local clock skew and on-die clock shrink circuits implemented in the Itanium processor.]

Yee, Wai Mun et al. "Laser Voltage Probe (LVP): A Novel Optical Probing Technology for Flip-Chip Packaged Microprocessors." *Proceedings of the 7th IPFA*, Singapore: 1999.

Glossary

1T cell   A dynamic memory cell made from a single transistor and capable of storing a single bit of data. This cell is the basis of DRAM memory.

1U server   A rack-mountable computer server 1.75 in in height.

2U server   A rack-mountable computer server 3.5 in in height.

Abort exception   An exception that prevents a computer program from continuing execution.

Accumulator   The name commonly used in early computer architectures for the register to which computational results are directed by default.

Active power   The power consumed by an integrated circuit due to the switching of voltages. This power is directly proportional to clock frequency.

ADL   Architectural description language.

Advanced graphics port (AGP)   A graphics bus standard based on the early PCI standard.

Advanced technology attachment (ATA)   A storage bus standard based on a 16-bit bus.

Advanced technology extended (ATX)   The most widely used form factor specifying the physical size and power connectors of a motherboard.

AGP   Advanced graphics port.

Amdahl's law   A relationship named after Gene Amdahl for calculating performance improvement of a computer system. Amdahl's law states that if speedup S is applied to a task occupying fraction F of the total execution time, the overall speed up will be $1/(1 - F + F/S)$.

AND   A logic function that is only true if all its inputs are true.

Anisotropic etch   An etch that proceeds in only one direction.

APCVD   Atmospheric pressure chemical vapor deposition.

Architectural description language (ADL)   A computer programming language designed to model computer architecture.

Architectural extension   Specialized instructions added to an already existing computer architecture.

Assembler   A program for converting assembly language code into machine language.

Assembly language Human-readable text mnemonics for the machine language instructions of a computer architecture. Unlike high-level programming languages, most assembly language instructions correspond to a single machine language instruction.

Associativity The number of locations where a single memory address may be stored in a cache memory.

ATA Advanced technology attachment.

ATE Automatic test equipment.

Atmospheric pressure chemical vapor deposition (APCVD) Deposition of material through the chemical reaction of gases at full atmospheric pressure. Full pressure allows high deposition rates but make it more difficult to create uniform high-quality layers.

ATPG Automatic test pattern generation.

ATX Advanced technology extended.

Automatic test equipment (ATE) Machines designed specifically to test integrated circuits by providing electrical stimulus and monitoring the response.

Automatic test pattern generation (ATPG) The generation by software of the minimal set of test vectors required to fully test a given circuit typically through the use of scan sequentials.

Balanced technology extended (BTX) A form factor specifying the physical size and power connectors of a motherboard. The BTX standard is designed to allow easier cooling of components than the older ATX standard.

Ball grid array (BGA) An integrated circuit package using multiple rows of solder balls on all four sides as leads. Common for chips to be used in portable applications.

Basic input output system (BIOS) Low-level routines programmed into a ROM chip on a PC motherboard. The BIOS initializes the PC upon start-up and provides subroutines that allow applications and the OS to interact with the motherboard hardware without knowledge of the details of its design.

BBUL Bumpless build-up layer.

BCD Binary coded decimal.

BESOI Bonded etched back silicon on insulator.

BGA Ball grid array.

Big endian A memory addressing format that assumes for numbers more than 1 byte in size, the lowest memory address is the most significant byte (the big end) of the number. See little endian.

BIM Binary intensity mask.

Binary coded decimal (BCD) A binary encoding scheme that represents each digit of a decimal number using four binary digits. This requires more bits than traditional binary encoding but makes conversion to and from decimal notation simpler.

Binary intensity mask (BIM)   A photolithography mask containing only two types of regions: those that transmit light and those that do not. See phase-shift mask.

BIOS   Basic input output system.

Bipolar junction transistor (BJT)   A transistor constructed of two P-N junctions. Charges emitted across one junction can be collected by the other. The voltage on the base in between the junctions can amplify or inhibit this effect.

BIST   Built-in self test.

Bit pitch   The fixed height of all the layout cells drawn to create repeated layout for several bits of data requiring the same logic. By drawing all the cells to have the same height, assembly of multiple rows of cells is made more efficient.

Bit   A single binary digit. A bit can only be one of two values.

BJT   Bipolar junction transistor.

Blue screen of death (BSOD)   The blue error screen shown by Microsoft Windows upon an irrecoverable error.

Bonded etched back silicon on insulator (BESOI)   A type of SOI wafer created by forming a layer of insulation on top of one wafer and bonding that wafer face-down to another. The wafer is then etched back to leave only a thin layer of silicon on top of the insulator.

Boundary scan   A scan chain connecting the input and output drivers of an integrated circuit. Connecting the boundary scan chains of different ICs on a motherboard allows tests of the board independent of the components.

Branch instruction   A computer instruction that specifies a test condition and can redirect program execution to a target address based on that test.

Branch prediction   A microarchitectural design feature that improves performance by predicting whether a branch instruction will redirect program flow or not based on its past behavior. If these predictions are usually correct, the time saved in executing branches can outweigh the added time required when a prediction is wrong.

Branch target buffer (BTB)   A memory cache supporting branch prediction by holding information on the past behavior of recent branches and their target addresses.

Braze   To solder two pieces of metal together using a hard high-temperature soldering material usually composed of copper and zinc.

BSOD   Blue screen of death.

BTB   Branch target buffer.

BTX   Balanced technology extended.

Glossary

Built-in self test (BIST) Circuits added to an integrated circuit capable of generating stimulus and comparing outputs of expected values in order to test a circuit block.

Bulk wafer A silicon wafer cut from a crystal ingot.

Bumping The process of adding solder bumps to finished die on a wafer before assembly into a flip-chip package.

Bumpless buildup layer (BBUL) packaging Packaging that uses the die itself as one of the core layers of the package. Package interconnects are built around and on top of the die, eliminating the need for die bumps and reducing package size.

Burn-in High-voltage and high-temperature stress testing designed to trigger latent defects and avoid shipping parts that would fail early in their lifetime.

Byte A group of 8 bits.

C4 Controlled collapse chip connections.

Cache coherency Protocols used to handle data sharing among multiple caches in a multiprocessor computer.

Cache hit When the memory address being searched for is currently held in cache memory.

Cache memory High-speed memory that holds recently accessed data or instructions in order to improve performance by reducing the average latency of accesses to memory.

Cache miss When the memory address being searched for is not currently held in cache memory.

Call instruction A computer instruction that redirects program execution to a new instruction after pushing the current instruction address onto the stack so that a return instruction will resume execution with the instruction immediately after the call. This allows the same block of code to be executed from many places and still return to the correct point in the program.

Canonical sum A function written as the logical sum of fundamental products each of which is true for only one combination of inputs.

Capacity miss A cache miss that would not have occurred in a larger cache.

Capillary underfill (CUF) Epoxy that can flow in between the C4 bumps of a die facedown on a package in order to hold the die in place under the stress caused by heating during use.

Carrier mobility The ability of free charge carriers in a semiconductor to move under the influence of an electric field. Higher mobility leads to higher currents and higher switching speeds. In silicon, N-type carriers typically have twice the mobility of P-type carriers, allowing NMOS transistors to produce the same current as PMOS transistors of twice the width.

CBD Cell-based design.

Cell-based design (CBD) A partial automated design method for creating layout by using a library of predesigned cells.

Ceramic dual in-line package (CERDIP) A ceramic package with rows of leads on two sides.

Chemical mechanical planarization (CMP) A manufacturing process to create a flat surface for the next processing step by spinning the wafer in an abrasive slurry.

Chemical vapor deposition (CVD) Deposition of material through the chemical reaction of gases.

Chip on board (COB) Attaching a bare integrated circuit die directly to a printed circuit board. Also called direct chip attach (DCA).

Chip scale package (CSP) An integrated circuit package with area less than 1.5 times the die area.

Chipset A pair of chips responsible for communication between the processor and other components on the motherboard. Typically one chip of the chipset is responsible for communication with high-speed components and the other with lower-speed components.

Circuit design The design step of changing logic (usually written into RTL) into a transistor implementation.

Circuit marginality A design problem that only appears at certain combinations of voltage, temperature, and frequency.

CISC Complex instruction set computing.

Cleanroom An integrated circuit fabrication facility where particles in the air are carefully controlled to limit manufacturing defects. The class of cleanroom is measured by the average number of particles above 0.5 µm in diameter per cubic foot of air.

Clock jitter Variation in clock frequency, causing some clock cycles to be longer or shorter than others.

Clock skew Variation in the arrival of a single clock edge at different points on the die, causing some sequentials to receive the clock signal before others.

CMOS Complementary metal oxide semiconductor.

CMP Chemical mechanical planarization.

COB Chip on board.

Coefficient of thermal expansion (CTE) The rate at which a material expands when heated. When assembling die into a package or packages onto a board it is important to try and minimize differences in CTE in order to limit the mechanical stress caused by heating.

Cold miss A cache miss caused by a memory address being accessed for the first time.

Compaction An integrated circuit design project that moves an earlier design onto a new manufacturing process while making few other changes.

Compiler A program that converts a high-level programming language into machine language.

Complementary metal oxide semiconductor (CMOS) A manufacturing process or logic gate that uses both NMOS and PMOS transistors.

Complex instruction set computing (CISC) Computer architectures that support memory operands in computation instructions, variable instruction size, and many addressing modes. Examples are the VAX and x86 architectures. See RISC.

Computer architecture The set of instructions that can be executed by a processor, their encodings, and their behavior. Computer architecture defines all the information needed to create machine language programs for a processor.

Conduction band The band of energy states available to electrons not bound to a particular atom in a crystal.

Conductor A material containing a large number of free charge carriers allowing it to conduct electricity easily.

Conflict miss A cache miss that would not have occurred in a cache of higher associativity.

Contact A vertical connection between the lowest level of interconnect and the semiconductor surface.

Control dependency The dependency caused when one instruction alters the program execution flow. If path of execution cannot be determined until an instruction is executed, following instructions must stall unless branch prediction is used to guess the proper instructions to execute.

Controlled collapse chip connection (C4) bumps Solder bumps used by die to be assembled in flip-chip packages. The C4 bumps are collapsed under heat and/or pressure to form electrical connections.

Cross talk Electrical noise on a wire caused by changing voltages on neighboring wires.

CSP Chip scale package.

CTE Coefficient of thermal expansion.

CUF Capillary underfill.

Cutoff region The MOSFET region of operation where gate voltage does not form an inversion layer at either end of the channel and there is almost no current flow. See linear and saturation regions.

CVD Chemical vapor deposition.

Czochralski (CZ) crystal growth A method for growing large crystals by gradually pulling a spinning seed crystal out of a crucible of molten material. This method is used to create silicon ingots, which are sliced into wafers for processing into integrated circuits.

Damascene   A process for creating interconnects by first etching trenches into a layer of insulation and then filling them with a conducting material. Commonly used for copper interconnects on integrated circuits.

Data dependency   A dependency between instructions using the same registers or memory locations. True data dependencies are caused by one instruction using the result of another, a read-after-write dependency. False data dependencies are caused by instructions waiting to write to locations being used by other instructions, write-after-read or write-after-write dependencies.

DCA   Direct chip attach.

Deadlock   When forward progress cannot be made in a program because multiple instructions are waiting for each other to complete.

Decoupling capacitors (Decaps)   Capacitors added to an integrated circuit die or package to reduce supply noise.

Defects per million (DPM)   The number of defects per million parts shipped.

DeMorgan's theorem   A theorem of boolean logic stating that any function is equal to the same function with all ANDs changed to ORs, all ORs changed to ANDs, and all terms inverted.

Design automation engineer   An engineer responsible for the creation and/or maintenance of CAD tools used in design.

Design for test (DFT) circuits   Circuits used specifically to help detect manufacturing defects. These circuits are also commonly used to help in silicon debug.

Device limited layout   Layout where the area required is determined primarily by the number of transistors and their sizes. See wire limited layout.

DFT   Design for test.

Die paddle   The portion of the leadframe for a plastic package to which the die will be attached. Also called a flag.

Die separation   The manufacturing step where a wafer is cut into individual die.

Die side cap (DSC)   Decoupling capacitors placed on the same side of the package substrate as the die.

Die yield   The percentage of functional die on each completed wafer.

Diffusion region   A region where N-type or P-type dopants have been added to a semiconductor to create free charge carriers.

Diode   A circuit element allowing current to flow in only one direction. Typically formed by a single P-N junction.

DIP   Dual in-line package.

Direct chip attach (DCA)   Attaching a bare integrated circuit die directly to a printed circuit board. Also called chip on board (COB).

Direct mapped cache   A cache memory that stores each memory address in a single location.

Doping Adding impurities to a semiconductor to create free charge carriers.

Double precision A binary floating point number format using 64 bits.

DPM Defects per million.

Drain terminal One of the diffusion terminals of a MOSFET.

DRAM Dynamic random access memory.

Dry etch An etch step performed using gases or plasma rather than liquid chemicals.

DSC Die side cap.

Dual damascene A damascene process where trenches are etched for vias and interconnects and both are filled with conducting material simultaneously.

Dual in-line package (DIP) A package with single rows of leads on two sides of the package. These packages typically have less than 50 total leads.

Dynamic random access memory (DRAM) Semiconductor memory using single transistor (1T) cells, which gradually leak away charge and must be refreshed approximately every 15 ms. DRAM chips provide the lowest cost per bit of the different types of semiconductor memory and therefore are used as the main memory store of almost all computers.

ECC Error correcting code.

Electrical test (E-test) Tests performed on a wafer, sometimes before processing is complete, to measure the electrical properties of the structures manufactured so far. Used to detect problems in the manufacturing process as early as possible.

Electromigration (EM) The tendency of high current densities to cause interconnects to fail over time by causing the movement of atoms out of thin spots in the wire.

Embedded processor A processor used in a product other than a computer.

EPIC Explicitly parallel instruction computing.

Epitaxy The deposition of a crystalline layer on top of a crystal substrate.

Epi-wafer A wafer made by depositing a crystalline epitaxial layer on top of a bulk wafer. This allows the creation of doping profiles that could not be created through traditional diffusion.

Errata Documented processor behaviors that are different from the design's intended behavior but are deemed unlikely or trivial enough that the processor design is still sold.

Error correcting codes (ECC) Schemes for adding redundant bits to memory stores to allow misread bits to be detected and in some cases corrected.

Etch rate The thickness of material, which can be removed by an etch step in a fixed length of time.

Etch selectivity The ratio of etch rates between the layer to be removed and the material beneath it.

Etching Physical and chemical means of removing a material layer.

E-test Electrical test.

EUV Extreme ultraviolet.

Exception A redirection of program flow caused by an unusual event during the execution of an instruction that would not normally alter the control flow. An example is redirection of the program to an error handling routine upon a division by zero.

Explicitly parallel instruction computing (EPIC) A computer architecture designed to allow more instructions per cycle by encoding the instructions with information showing which can be executed in parallel. The architecture used by the Itanium family of processors.

Extended precision A binary floating-point number format using 80 bits.

External lead The part of a package lead outside the die encapsulation.

Extreme ultraviolet (EUV) lithography Photolithography using wavelengths as short as 13.5 nm.

Fanout The ratio between the total capacitive load being driven by a gate and the capacitance of one of that gate's inputs. The higher the fanout of a gate, the larger its delay.

Fault exception An exception that after being called retries the instruction that caused the exception.

Fault model The type of defect being assumed during automatic test pattern generation. The most common fault model is the stuck-at model, which assumes inputs of each gate may be permanently fixed at high or low voltages.

FC Flip-chip.

FF Flip-flop.

FIB Focused ion beam.

Field programmable gate array (FPGA) An integrated circuit designed as an array of generic logic gates and programmable storage sequentials, which allows the logic function being performed to be dynamically reconfigured. In processor design, FPGAs are commonly used to allow high-speed simulation of designs before they are fabricated.

FireWire A peripheral bus standard common among digital video cameras and recorders. Also called IEEE 1394.

First silicon The arrival of the first finished chips of a new integrated circuit design.

Flame retardant material type 4 (FR4) A fiberglass material commonly used to make printed circuit boards.

Flip-chip (FC) A method for creating electrical connections between a die and package by placing solder bumps on the face of the die and then turning the die facedown on the package substrate and applying heat and/or pressure. See wire bond package.

Flip-flop (FF) A storage sequential circuit that captures the data on its input at each new rising edge of a clock signal.

Floating-point execution unit (FPU) The portion of a microprocessor responsible for computation involving floating-point numbers.

Floating point Numbers formatted in scientific notation as a fraction and an exponent.

Flux Chemicals that help solder flow and create good electrical connections by preventing the formation of oxides.

Focused ion beam (FIB) edit Altering an integrated circuit design by cutting or depositing structures using a FIB process. This allows design changes to be tested before new masks are made.

FPGA Field programmable gate array.

FPU Floating-point execution unit.

FR4 Flame retardant material type 4.

Front-side bus (FSB) A bus connecting the processor and the chipset.

Fully associative cache A cache memory where the data from any memory address can be stored in any location in the cache.

Gate leakage Leakage current through the gate oxide of a MOSFET.

Gate oxide The layer of insulation separating the gate terminal and the channel of a MOSFET.

Gate terminal The MOSFET terminal controlling whether current will flow between the source and drain.

Gigabyte (GB) 2^{30} or approximately 1 billion bytes.

GMCH Graphics and memory controller hub.

GPU Graphics processor unit.

Graphics memory controller hub (GMCH) A chipset component controlling communication between the processor and main memory and containing a built-in graphic processor. Also called the Northbridge.

Graphics processor unit (GPU) A processor specifically designed to create graphics images.

Grid array package An integrated circuit package using multiple rows of leads on all four sides of the package.

Gull-wing pin A package pin that bends outward.

Hardware description language (HDL) A computer programming language specifically intended to simulate computer hardware.

HDI High-density interconnect.

HDL Hardware description language.

Heat sink A metal structure attached to an integrated circuit package to improve heat dissipation. Passive heat sinks work by simply providing increased surface area. Active heat sinks include a built-in fan.

High-density interconnect (HDI) Package interconnects created using thin-film process technologies.

High-K gate dielectric A layer of high permittivity (high-K) material created under the gate of a MOSFET. Using high-K materials gives the gate more control over the channel without needing to further thin the gate oxide layer.

High-level programming language A programming language that is independent of a particular computer architecture. Examples are C, Perl, and HTML.

Hold noise Noise on a sequential node created by inputs switching just after the sequential has captured a new value.

HyperThreading An x86 architectural extension that allows software to treat a single processor as if it were two processors executing separate program threads. This improves performance by allowing one thread of execution to make forward progress while another thread is stalled.

IC Integrated circuit.

ICH Input/output controller hub.

IHS Integrated heat spreader.

ILD Interlevel dielectric.

ILP Instruction level parallelism.

Infrared emissions microscope (IREM) A microscope that takes pictures of infrared light. Silicon is transparent to infrared and transistors will emit at this frequency when drawing current, allowing images to be taken showing where power is being consumed on the die.

Input output controller hub (ICH) A chipset component controlling communication the between the Northbridge of the chipset and all the lower bandwidth buses. Also called the Southbridge.

Instruction level parallelism (ILP) The ability of individual machine language instructions from a single computer program to be executed in parallel. The amount of ILP determines the maximum instructions per cycle possible for a particular program.

Instruction pointer (IP) A register containing the address of the next instruction to be executed.

Instructions per cycle (IPC) The average number of instructions completed per cycle. Overall processor performance is determined by the product of clock frequency and IPC.

Insulator A material containing few or no free charge carriers preventing it from conducting electricity under normal conditions.

Integrated circuit (IC) An electronic circuit containing multiple components all made from a single piece of semiconductor. Modern integrated circuits can contain more than a billion transistors.

Integrated heat spreader (IHS) A metal plate in an integrated circuit package designed to spread out heat generated by the die.

Interlevel dielectric (ILD) Insulating material deposited between levels of interconnect in an integrated circuit.

Internal lead The part of a package lead inside the die encapsulation.

Interrupt Temporary suspension of the execution of a program to run an interrupt handler pointed to by a global address table. Software interrupts are caused by the execution of an interrupt instruction. Hardware interrupts are triggered by signals from external devices. Interrupts allow subroutines that are independent of the program currently being run to be called to handle special events.

Inversion layer A layer of free charge carriers beneath the gate oxide of a MOSFET of the opposite type of the transistor's well. This layer is created by the gate voltage in the linear and saturation regions of operation.

Ion implantation The addition of dopant atoms to a semiconductor by ionizing the element to be added and accelerating it through an electric field at the surface of a wafer.

IP Instruction pointer.

IPC Instructions per cycle.

IREM Infrared emissions microscope.

Isotropic etch An etch that proceeds at the same rate in all directions.

J-bend pin A package pin that bends inward.

Jump instruction An instruction that always redirects program flow to new instruction address.

Karnaugh map (K-map) An aid to by hand logic minimization.

kB kilobyte

Keeper device A transistor used only to prevent a wire's voltage from changing when no other transistor is driving it. Keeper devices are sized very small to allow other transistors to easily override them when the wire's voltage needs to be changed.

Keyhole void A hole in the material of a via caused by a deposition step that fails to fill narrow gaps fully.

KGD   Known good die.

Kilobyte (kB) 2^{10} or approximately 1000 bytes.

K-map   Karnaugh map.

Known good die (KGD)   A die that has been fully tested and proven to be functional before packaging.

Land grid array (LGA)   An integrated circuit package using multiple rows of flat circular metal lands on all four sides as leads.

Land side cap (LSC)   Decoupling capacitors placed on the opposite side of the package substrate as the die.

Laser voltage probing (LVP)   Reading the transient voltage waveforms of an integrated circuit using the reflection of a laser beam.

Latch   A storage sequential circuit that based on a clock signal either passes its input directly to its output or holds a previous value.

Latchup   A positive feedback between N-type and P-type wells, causing large amounts of current to flow beneath the surface of a chip. The only way to return the chip to normal operation is to switch its power off. The possibility of latchup is reduced by frequent well taps, or using epi or SOI wafers.

Layout density   The number of components per area drawn in layout. Because large area means larger die cost, it is important to achieve the highest layout density possible.

Layout design rules   Rules restricting the widths and spacings of the different materials drawn in layout to allow for reliable manufacturing.

Layout   The task of drawing the individual layers of material that will make up the components needed for an integrated circuit.

Lead   An electrical connection from a packaged integrated circuit to the printed circuit board.

Lead design   A design project that reuses almost nothing from previous designs.

Lead pitch   The spacing between leads on an integrated circuit package, commonly between 1 and 0.4 mm.

Leadframe   A strip of leads for multiple plastic packages. After die are attached and encapsulated the individual packages will be cut free of the leadframe.

Leakage power   The power consumed by an integrated circuit due to sub-threshold and gate leakage. This power is independent of clock frequency.

LGA   Land grid array.

Linear region   The MOSFET region of operation where gate voltage forms an inversion layer at both ends of the channel. Current is determined by the voltage at the source, gate, and drain terminals. See cutoff and saturation regions.

Little endian A memory addressing format that assumes for numbers more than 1 byte in size, the lowest memory address is the least significant byte (the little end) of the number. See big endian.

Livelock When forward progress cannot be made in a program because multiple instructions repeatedly cause each other to be executed again.

Load instruction An instruction that moves data from a memory location to a register.

Local oxidation of silicon (LOCOS) The formation of thermal oxide on only parts of a silicon wafer by first covering regions where oxide is not wanted in a layer of silicon nitride.

Logic bug A design flaw that causes an integrated circuit to produce a logically incorrect result.

Logic design The task of converting a microarchitectural specification into a detailed simulation of cycle-by-cycle behavior using a hardware description language.

Low-pressure chemical vapor deposition (LPCVD) Deposition of material through the chemical reaction of gases at low pressure. Low pressure allows good control and creation of high-quality films but slows the deposition rate.

LSC Land side cap.

LVP Laser voltage probing.

Machine language The encoded values representing the instructions of particular computer architecture. Also called binary code. Any software must be converted to machine language before being run on a processor.

MB megabyte.

MCH Memory controller hub.

MCM Multichip module.

MCM-C Multichip module with ceramic substrate.

MCM-D Multichip module with deposited substrate.

MCM-L Multichip module with laminate substrate.

Megabyte (MB) 2^{20} or approximately 1 million bytes.

Memory controller hub (MCH) A chipset component controlling communication between the processor and main memory. Also called the Northbridge.

Memory hierarchy The use of progressively larger but slower levels of cache memory to provide high storage capacity with low average latency.

MESI Modified, exclusive, shared, and invalid. The states of a typical cache coherence protocol keeping track of how memory addresses are shared among multiple cache memories.

Metal oxide semiconductor field-effect transistor (MOSFET) A transistor created by a conducting wire resting on a thin layer of insulation deposited on top of a semiconductor. This gate wire separates two doped regions in the semiconductor called the source and the drain. The electric field from the gate penetrates the semiconductor and determines whether current will flow between the source and the drain.

MFLOPS Million floating-point operations per second.

Microarchitecture The overall design details of a processor that are not visible to the software. Microarchitectural choices largely determine the performance of a processor. Some examples of these design choices are pipeline length, instruction issue width, methods to resolve control and data dependencies, and the design of the memory hierarchy.

Microcode Programs of low-level steps performed by the processor to execute complex machine language instructions or handle special events such as exceptions or processor reset. These programs are stored in a read-only memory array, the microcode ROM.

Micrometer (Micron, μm) One millionth (10^{-6}) of a meter.

Microprocessor An integrated circuit capable of executing a stored series of instructions from memory and writing results to memory.

Microvias Small laser-drilled vias connecting levels of package interconnect.

Million floating-point operations per second (MFLOPS) A measure of floating-point performance. This metric can be deceiving since different floating-point operations require different amounts of computation and peak MFLOPS may be far higher than the rate achieved by any real program.

Million instructions per second (MIPS) A measure of computer performance. This metric can be deceiving since different computers support different machine language instructions, different instructions require different amounts of computation, and peak MIPS may be far higher than the rate achieved by any real program. Also referred to as *meaningless indicator of processor speed*.

Minimal sum The simplest sum-of-products form of a logic function.

MIPS Million instructions per second.

Moore's law The trend that maximum number of components that can be fabricated on an integrated circuits typically doubles every 2 years. This trend has allowed the semiconductor industry to quickly increase performance while at the same time reducing cost.

MOSFET Metal oxide semiconductor field-effect transistor.

Motherboard The main printed circuit board of a computer.

Multichip module (MCM) An integrated circuit package containing multiple die.

Mux A multiplexer circuit that routes one of multiple inputs to its output.

NAND A logic function that is only false if all its inputs are true.

Nanometer (nm) One billionth (10^{-9}) of a meter.

NMOS N-type metal oxide semiconductor field-effect transistor.

NOR A logic function that is false if any of its inputs are true.

Northbridge The chipset component responsible for communicating with the processor and the highest bandwidth components, typically main memory and the graphics adapter. Also called the memory controller hub (MCH) or graphics memory controller hub (GMCH).

NOT A logic function that is true if its input is false.

nm Nanometer.

N-type metal oxide semiconductor (NMOS) A MOSFET with N-type source and drain terminals in a P-well. An NMOS transistor will conduct when the gate voltage is high.

N-type semiconductor Semiconductor with donor dopants added to create negative free charge carriers.

OPC Optical proximity correction.

Opcode The bits of an encoded machine language instruction determining the operation to be performed.

Operating system (OS) The program responsible for loading and running other programs.

Optical proximity correction (OPC) Changes made to a photolithography mask to compensate for the wave nature of light.

OR A logic function that is true if any of its inputs is true.

OS Operating system.

Out-of-order execution Executing instructions in a different order than specified by the software in order to improve performance. The processor must make sure that the program results are unchanged by this reordering.

PA-RISC Precision architecture reduced instruction set computing. A RISC computer architecture created by Hewlett Packard.

Parity bits Redundant bits added to memory stores to allow misread bytes to be detected.

PCA Printed circuit assembly.

PCB Printed circuit board.

PCI Peripheral component interconnect.

PDIP Plastic dual in-line package.

PECVD Plasma-enhanced chemical vapor deposition.

Performance bug A design flaw that causes an integrated circuit to spend more cycles than intended executing a program but still produce the logically correct result.

Peripheral component interconnect (PCI) An expansion bus standard. The most common form is 32 bits wide and runs at 33 MHz.

PGA Pin grid array.

Phase-shift mask (PSM) A photolithography mask that allows light to pass through with different phases in order to use destructive interference to create sharper shadows on the photoresist and to allow patterning of smaller features.

Photolithography mask A patterned plate through which light is projected during photolithography to determine the pattern of a particular layer. Masks are usually made of quartz with the pattern drawn in chrome. Also called a reticle. A complex integrated circuit may require as many as 40 masks.

Photolithography The process of creating extremely small features from masks containing the desired pattern. The wafer is coated with a light sensitive chemical called photoresist. Light is then shown through the mask causing a chemical reaction in the photoresist that is exposed. The photoresist is developed leaving behind a pattern determined by the mask and allowing etching to be performed on the layers beneath only where desired. Repeated steps of deposition and photolithography are used to create integrated circuits.

Photoresist (PR) A light sensitive chemical used during photolithography. Positive resists become more soluble when exposed to light, and negative resists become less soluble.

Pin grid array (PGA) An integrated circuit package using multiple rows of pins on all four sides as leads.

Pipeline The series of steps required to fetch, decode, execute, and store the results of a processor instruction. Pipelines improve performance by allowing the next instruction to begin before the pervious instruction has completed.

Planarization A manufacturing process to create a flat surface for the next processing step.

Plasma-enhanced chemical vapor deposition (PECVD) Deposition of material through the chemical reaction of plasma. Microwaves are used to turn low-pressure gases into a plasma allowing the high-rate deposition of high-quality films.

Plasma oxide (PO) Silicon dioxide deposited using PECVD.

Plastic dual in-line package (PDIP) A plastic package with rows of leads on two sides.

Plastic leaded chip carrier (PLCC) A plastic package with a single row of J-bend pins on all four sides.

Plastic quad flat pack (PQFP) A plastic package with a single row of gull-wing pins on all four sides.

Plated through hole (PTH) A package via created by drilling through the substrate and filling the hole with a conductor.

PLCC Plastic leaded chip carrier.

Plug and Play (PnP) A hardware and software specification that allows a computer system to automatically detect and configure new components.

PMOS P-type metal oxide semiconductor field-effect transistor.

P-N junction The boundary formed between regions of P-type and N-type semiconductors.

PnP Plug and Play

Pollack's rule The observation that for processors using the same fabrication technology, performance improves roughly with the square root of added die area.

Polycrystalline silicon (poly) Microscopic silicon crystals randomly oriented and fused into a single piece. Commonly heavily doped and used as the gate terminal of a MOSFET.

POST Power on self test.

Post-silicon validation The task of proving an integrated circuit design is logically correct and meets all the product specifications by testing manufactured chips.

Power bug A design flaw that causes an integrated circuit to use more power than intended in executing a program but still produce the logically correct result.

Power on self test (POST) Initialization and diagnostics routines run when a computer system is first turned on.

PowerPC A RISC computer architecture created by IBM and Motorola.

PQFP Plastic quad flat pack.

PR Photoresist.

Prepreg Fiberglass sheets that have been impregnated with epoxy resin but not yet fully cured. These flexible sheets are used to separate interconnect layers in multilayer packages and PCBs.

Pre-silicon validation The task of proving an integrated circuit design is logically correct and meets all the product specifications by performing simulations before the design is manufactured.

Printed circuit assembly (PCA) A printed circuit board with components attached.

Printed circuit board (PCB) Rigid sheets of insulator and conducting wires (also called traces) used to physically support and provide electrical connections between electronic components. Most PCBs uses sheets of fiberglass and copper traces.

Proliferation design An integrated circuit design that reuses some elements of a previous design but makes large logic changes and uses a new manufacturing process.

PSM Phase-shift mask.

PTH Plated through hole.

P-to-N ratio The ratio of the PMOS transistor widths to NMOS transistor widths in a CMOS logic gate. This ratio will determine what voltage level on the input must be reached to switch the output of the gate.

P-type metal oxide semiconductor (PMOS) A MOSFET with P-type source and drain terminals in an N-well. A PMOS transistor will conduct when the gate voltage is low.

P-type semiconductor Semiconductor with acceptor dopants added to create positive free charge carriers (holes).

Quad package A package with single rows of leads on four sides of the package. These packages typically have less than 250 total leads.

Random access memory (RAM) Semiconductor memory that allows stored data to be read or written in any order. The mechanical movement of hard drives or magnetic tapes requires data to be accessed in a specific order. Solid-state memories do not have this limitation. The two basic types of RAM are dynamic RAM (DRAM) and static RAM (SRAM).

Random access memory digital-to-analog converter (RAMDAC) The circuit in a video adapter that scans the frame buffer RAM containing an encoded image to be displayed and converts the binary values to analog voltages that drive the monitor controls.

RAT Register alias table.

Reactive ion etch (RIE) An etch process that uses ions that will chemically react with the material to be removed and an electric field to accelerate the ions toward the surface of the wafer.

Read only memory (ROM) Semiconductor memory that can be read from but not written to.

Real time clock (RTC) The computer circuit that tracks the passage of time when the system is on or off. This is separate from the clock signals used to synchronize communications between components.

Reduced instruction set computing (RISC) Computer architectures that support only register or immediate operands in computation instructions, fixed instruction size, and few addressing modes. Examples are SPARC and PA-RISC. See CISC.

Register alias table (RAT) A table listing which physical registers are currently holding the values of the architectural registers.

Register renaming The microarchitectural algorithm assigning architectural registers to different physical registers in order to eliminated false data dependencies and improve performance.

Register transfer level (RTL) The level of abstraction of hardware description language code, which models the cycle by cycle logic state of an integrated circuit but does not model the relative timing of different events within a cycle.

Register On-die storage for a single value on a processor.

Rent's rule The observation that the number of package leads required by an integrated circuit increases with the total number of transistors according to the formula: Num leads $\propto$ (num components)^R. The value of the Rent's exponent R varies for different types of integrated circuits but is usually estimated for microprocessors as approximately 0.5.

Reorder buffer (ROB) The functional unit in an out-of-order processor responsible for committing results in the original program order to guarantee that the results are unaffected by the instruction reordering.

Repackage design An integrated circuit product that reuses an existing silicon design in a different package.

Required window The window of time during which the input to a sequential must be stable in order to guarantee the correct capture of the signal value by the sequential. See valid window.

Resource conflict When two instructions cannot be executed in parallel because they both require use of a same processor functional area.

Reticle A photolithography mask.

Return instruction A computer instruction that pops the address from the top of the stack and redirects program execution to that instruction, returning the program flow to the point after the last call instruction.

RIE Reactive ion etch.

RISC Reduce instruction set computing.

ROB Reorder buffer.

ROM Read only memory.

RTC Real time clock.

RTL Register transfer level.

SATA Serial advanced technology attachment.

Saturation region The MOSFET region of operation where gate voltage forms an inversion layer at one end of the channel. Current is determined by the voltage at the source and the gate but is largely unaffected by the drain voltage. See cutoff and linear regions.

Scan sequentials Sequential circuits that contain extra circuitry to allow the sequential value to be read and/or written by shifting values along a serial chain connecting all the scan sequentials. Scan sequentials make it easier to test integrated circuits by providing more points of control and observability.

SCSI Small computer systems interface.

SDRAM Synchronous dynamic random access memory.

Semiconductors Materials such as silicon that conduct electricity easily under some conditions. The conductivity of semiconductors can be dramatically altered by the addition of impurities and by the variation of electric fields.

Sequential   A circuit that stores a value.

SER   Soft error rate.

Serial advanced technology attachment (SATA)   A storage bus standard. SATA is software compatible with the older ATA standard but uses a 1-bit-wide bus instead of 16 bits. SATA achieves higher bandwidth than ATA by using much higher clock rates.

Setup noise   Noise on a sequential node created by inputs switching just before the sequential captures a new value.

Shallow trench isolation (STI)   Separating MOSFETs by etching trenches into the semiconductor surface and filling the trenches with an insulator. STI allows tighter spacing between transistors than earlier LOCOS methods of isolation.

Shmoo diagram   A diagram formed by repeatedly running the same test under different conditions and marking each point on a graph according to whether the test passes or fails. The most common conditions varied are clock frequency and voltage. Shmoo diagrams are used to help determine the cause of design bugs.

Sidewall spacer   Insulating material on either side of a MOSFET gate terminal used to separate the deep source and drain implant from the shallow source and drain extensions.

Silicide   A compound of silicon and a refractory metal such as nickel, cobalt, or tungsten. Silicide layers are formed on top of MOSFET poly gates and source and drain regions in order to dramatically reduce their resistance and therefore improve transistor performance.

Silicon bugs   Design flaws that are discovered after an integrated circuit has been manufactured.

Silicon debug   The task of finding silicon bugs, determining their root cause, and identifying a fix.

Silicon implanted with oxygen (SIMOX)   A process for creating SOI wafers by using ion implantation to deposit oxygen beneath the surface of a silicon wafer and then heating the wafer to cause a layer of silicon dioxide to form beneath the surface.

Silicon-on-insulator (SOI) wafer   A silicon wafer with a layer of silicon dioxide just beneath the surface. This layer of insulation reduces the capacitance of the MOSFET source and drain regions, which improves switching speed.

SIMD   Single instruction multiple data.

SIMOX   Silicon implanted with oxygen.

Single instruction multiple data (SIMD)   Instructions that simultaneously perform the same operation on multiple pieces of data. For example, a SIMD Add might add two vectors of four numbers each to produce a four-number result vector.

Single precision   A binary floating-point number format using 32 bits.

Small computer systems interface (SCSI) A storage bus standard. Pronounced "scuzzy."

SMT Surface mount technology.

Soft error rate (SER) The rates of processor errors caused by memory bits changing values due to cosmic rays.

SOI Silicon on insulator.

Solder Metal alloys that are melted and applied to metallic parts to form an electrical connection. Usually a mixture of lead and tin.

Solid state diffusion The diffusion of impurities through a solid material. Used to distribute dopant atoms through semiconductors.

Source terminal One of the diffusion terminals of a MOSFET.

Southbridge The chipset component responsible for communicating with the Northbridge and the lower bandwidth components. Also called the input/output controller hub (ICH).

SPARC A RISC computer architecture created by Sun Microsystems.

Spatial locality The tendency of memory accesses to a particular address to be more likely if nearby addresses have recently been accessed.

Speedpath The slowest circuit path through block of logic, which will determine the maximum clock frequency at which that logic can operate.

Spin-on deposition The deposition of material to a wafer by applying a liquid solution while spinning the wafer and later curing the liquid to form a solid layer.

Spin-on dielectric (SOD) Insulating material that can be applied through spin-on deposition.

Spin-on glass (SOG) Silicon dioxide applied through spin-on deposition.

Sputtering The deposition of material by placing a target of the material to be deposited inside a vacuum chamber and striking it with a beam of ions causing bits of the material to be vaporized. This vapor will condense on any substrates placed inside a vacuum chamber, gradually forming a new layer of material.

SRAM Static random access memory.

Stack A first-in last-out queue of stored values. A push instruction is used to put a new value on the top of the stack, and a pop instruction is used to remove the top value on the stack. Memory stacks are commonly used to hold instruction addresses for return instructions allowing them to automatically return execution to the point after the last call instruction.

Static random access memory (SRAM) Semiconductor memory that uses feedback to maintain its values as long as power is applied. This typically requires six transistors per cell rather than the single transistor (1T) cells of DRAM. SRAM is therefore more expensive per bit than DRAM, but does not require refreshing and can provide lower latencies. For these reasons, SRAM is commonly used as cache memory.

Stepper A machine for performing photolithography. Steppers are named for the fact that the wafer must be carefully repositioned after each exposure to serially expose the entire wafer.

STI Shallow trench isolation.

Store instruction An instruction that moves data from a register to a memory location.

Stuck-at fault A manufacturing or design flaw causing a particular circuit node to always be at a high voltage or a low voltage. Automatic test pattern generation must assume how faults will logically appear (a fault model) in order to create input vectors to find these faults.

Subthreshold leakage Leakage current that flows between a MOSFET's source and drain even in the cutoff region. Subthreshold currents are typically orders of magnitude smaller than the current that flows when the MOSFET is fully on, but for integrated circuits containing millions of transistors subthreshold leakage can contribute significantly to total power.

Super I/O A component of some chipsets responsible for communicating with the lowest bandwidth components using the oldest standards least likely to change, such as buses to the keyboard, mouse, and floppy disk.

Superscalar Processors capable of completing more than one instruction per cycle.

Supply noise Electrical noise on the high-voltage or low-voltage supply lines caused by sudden changes in current draw.

Surface mount technology (SMT) Technology for attaching components to a printed circuit board without requiring holes to be drilled through the board. Mounting components on the surface allows for components to be attached to both sides of the board but does not give as good a physical connection as pins inserted through the board. See through hole technology (THT).

Synchronous dynamic random access memory (SDRAM) DRAM chips that use clock signals to improve bandwidth.

Tapeout The completion of the entire layout for an integrated circuit. After tapeout photolithography masks can be created and the first chips manufactured.

TB terabyte.

Temporal locality The tendency of memory accesses to a particular address to be more likely if that address has recently been accessed.

Terabyte (TB) 2^{40} or approximately 1 trillion bytes.

Thermal interface material (TIM) A thin layer of material added between two objects to aid in the transmission of heat. Usually refers to the material added between a computer chip and the package integrated heat spreader (IHS) or between the heat spreader and a heat sink.

Thermal oxidation The formation of silicon dioxide by exposing a silicon wafer to heat in the presence of oxygen or steam.

Thermal resistance The ratio of the temperature difference at thermal equilibrium between two points and the amount of power released as heat at one of the points. A processor with a thermal resistance of 2°C/W running at 25 W will reach thermal equilibrium at 50°C over its surrounding temperature.

Threshold voltage The gate voltage at which a MOSFET turns on by leaving the cutoff region. Threshold voltage is determined by the thickness of the gate oxide and the doping in the MOSFET channel.

Through hole technology (THT) Technology for attaching components to a printed circuit board by inserting pins through holes drilled in the board and plated with metal. This provides a good physical connection to the board but does not allow components to be attached to both sides of the board. See surface mount technology (SMT).

Tie bar The part of a leadframe for a plastic package connecting the die paddle to the leadframe. The tie bar will be cut after die encapsulation when the finished packages are separated from the leadframe.

TIM Thermal interface material.

TLB Translation lookaside buffer.

Trace cache A cache memory containing already partially decoded instructions, which may be stored in program execution order rather than the order they were original stored in main memory.

Transfer curve The curve showing how the output voltage of a logic gate changes with a changing input voltage.

Transistor legging The layout practice of drawing wide transistors as multiple smaller transistors connected in parallel. This keeps poly gate terminals short, which improves performance.

Transistor length The distance between a MOSFET source and drain. This can be measured as the drawn length, which is the width of the gate terminal in layout, the final length, which is the width of the poly gate line after processing, or the effective length, which is the separation between the source and drain diffusion regions underneath the gate oxide.

Transistor width The width of a MOSFET source and drain. Drawing a transistor with a larger width allows it to draw more current and switch more quickly but consumes more die area and more power.

Transistor A semiconductor component where the current between two terminals is controlled by the voltage on a third terminal.

Translation lookaside buffer (TLB) A memory cache containing recent virtual to physical memory address translations to improve the performance of processors supporting virtual memory.

Trap exception An exception that after being called continues execution with the instruction after the one that caused the exception.

Two's complement A binary encoding scheme for integers where negative values are designed by negating all the bits and then adding one. This scheme makes sure there is only one way of representing a zero.

UBM Under-bump metallization.

UF Underfill.

Ultraviolet (UV) lithography Photolithography using wavelengths shorter than the visible spectrum (<380 nm).

μm Micrometer or micron.

Under-bump metallization (UBM) The layers of metal applied beneath solder bumps on a flip-chip integrated circuit. UBM typically comprises an adhesion layer to provide a good mechanical connection to the die, a diffusion barrier to prevent contamination, and a solder wettable layer to provide a good electrical connection to the C4 bump.

Underfill (UF) Epoxy applied underneath a die in a flip-chip package to hold the die to the package substrate.

Unity gain points The points on a logic gate's transfer curve with a slope of negative one. These points mark the boundary between where electrical noise on the inputs will be dampened or amplified.

Universal serial bus (USB) A 1-bit-wide peripheral bus standard.

UV Ultraviolet.

Vacuum tube Electronic components formed from metal filaments inside a glass enclosure from which the air has been removed. With different types and numbers of filaments vacuum tubes can form diodes or amplifiers. Today, vacuum tubes have been almost entirely replaced by transistors.

Valence band The band of energy states available to electrons that are bound to a particular atom in a crystal.

Valid window The window of time during which an output signal is stable. This time occurs after the latest possible transition of one cycle and before the earliest possible transition of the next cycle. See required window.

Validation The task of proving an integrated circuit design is logically correct and meets all the product specifications. Pre-silicon validation uses simulations before the design is manufactured, and post-silicon validation tests real chips.

VAX A computer architecture developed by Digital Equipment Corporation in the mid-1970s. Because of its very large number of instructions and addressing modes, the VAX architecture is commonly considered the ultimate example of a CISC architecture.

V_{dd} Common labeling of a wire connected to the high-voltage rail of a power supply. See V_{ss}.

Verilog A hardware description language created in 1984 by Gateway Design Automation, which was later purchased by Cadence Design Systems.

Very high-speed integrated circuit hardware description language (VHDL) A hardware description language whose development was started in 1981 by the U.S. Department of Defense to document the logical behavior of military electronics.

Very long instruction word (VLIW) A type of computer architecture where multiple instructions are encoded together for parallel execution.

Via A vertical connection between two different interconnect layers.

VID Voltage identification.

Virtual memory The translation of memory addresses according to a page table maintained by the operating system. This translation prevents memory conflicts between multiple programs being run at the same time and allows pages of memory to be swapped to and from the hard drive, creating the illusion of more memory than the computer truly has.

VLIW Very long instruction word.

Voltage identification (VID) A signal to a voltage regulator that determines what supply voltage the regulator should provide. This signal allows a single board design to automatically support different products using different voltage levels.

Voltage regulator (VR) Components on a printed circuit board that transform a power supply from one voltage to another to provide power for other components on the board.

V_{ss} Common labeling of a wire connected to the ground voltage rail of a power supply. See V_{dd}.

Wafer probe test Testing of finished die while still on the wafer before die separation or packaging.

Wafer yield The percent of wafers started that successfully finish processing in a fab.

Wafer A flat round piece of semiconductor used as the starting material for making integrated circuits.

Well tap An electrical contact to a well. Well taps help prevent latchup by holding wells at fixed voltages.

Well A deep diffusion area typically formed as the first step in creating a MOSFET.

Wet etch An etch step performed using liquid chemicals, usually acids mixed with water.

Wettable Material to which molten solder will cling and form a good electrical connection.

Wire bond package A method for creating electrical connections between a die and a package through individual wires between metal pads on the die and the package substrate. See flip-chip.

Wire limited layout Layout where the area required is determined primarily by the number of wires and their connections. See device limited layout.

x86 The computer architecture created by Intel and supported by the Intel 286, 386, and 486 processors. The Intel Pentium and AMD Athlon processors support extensions to this computer architecture.

XNOR A logic function that is true only if an even number of its inputs are true.

XOR A logic function that is true only if an odd number of its inputs are true.

Index

1T DRAM cell, 217, 218, 357
1U server, 357
2-bit prediction, 150
2-operand instruction, 117
2U server, 357
3D imaging, 51–52
3DNow!, 105
3DNow! Professional, 82–84, 106
3-operand instruction, 117
64-bit address(es), 97, 110
64-bit register, 105
90-nm fabrication generation, 30, 32
90nm-process, 247
"400-MHz" buses, 45
2001 Design Automation Conference, 79
2003 ITRS (*see* International Technology Roadmap for Semiconductors)

Abort exceptions, 115, 357
Absolute address mode, 107
Accelerated Graphics Port (AGP), 52
Accumulator, 100, 357
Accumulator architecture, 101
Active power (P_{active}), 232, 357
Address(es):
  64-bit, 97, 110
  memory, 106–108
  registers vs., 111
  supported by EPIC, 111
  virtual, 109
Addressing:
  absolute address mode, 107

Addressing (*Cont.*):
  control flow addressing model, 113
  displacement addressing mode, 107, 108
  indexed addressing mode, 107, 108
  IP relative addressing, 113
  modes for RISC, 108
  modes for VAX architecture, 108
  register indirect addressing, 107, 108
  stack addressing, 113
ADL (*see* Architectural description languages)
Advanced graphics port (AGP), 357
Advanced Micro Devices (AMD):
  90-nm fabrication generation of, 31
  AMD64 Technology, 123
  architecture and processors used by, 96
  Cool'n'Quiet, 233
  processors by, 82–84, 97 (*See also specific types, e.g.*: Athlon processors)
  terminology of, vs. Intel, 123
  and x86 architecture, 124
  (*See also specific products of, e.g.*: PowerNow!)
Advanced technology attachment (ATA), 53, 54, 357
Advanced technology extended (ATX), 59, 60, 357 (*See also specific types, e.g.*: Flex-ATX)
AGP (Accelerated Graphics Port), 52
AGP (advanced graphics port), 357

Allocation (instructions), 164–165
Alpha architecture, 119
AltiVec (PowerPC), 106
Alumina, 314
AMD (*see* Advanced Micro Devices)
AMD processors, 82–84 (*See also specific types, e.g.*: Athlon processors)
AMD64 Technology, 123
Amdahl's law, 136, 357
AND gates, 183, 357
Anisotropic etch, 286, 357
APCVD (atmospheric pressure chemical vapor deposition, normal atmospheric pressure), 275, 358
Apple (company), 57–58
Applebred (processor), 82–84
Arbitration, 39
Architectural description languages (ADL), 182, 357
Architectural extension, 357
Architecture(s):
 load/store, 102
 microarchitecture, 127–168, 371
 oldest form of, 96
 processor, 72
 Very Long Instruction Word, 119
 (*See also specific types, e.g.*: VAX architecture)
Arithmetic operations, 99
Arrays, 365
 ball grid, 358
 data, 145
 field programmable gate, 179, 365
 grid, 310, 366
 land grid, 369
 pin grid, 373
 tag, 145
Assembler (program), 357
Assemblers, 98
Assembly flow, 325–328
Assembly language, 358
Assembly language instructions, 98
Associativity, 358
Associativity of the cache, 145
ATA (*see* Advanced Technology Attachment)

Atalla, John, 13
ATE (*see* Automatic test equipment)
Athlon 64, 42, 119
Athlon MP, 82–84
Athlon processors, 44, 82–84
Athlon XP, 82–84
 bus ratios for, 45
 computer architectures using, 96
 data transfers for, 44–45
 and FSB400, 46
Atmospheric pressure chemical vapor deposition (*SEE* APCVD)
ATPG (*see* Automatic test pattern generation)
AT&T Bell Laboratories, 3, 6, 8, 10, 11, 13
Attach:
 die, 318–319
 direct chip, 304
ATX (*see* Advanced technology extended)
Automated logic synthesis, 176
Automatic test equipment (ATE), 344, 358
 defined, 358
 stored-response test for, 351–352
Automatic test pattern generation (ATPG), 335, 358
Average instructions per cycle (IPC), 134 (*See also* Instructions per cycle)

Back-side bus, 44
Backward branches, 150
Balanced Technology Extended (BTX), 60, 358 (*See also specific types, e.g.*: Micro-BTX)
Ball grid array (BGA), 358
Ball leads, 312
Bandwidth, 40
Bardeen, John, 7–8, 10
Barton (processor), 82–84
Basic Input Output System (BIOS), 62–63, 155
 BIOS ROM, 62, 155
 defined, 358
BBUL package (*see* Bumpless buildup layer package)

BCD (*see* Binary coded decimal)
Beckman, Arnold, 11
Behavior level model, 174
Bell Laboratories (*see* AT&T Bell Laboratories)
Benchmarks:
   and processor performance, 141–142
   SPEC integer benchmark program, 141
   SPEC89, 151
   SPECfp2000, 139, 140
   SPECint2000, 138–139
Berkeley University, 96, 97
BESOI (*see* Bonded etched back silicon on insulator wafers)
BGA (ball grid array), 358
Big endian, 106–107, 358
BIMs (*see* Binary intensity masks)
Bin frequency, 46
Binary coded decimal (BCD), 104, 358–359
Binary intensity masks (BIMs), 284, 359
BIOS (*see* Basic Input Output System)
BIOS ROM, 62, 155
Bipolar junction transistors (BJTs), 10, 200
  defined, 359
  MOSFETS vs., 14
BIST (*see* Built-in self-test)
Bit pitch, 252, 359
Bits, parity, 74
BJTs (*see* Bipolar junction transistors)
Blade servers, 74
Blanking, 314, 315
Blind vias, 306
Blue screen of death (BSOD), 342, 359
Bonded etched back silicon on insulator (BESOI) wafers, 31, 267, 359
Boundary scan, 336, 359
Branch history table, 150
Branch instruction, 359
Branch prediction, 149–152, 162–163, 359
Branch prediction buffer, 150

Branch target buffer (BTB), 152, 162, 359
Branches, 112
Branch/jump instruction, 117
Brattain, Walter, 7–8
Braze, 359
Breakdown, dielectric, 22
Breakpoints, 337
BSOD (*see* Blue screen of death)
BTB (*see* Branch target buffer)
BTX (*see* Balanced Technology Extended)
Buffer(s):
   branch prediction buffer, 150
   branch target, 152, 162
   data translation lookaside, 166
   instruction translation lookaside, 161
   translation lookaside, 146
Buffered latch, 218
Buffering, double, 51
Bugs:
   logic, 338, 339, 341, 370
   Pentium FDIV, 180–181, 338
   performance, 339, 372
   post-silicon, 342
   power, 339, 374
   silicon, 338, 347, 376
   software, 342, 349
   sources of, 341–344
Built-in self-test (BIST), 337, 360
Bulk metal oxide semiconductor field-effect transistors, 31
Bulk wafers, 266, 267, 360
Bumping, 325–327, 360
Bumpless buildup layer (BBUL) package, 329, 360
Bunny suits, 265
Buried vias, 306
Burn-in, 360
Bus(es), 38
   400-MHz, 45
   back-side, 44
   differential, 39
   double-pumped, 40
   expansion, 55
   for expansion, 56
   front-side, 366

388   Index

Bus(es) *(Cont.)*:
  Front-Size Bus, 44
  FSB400, 45
  I/O, 55
  peripheral, 57–58
  point-to-point, 39
  processor, 44–47
  quad-pumped, 40
  single ended, 39
  snooping of, 147
  standards for, 38–40
  synchronous, 39
  USB, 57
  USB 2.0, 58
  wires in, 38–39
Bus arbitration, 39
Bus multiplier, 45
Bus ratios, 45, 46
  for Athlon XP, 45
  for Intel Pentium 4, 45
Busicom, 15, 16
Byte swap instruction, 106–107
Bytes, 106, 360

C (programming language), 98, 173, 174
C4 (controlled collapse chip connection), 318
C4 bumps, 319, 362
Cache:
  associativity of the cache, 145
  fully associative, 366
  L2 cache, 162–164
  multilevel, 65
  size increases of, 65
  trace, 380
  trace cache, 157
  tree cache, 163
  two-level, 65
  unified, 162
  (*See also specific types*, e.g.: Multilevel cache)
Cache coherency, 147–149, 360
Cache hit, 360
Cache line, 145
Cache memory(-ies), 48, 143–146
  defined, 360
  first used in IBM PC, 67

Cache memory(-ies) *(Cont.)*:
  high-speed, 44
  latency of, 64
Cache miss(es), 65, 144, 360
CAD, 220
Call instructions, 112, 360
Canonical sum, 184, 360
Capacitance:
  of interconnects, 32
  per length, 25
  of transistors, 19
  of wires, 25, 26
Capacitors:
  die side capacitors, 320
  land side, 320, 369
Capacity miss, 144, 360
Capillary underfill (CUF), 319, 360
Carrier mobility, 360
CAS (column access strobe), 48–49
CAS latency (T_{CL}), 48
Cat whisker diodes, 7, 8
CBD (*see* Cell-based design)
CD (compact disks), 54
CD-R standards, 54
CD-RW, 54
Celeron (processor), 78
Cell-based design (CBD), 177, 361
Central Processing Unit (CPU), 52
Ceramic dual in-line package (CERDIP), 361
Ceramic packages, 316
Ceramic substrate multichip modules (MCM-C), 325
Ceramic substrates, 314
CERDIP (ceramic dual in-line package), 361
C_{GATE} (capacitance of the gate), 19
Channel length, 20
Chemical mechanical planarization (CMP), 361
Chemical vapor deposition (CVD), 272, 274
  APCVD, 358
  defined, 361
  plasma-enhanced, 275, 288
Chemical-mechanical polishing (CMP), 278
Chip on board (COB), 304, 361

Chip scale package (CSP), 329, 361
Chipsets, 38, 41–44
   defined, 361
   Northbridge chips, 41
   Southbridge, 41, 43
   super I/O chip, 43
Circuit(s):
   integrated (*see* Integrated circuits)
   noise affecting, 226–231
Circuit checks, 220–221
Circuit design, 72, 73, 199–236
   and circuit checks, 220–221
   and circuit timing, 221–226
   and CMOS logic gates, 207–212
   defined, 361
   and MOSFET behavior, 200–207
   noise affecting, 226–231
   power concerns in, 231–235
   prelayout, 235
   and sequentials, 216–220
   transistor sizing for, 212–216
Circuit marginality(-ies), 339, 342, 361
Circuit timing, 221–226
Circuits, sequential, 195
CISC (*see* Complex instruction set computing)
C_L (capacitance per length), 25
Cleanrooms, 265, 361
Clock frequency, 40
Clock jitter, 225, 361
Clock signals, 129
Clock skew, 222, 224, 225, 361
CMOS (*see* Complementary metal oxide semiconductor)
CMOS logic gates, 207–212
CMOS pass gate, 218
CMOS process flow, 289–298
CMOS RAM, 61
CMP (chemical mechanical planarization), 361
CMP (chemical-mechanical polishing), 278
COB (*see* Chip on board)
Cocke, John, 96
Coefficient of thermal expansion (CTE), 308, 361

Coherency, cache, 360
Cold misses, 144, 146, 361
Column access strobe (CAS), 48–49
Combinational logic, 182–192
Compact disks (*see under* CD)
Compaction designs, 28
Compactions, 28, 79, 362
Compaq, 56
Compilers, 98, 362
Complementary metal oxide semiconductor (CMOS), 202
   defined, 362
   logic gates, 207–212
   pass gate, 218
   process flow, 288–298
   RAM, 61
Complex instruction set computing (CISC), 97
   defined, 362
   macroinstructions for, 155
   RISC vs., 118–119, 157
Computational instructions, 99–103
Computer architecture, 95–125
   and CISC vs. RISC question, 118–119
   defined, 362
   instructions in, 98–117
   and RISC vs. EPIC question, 119–122
   x86 extensions affecting, 122–124
Computer components, 37–69
   BIOS, 62–63
   bus standards, 38–40
   chipsets, 41–44
   expansion cards, 55–56
   memory as, 47–51
   memory hierarchy, 63–68
   motherboard, 38, 58–61
   peripheral bus, 57–58
   processor bus, 44–47
   storage devices, 53–55
   video adapters, 51–53
Computing address (*see specific types, e.g.*: Virtual addresses)
Conduction bands, 5, 362
Conductor printing, 315

Conductors, 4
  defined, 362
  N-type, 372
  PMOS, 375
  semi-, 372
Conflict miss, 144, 362
Constant field scaling, 23
Constant R-scaling, 26
Contacts, 242, 362
Control dependency(-ies), 131, 149, 152, 362
Control flow addressing model, 113
Control flow changes, 114
Control flow instructions, 111–115
Controlled collapse chip connection (C4), 318
Controlled collapse chip connection (C4) bumps (*see* C4 bumps)
Cooling system(s):
  for chipsets and graphics cards, 60
  for high-performance processors, 321
  in server farms, 74
Cool'n'Quiet, 233
Copper, 296, 314
Copper wires, 26
Coppermine (processor), 76, 77
Core layers, FR4, 329
Corner cases, 341
Costs:
  of packages, 88.
  test, 89
  transistor size affecting, 18
CPU (Central Processing Unit), 52
Cross talk, 227, 362
Crystal growth, Czochralski, 266
Crystal oscillator, 61
CSP (*see* Chip scale package)
CTE (*see* Coefficient of thermal expansion)
CUF (*see* Capillary underfill)
Currents:
  drawn by MOSFETs, 19
  gate leakage, 32
  gate tunneling, 22
  gate tunneling current, 235
  tunneling, 22
Custom designs, 177

Cutoff region, 203, 362
CVD (*see* Chemical vapor deposition)
CVD at low pressures (LPCVD), 275
CZ (Czochralski) crystal growth, 266
Czochralski, J., 9
Czochralski (CZ) crystal growth, 266, 362

Daisy chaining, 57
Damascene, 297
  defined, 363
  dual, 297, 364
Data array, 145
Data dependencies, 152, 363
Data transfer instructions, 103–111
  for memory addresses, 106–108
  for virtual memory, 108–111
Data transfers:
  for Athlon XP, 44–45
  for Intel Pentium 4, 44–45
Data translation lookaside buffer (DTLB), 166
Daughterboards (*see* Expansion cards)
DCA (*see* Direct chip attach)
DDR SDRAM (double data rate SDRAM), 50
DDR standard, 50
DDR2 standard, 50
DDR266 standard, 50
Deadlock, 342, 363
Debug (*see* Silicon debug)
DEC (*see* Digital Equipment Corporation)
Decaps (*see* Decoupling capacitors)
Decode (instruction), 129
Decoding, 162
Decoupling capacitors (decaps), 319–320, 363
Defects:
  and DFT circuits, 337
  on die area, 88
  in gate oxide, 22
  manufacturing, 338, 343
Defects per million (DPM), 353, 363
Delay(s), 27
  interconnect, 27
  of MOSFETs, 20

Delay(s) (*Cont.*):
  RC, 25
  T_{DELAY}, 19
DeMorgan's theorem, 184, 363
Dennard, Robert, 23
Density targets, 288
Denver, Colorado, 230
Dependency(-ies):
  control, 149, 152, 362
  data, 152, 363
  read-after-write, 153
  write-after-read, 153
  write-after-write, 153
Deposited substrate multichip module (MCM-D), 325
Deposition, 272–276, 358
Design:
  circuit (*see* Circuit design)
  lead designs, 79
  logic (*see* Logic design)
  processor, 79, 81
  proliferation, 374
  repackage, 376
  of semiconductors, 353
Design automation, 175–178
Design Automation Conference (2003), 79
Design automation engineer, 363
Design for test (DFT) circuits, 333–338
  and defects, 337
  defined, 363
  trade-offs of, 333
Design planning, 71–92
  design types/time affecting, 79–85
  processor roadmaps for, 74–78
  product cost affecting, 85–91
Design verification, 179, 180
Desktop computers:
  architecture for, 96
  processors for, 74–75
Destructive scan, 334
Device count:
  drawn, 249–250
  schematic, 250
Device drivers, 62
Device limited layout, 363
DFT (*see* Design for test circuits)

Die attach, 314, 318–319
Die paddle, 363
Die separation, 327, 363
Die side capacitors (DSC), 320, 363
Die yield, 88, 363
Dielectric breakdown, 22
Dielectrics, low-K, 26
Differential buses, 39
Differential signaling, 39
Diffusion:
  and copper wires, 26
  N+, 254
  P+, 254
  solid-state, 269, 378
Diffusion region, 268, 363
Digital Equipment Corporation (DEC), 96, 119, 137
Digital versatile disk (*see under* DVD)
Digital videodisk (*see under* DVD)
Diodes, 3
  cat whisker, 7, 8
  defined, 363
  P-N junction diode, 6, 374
  vacuum tube, 3
DIP (*see* Dual in-line package)
Direct chip attach (DCA), 304, 363
Direct mapped cache, 363
Dispatch, 165, 166
Displacement addressing mode, 107, 108
Divide by zero error, 350
DLTB (data translation lookaside buffer), 166
Doping, 268–272, 364
Double buffering, 51
Double data rate SDRAM (DDR SDRAM), 50
Double precision, 364
Double-pumped bus, 40
Down binning, 46, 352
DPM (*see* Defects per million)
Drain (MOSFET), 200
Drain terminal, 364
DRAM (*see* Dynamic random access memory)
Drawn device count, 249–250
Drivers, device, 62

Dry etches, 287, 364
Dry oxidation, 276
DSC (die side capacitors), 320
DTLB (data translation lookaside buffer), 166
Dual damascene, 297, 364
Dual in-line package (DIP), 310, 364
Duron, 65, 82–84
DVD (digital videodisk, digital versatile disk), 54
DVD-R standards, 54
DVD-RW, 54
Dynamic memory, 47
Dynamic prediction, 150
Dynamic random access memory (DRAM), 47
   1T, 218, 357
   defined, 364
   double data rate SDRAM, 50
   Moore's Law applied to, 118
   synchronous, 379

E (exclusive) ownership state, 148
E-beam lithography, 281
ECC (*see* Error Correcting Codes)
Effective channel length (L_{EFF}), 21
EIDE (Enhanced IDE), 54
EISA (Extended ISA), 56
Electrical oxide thickness ($T_{OX\text{-}E}$), 21
Electrical test (E-test), 350, 351, 364
Electromigration, 27, 257–258, 364
Electrons, 4
EM64T (Extended Memory 64-bit Technology), 123
Embedded processor(s), 75, 364
Encoding, 115–117
Enhanced IDE (EIDE), 54
Enhanced SpeedStep, 61
Enhanced Virus Protection, 123
EPIC (*see* Explicitly Parallel Instruction Computing)
Epitaxy, 364
Epi-wafer, 267, 364
Errata, 349, 364
Error(s):
   divide by zero, 350
   soft, 230–231
Error correcting codes (ECC), 74, 364

Etch rate, 364
Etch selectivity, 365
Etches (*see specific types, e.g.*: Dry etches)
Etching, 280, 286–288
   defined, 365
   ion beam, 287, 288
   limits of, 288
   sputter, 287, 288
   (*See also specific types, e.g.*: Reactive ion etching)
E-test (*see* Electrical test)
EUV (*see* Extreme ultraviolet lithography)
Exception handler, 115
Exceptions, 112
   abort, 115
   defined, 365
   fault, 114
   trap, 115
Exclusive (E) ownership state, 148
Execute (instruction), 129
Execute Disable Bit, 123
Execution pipelines, 160, 161
Expansion bus(es), 55
   developmental history of, 56
   for IBM PC, 55
Expansion cards (daughterboards), 55–56, 305
Explicitly Parallel Instruction Computing (EPIC), 97
   addresses supported by, 111
   defined, 365
   features of, 121
   processors for, 96
   RISC vs., 119–122
   special architecture modes of, 107
Extended ISA (EISA), 56
Extended Memory 64-bit Technology (EM64T), 123
Extended precision, 365
External leads, 365
Extreme ultraviolet (EUV) lithography, 285, 365

Fab utilization, 86
Faggin, Fedrico, 14, 16
Fairchild Camera and Instruments, 11

Fairchild Semiconductor, 11, 12, 14, 15
Fanout, 214, 365
Fault exceptions, 114, 365
Fault model, 365
FC (*see* Flip-chip)
FDIV bug (*see* Pentium FDIV bug)
Fetch (instruction), 129
FF (*see* Flip-flops)
FIB (*see* Focused ion beam edits)
Fiducials, 279
Field programmable gate array (FPGA), 179, 365
Field-effect transistor, 7, 8, 13
FireWire, 57, 58
  Apple development of, 57–58
  defined, 365
  Intel licensing of, 58
  Plug-n-Play capability of, 57
  Universal Serial Bus vs., 58
  use and licensing by Intel, 58
First silicon, 73, 331, 365
Fixed bus ratio, 46
Fixed length instructions, 116, 117
Flag values, 167
Flame retardant material type four (FR4), 305, 329, 365
Flash memory, 62
Flaws:
  hardware design, 342
  microarchitectural, 341
Flex-ATX, 59
Flip-chip (FC), 318
  assembling, 325–328
  defined, 366
Flip-flops (FF), 191, 192, 219, 366
Floating point, 104, 366
Floating-point numbers, 105
Floating-point unit (FPU), 166, 366
Floppy disks, 54
Flux, 366
Focused ion beam (FIB) edits, 347–348, 366
Fortran, 173
Forward branches, 150
FPGA (*see* Field programmable gate array)
FPU (*see* Floating-point unit)

FR4 (*see* Flame retardant material type four)
FR4 core layers, 329
Fracture, 281
Frame buffer, 51
Frequency:
  bin, 46
  clock, 40
  measuring, 138
Frequency bin splits, 46
FR4 (flame retardant material type four), 305
Front-end pipeline, 160
Front-side bus (FSB), 44
  defined, 366
  transfer rate of, 76
FSB400 bus, 45, 46
Full scan, 335
Full shielding, 257
Fully associative cache, 366
Fundamental products, 184
Fusing, 352

G5, 96
Gallatin (processor), 78
Gap fill, 274
Gate(s), 200
  capacitance of, 19
  CMOS, 207–212, 218
  drawn length of, 20
  length, 21
  length of, 29
  triple, 32–33
  See also Logic gates; *specific types, e.g.:* AND gates
Gate dielectric, high-K, 32
Gate leakage, 234, 235, 366
Gate leakage currents, 32
Gate oxide (T_{OX}), 19
  defects in, 22
  defined, 366
Gate oxide thickness, 21
Gate terminal, 366
Gate tunneling current, 22, 235
GB (gigabyte), 366
Germanium, 7, 10
Gigabyte (GB), 366
Global lookup, 114

GMCH (*see* Graphics Memory Controller Hub)
Goofs, 341
GPU (*see* Graphics Processor Unit)
Graphics accelerators, 52
Graphics cards (video adapters), 51–53
Graphics memory controller hub (GMCH), 41, 366
Graphics Processor Unit (GPU), 52, 366
Graphics tunnel, 42
Grid array packages, 310, 366
Gulliver's Travels, 106
Gull-wing pin, 311, 312, 366

Half-shielding, 257
Hardware description language (HDL), 172–182
  defined, 367
  design automation for, 175–178
  model, 172–182
  and pre-silicon validation, 178–182
Hardware design flaws, 342
Hardware interrupts, 114
HDI (*see* High-density interconnect)
HDL (*see* Hardware description language)
Heat sinks, 322, 367
Hennessey, John, 96
Hewlett Packard (HP):
  architecture and processors used by, 96
  and PA-RISC, 97
HF (hydrofluoric acid), 287
High-density interconnect (HDI), 316, 367
High-K gate dielectric, 32, 367
High-level programming languages, 98, 173
  defined, 367
  instruction for, 98
High-performance processors, 321
High-speed memory:
  of cache, 44
  latency of, 64
High-speed SRAM, 67
Hit, cache, 360

Hit rate, 64
Hoff, Ted, 15
Hold noise, 230, 367
Holes (silicon), 4
Hot swapping, 57
HP (*see* Hewlett Packard)
HT (*see* HyperThreading)
HTML (hypertext markup language), 98
Huck, Jerome, 102
Hung system, 342
Hydrofluoric acid (HF), 287
Hypertext markup language (HTML), 98
HyperThreading (HT), 123, 133, 367

I (invalid) ownership state, 148
IBM (International Business Machines), 14
  architecture and processors used by, 96
  architecture simplification by, 96
  computers produced by, 95
  and POWER architecture, 97
  and Rent's Rule, 309
IBM 360, 95
IBM 801, 96
IBM PC:
  cache memories first used in, 67
  expansion bus for, 55
  Macintosh vs., 97
  processors used in, 56
  quartz crystal oscillator for, 61
ICH (Input output Controller Hub), 367 (*See also* Southbridge chip)
ICs (*see* Integrated circuits)
IDE (Integrated Drive Electronics), 54
Ideal interconnect scaling, 29, 32
Ideal scaling, 26
IEEE standard, 104
IEEE standard #1394, 57
IEEE-1394b, 58
IEU (integer execution unit), 166
IHS (*see* Integrated heat spreader)
ILD (interlevel dielectric), 368
ILP (*see* Instruction level parallelism)
Imaging, 3D, 51–52

Immediate value, 101
Implementation verification, 179
Implicit operands, 101
Indexed addressing mode,
    107, 108
Industry Standard Architecture
    (ISA), 55, 56
Infinity (number), 104
Infrared emissions microscope
    (IREM), 346, 367
Initialization, 341
In-order pipelines, 131
Input/output controller hub (ICH),
    367 (*See also* Southbridge chip)
Input/ouput (I/O) bus, 55
Instruction(s), 98–117
    allocation step of, 164–165
    and branch prediction, 162–163
    decoding, 162
    and dispatch, 165, 166
    encoding, 115–117
    for high-level programming
        languages, 98
    IPC of, 134
    and L2 cache, 162–164
    life of, 160–167
    and load queue, 165
    macroinstructions, 154
    and microbranch prediction, 163
    prefetch of, 161
    and register files, 166
    and register renaming, 165
    retirement of, 158–159, 167
    for SIMD, 105
    single instruction multiple data, 105
    and tree cache, 163
    and uops, 164, 166, 167
    (*See also specific types, e.g.*: Byte
        swap instruction)
Instruction level parallelism (ILP),
    128–129, 367
Instruction pointer (IP), 112, 161, 367
Instruction translation lookaside
    buffer (ITLB), 161
Instructions per cycle (IPC), 131, 134
    defined, 368
    of Itanium 2, 140
    of superscalar processors, 133

Insulators, 4, 368
Integer benchmark program, 141
Integer execution unit (IEU), 166
Integrated circuits (ICs), 10–14
    defined, 368
    history of, 12, 14
Integrated Drive Electronics
    (IDE), 54
INTegrated ELectronics, 15
Integrated heat spreader (IHS),
    322, 368
Intel, 15
    architecture and processors used
        by, 96
    and FireWire licensing, 58
    lead and compaction designs of, 28
    and Pentium FDIV bug, 180–181
    SpeedStep technology, 233
    terminology of, vs. AMD, 123
    USB support by, 57
    and x86 architecture, 124
    (*See also specific products of, e.g.*:
        Enhanced SpeedStep)
Intel 90-nm fabrication generation,
    30, 32
Intel 268, 56
Intel 386, 56
Intel 4004, 16–18
Intel 8088, 56
Intel MXX, 105
Intel processors, 76–77 (*See also
    specific types, e.g.*: Pentium 4)
Interconnect(s):
    capacitance improvements of, 32
    scaling, 23–27
    substrate types affecting, 313–318
Interconnect delays, 27
Interconnect scaling, quasi-ideal,
    29, 32
Interference, 39
Interlevel dielectric (ILD), 368
Internal leads, 368
International Business Machines
    (*see* IBM)
International Technology Roadmap
    for Semiconductors (ITRS)
    (2003), 80, 87
Internet Explorer, 62

396　Index

Internet protocol (IP) relative addressing, 113, 116
Interrupts, 112
　defined, 368
　hardware, 114
Invalid (I) ownership state, 148
Inversion layer, 368
Inverter, 183, 209
I/O (input/ouput) bus, 55
Ion beam etching, 287, 288
Ion etching, reactive, 288
Ion implantation, 270–272, 368
Ion milling, 287, 288
IP (*see* Instruction pointer)
IP (Internet protocol) (*see* Internet protocol relative addressing)
IPC (*see* Instructions per cycle)
IREM (*see* Infrared emissions microscope)
ISA (Industry Standard Architecture), 55
ISA expansion cards, 56
Isotropic etch, 286, 368
Itanium 2, 96, 140
Itanium processor, 97
ITLB (instruction translation lookaside buffer), 161
ITRS (*see* International Technology Roadmap for Semiconductors)

Jam latch, 217
Java, 96, 97
Java Virtual Machine (JVM), 100
　and computations, 101
　processors for, 96
　on Web sites, 97
J-bend pin, 311, 312, 368
Jitter, clock, 225
Jump instruction, 368
Jumps, 112
Junction diode, 4, 6
Junction transistors, 8–10
JVM (*see* Java Virtual Machine)

K6 (processor), 179
K7 (processor), 82–84
K75 (processor), 82–84

Karnaugh maps (K-maps), 185–191, 368
Katmai (processor), 76, 77
kb (kilobyte), 369
Keeper device, 368
Kennedy, John F., 14
Keyhole gaps, 274
Keyhole voids, 274, 368
KGD (*see* Known good die)
Kilby, Jack, 12
Kilobyte (kB), 369
K-maps (*see* Karnaugh maps)
Known good die (KGD), 305, 369

L2 (*see* Level 2 cache)
Laminate substrate multichip module (MCM-L), 325
Land (lead), 311
Land grid array (LGA), 369
Land leads, 312
Land pad, 312
Land side capacitors (LSC), 320, 369
Laser voltage probing (LVP), 346, 369
Latch, 217, 369
Latch circuit, 218
Latchup, 266–267, 369
Late changes, 341
Layering, 268–279
Layering (semiconductor manufacturing):
　deposition, 272–276
　doping, 268–272
　and planarization, 278–279
　and thermal oxidation, 276–278
Layout, 73, 239–260
　creating, 240–246
　defined, 369
　density of, 245–253
　device limited, 363
　quality of, 253–259
　wire limited, 383
Layout density, 245–253, 369
Layout design rules, 247, 369
Layout transistor count, 250
L_{DRAWN} (drawn gate length), 20

Lead(s), 369
  ball, 312
  external, 365
  internal, 368
  land, 312
  number/configuration of, 309–311
  package, 311
  package on board, 311
  pin, 311
  power, 310
  SMT as, 311
  substrate types affecting, 313–318
  THT as, 311
  types of, 311–312
Lead designs, 27–28, 79
  defined, 369
  of Intel, 28
Lead pitch, 369
Leadframe, 369
Leadless packages, 312
Leakage:
  gate, 234, 235, 366
  leakage power, 233
  subthreshold, 21, 234, 379
Leakage currents, 32
Leakage power ($P_{leakage}$), 233, 369
L_{EFF} (effective channel length), 21
Legging a transistor, 248–250
Level 1 cache, 67–68
Level 2 (L2) cache, 68, 162–164
LGA (land grid array), 369
L_{GATE} (gate length), 21, 29
Light wavelengths, 282–286
Li'l Abner, 343
Lilienfield, Julius, 8
Lilliput, 106
Linear region, 203, 369
Lithography:
  e-beam, 281
  extreme ultraviolet, 365
Little endian, 106–107, 370
Livelock, 342, 370
Load instruction, 370
Load queue, 165
Loads, 103
Load/store architecture, 102

Local oxidation of silicon (LOCOS), 277
  defined, 370
  and thermal field oxide, 289, 290
Locality, 143
Logic:
  combinational, 182–191
  sequential, 191–196
Logic analyzer, 339
Logic bugs, 338, 339
  defined, 370
  sources of, 341
Logic design, 73, 171–196
  defined, 370
  and HDL model, 172–182
  and logic minimization, 182–196
Logic gates, 183, 210
Logic minimization, 182–196
  and combinational logic, 182–191
  Karnaugh maps for, 185–191
  and sequential logic, 191–196
Logic synthesis (*see* Logic minimization)
Logical operations, 99
Logical shielding, 257
LongRun, 61, 233
Low-*K* dielectrics, 26
Low-pressure chemical vapor deposition (LPCVD), 275, 370
LSC (*see* Land side capacitors)
LVP (*see* Laser voltage probing)
L_{WIRE} (length of wire), 25

M (modified) ownership state, 148
M1 (*see* Metal 1 wires)
Machine language, 98, 370
Macintosh computers, 97
Macroinstructions, 154–155
Magneto-optic (MO) discs, 54
Main memory, 47–51 (*See also* Memory)
Manufacturing, semiconductors, 263–301
Manufacturing defects, 338, 343
Marginality, circuit, 361
Markets, for microprocessors, 75

Mask designers, 240
Masks:
  binary intensity masks, 284
  phase-shift masks, 283
  photolithography, 373
  and photolithography, 280–282
Maximum delay (maxdelay), 200
Mazor, Stan, 15, 16
MB (megabyte), 370
MCA (Micro Channel Architecture), 56
MCH (*see* Memory controller hub)
MCM (*see* Multichip modules)
MCM-C (ceramic substrate multichip module), 325
MCM-D (deposited substrate multichip module), 325
MCM-L (laminate substrate multichip module), 325
MCP (multichip processors), 325
Meaningless Indicator of Processor Speed (MIPS), 137
Megabyte (MB), 370
Memory, 47–51
  cache, 360
  core, 15
  flash, 62
  hierarchy of, 63–68
  high-speed, 64
  interleaving, 42
  virtual, 65, 154, 382
Memory addresses, 106–108
Memory architectures, 101, 102
Memory bus standards, 107
Memory cell, 1T, 217
Memory controller hub (MCH), 41, 370
Memory execution unit (MEU), 166
Memory gap, 67, 68
Memory hierarchy:
  defined, 370
  latency of, 66
  need for, 66
  processors using, 66
Mendocino (processor), 76, 77
MESI protocols, 148, 370
Metal 1 (M1) wires, 27, 242, 248, 289, 297

Metal oxide semiconductor field-effect transistors (MOSFETs), 13, 200
  behavior of, 200–207
  bipolar junction transistors vs., 14
  bulk, 31
  current drawn by, 19
  defined, 370–371
  delay of, 20
  silicon-on-insulator, 31
  threshold voltages of, 202, 203
Metallization, under-bump, 381
Metal-only stepping, 347
MEU (memory execution unit), 166
MFLOPS (*see* Million floating-point operations per second)
Micro Channel Architecture (MCA), 56
Microarchitectural flaws, 341
Microarchitecture, 73, 127–168
  and branch prediction, 149–152
  and cache coherency, 147–149
  and cache memory, 143–146
  concepts of, 142–143
  defined, 371
  and instruction life, 160–167
  and macro/microinstructions, 154–155
  and microcode, 155–157
  of Pentium 4, 157
  performance design, 134–137
  performance measuring, 137–142
  pipelining for, 128–134
  and register renaming, 152–154
  and reorder buffer, 157–159
  and replay, 160
  and retiring an instruction, 158–159
Micro-ATX, 59
Microbranch prediction, 163
Micro-BTX, 59
Microcode, 155–157
  and BIOS ROM, 155
  defined, 371
Microcode patch, 348

Microinstruction pipeline, 160
Microinstructions, 154–155
Micrometer (micron, μm), 371
Microprocessor(s), 2–3, 14–17
  defined, 371
  first, 15
  historical trends in fabrication of, 29
  instructions for, 96
  markets for, 75
  power of, 60
  scaling, 27–30
Microsoft Windows, 62
MicroVax 78032, 96
Microvias, 317, 371
Milling, ion, 287, 288
Million floating-point operations per second (MFLOPS), 138, 371
Million instructions per second (MIPS), 96, 137, 371
Mindelay (minimum delay), 200
Minesweeper, 62
Mini-ATX, 59
Minimal sum, 185, 371
Minimization, logic (*see* Logic minimization)
Minimum delay (mindelay), 200
MIPS (*see* Million instructions per second)
Misses (*see specific types, e.g.*: Cache miss(es))
MMX (Multi-Media Extension), 96
MO (magneto-optic) discs, 54
Mobile processors, 75
Modified (M) ownership state, 148
Moore, Gordon, 11, 15, 17–18, 309
Moore's Law, 17–18
  applied to DRAM and processor development, 118
  defined, 371
  future of, 30–33
  and high-K gate dielectric, 32
  and interconnect improvements, 32
  and multiple threshold voltages, 30
  and program size, 110
  and SOI, 31
  and strained silicon, 31, 32
  and triple gates, 32–33
Morgan (processor), 82–84

MOSFETs (*see* Metal oxide semiconductor field-effect transistors)
Motherboards, 38, 58–61, 304, 371
Multichip modules (MCM), 305, 323–325
  with ceramic substrate, 325
  defined, 371
  with deposited substrate, 325
  with laminate substrate, 325
Multichip processors (MCP), 325
Multilevel cache, 65
Multi-Media Extension (MMX), 96
Multiple threshold voltages, 30
Multitasking, 109, 123
μm (micrometer, micron), 371
Mux, 371

N+ diffusion, 254
NAND gates, 183, 209, 371
Nanometer (nm), 371
Nanotechnology, 33
NASA, 14
Nickel-iron alloy, 314
nm (nanometer), 371
NMOS (*see* N-type metal oxide semiconductor)
NMOS transistors, 200–201
  for CMOS RAM, 61
  and voltage, 217
Noise, 211, 212
  affecting circuit design, 226–231
  hold, 230, 367
  setup, 376
  supply, 379
NOR logic gates, 183, 210
Normal atmospheric pressure (*see* APCVD)
Northbridge chip(s), 41
  defined, 372
  and video adapters, 51
Northwood (processor), 76–78
NOT gates, 183, 372
Not-A-Number, 104
Notebook computers, 75
Noyce, Robert (Bob), 11, 12, 15
N-type metal oxide semiconductor (NMOS), 372

N-type semiconductors, 6, 372
N-type silicon, 4

Ohl, Russell, 3–4
OPC (*see* Optical proximity correction)
Opcode, 115, 372
Operating system (OS), 155, 372
Opteron (processor), 65, 140
Optical drives, 54
Optical proximity correction (OPC), 283, 372
OR gates, 183, 372
Organic laminate packages, 316
OS (*see* Operating system)
Out-of-order execution, 157, 372
Out-of-order pipeline, 133
Out-of-order processors, 132
Overclocking, 46–47
Ownership states, 147–148
Oxidation:
  local, 277
  thermal, 276–278, 379
Oxide, 277

P+ diffusion, 254
PA 8800, 96
Package(s):
  BBUL, 329, 360
  ceramic, 316
  CERDIP, 361
  chip scale, 329
  cost of, 88
  DIP, 310, 364
  grid array, 310, 366
  leadless, 312
  organic laminate, 316
  PDIP, 373
  plastic, 313–314
  quad, 310, 375
  wire bond, 314
Package costs, 88
Package leads, 311
Package on board leads, 311
Packaged part testing, 352
Packaging, 303–329
  assembly flow example for, 325–328

Packaging (*Cont.*):
  design choices for, 308–325
  hierarchy for, 304–308
  and multichip modules, 323–325
  and thermal resistance, 320–323
P_{active} (active power), 232
Page fault, 110
Palo Alto, California, 11
Palomino, 82–84
Parallel port, 57
PA-RISC (*see* Precision Architecture RISC)
PA-RISC MAX2, 106
Parity bits, 74, 372
Part testing, 352
Partial scan, 335
Patterson, Dave, 96, 97
PC (*see* IBM PC)
PC programs, 97
PCA (*see* Printed circuit assembly)
PCB (*see* Printed circuit boards)
PCI (*see* Peripheral component interconnect)
PCI-Express, 56
PDAs (Personal Digital Assistants), 75
PDIP (plastic dual in-line package), 373
PECVD (*see* Plasma-enhanced chemical vapor deposition)
Pellicle, 282
Pentium 4, 30
  bus ratios for, 45
  computer architectures using, 96
  data transfers for, 44–45
  development of, 76, 77
  L2 cache of, 162
  as lead design, 79, 80
  microarchitecture of, 157
  performance comparisons of, 97
  pipelines of, 140, 160–161
  reorder buffer of, 164
  two-level cache incorporated in, 65
  uop retirement in, 167
  and x86, 119
Pentium 4 Extreme Edition, 78
Pentium FDIV bug, 180–181, 338
Pentium II, 65

Pentium III, 44
  and 32-bit virtual addresses, 110–111
  microarchitecture of, 157
  two-level cache incorporated in, 65
Pentium processor(s), 97 (*See also specific processors, e.g.*: Pentium 4)
Performance bugs, 338, 339, 372
Performance design, 134–137
Performance measuring, 137–142
Peripheral bus, 57–58
Peripheral component interconnect (PCI), 56, 373
Perl, 98
Personal Digital Assistants (PDAs), 75
PGA (pin grid array), 373
Phase-shift masks (PSMs), 283, 373
Photolithography, 264, 279–286
  defined, 373
  light wavelengths affecting, 282–286
Photolithography masks, 280–282, 373
Photoresist (PR), 264, 279, 280, 373
Physical addresses, 111
Physical gate oxide thickness ($T_{OX\text{-}P}$), 21
Physical shielding, 257
Pico-BTX, 59
Pico-Java, 96
Pin (lead), 311
Pin brazing, 316
Pin grid array (PGA), 373
Pipeline breaks, 131
Pipelined processors, 129–130
  combinational logic in, 191, 192
  sequential processor vs., 129, 130
Pipelines:
  defined, 373
  execution, 160, 161
  front-end, 160
  in-order, 131
  microinstruction, 160
  out-of-order, 133
  of Pentium 4, 140, 160–161
Pipelining, 128–134
Planarization, 278–279, 373

Plasma oxide (PO), 275, 373
Plasma-enhanced chemical vapor deposition (PECVD), 275, 288, 373
Plastic dual in-line package (PDIP), 373
Plastic leaded chip carrier (PLCC), 373
Plastic molding, 314
Plastic packages, 313–314
Plastic quad flat pack (PQFP), 373
P-latch, 217–218
Plated through hole (PTH), 373
PLCC (plastic leaded chip carrier), 373
$P_{leakage}$ (*see* Leakage power)
Plug-n-Play (PnP), 57
  defined, 374
  FireWire capability for, 56
PMOS (P-type metal oxide semiconductor), 375
PMOS transistors, 200–201
P-N junction diode, 6, 374
PnP (*see* Plug-n-Play)
PO (*see* Plasma oxide)
Point-contact transistor, 8
Point-to-point buses, 39
Pollack, Fred, 136
Pollack's rule, 136, 374
Poly gate layer, 241
Polycrystalline silicon (poly), 15, 374
Polysilicon, 15, 277
Polysilicon wires, 248
Poor step coverage, 274
Pops, 100
Ports, 39
POST (*see* Power on self test)
Post-silicon bugs, 342
Post-silicon validation, 338–344
  defined, 374
  platforms/tests for, 339–340
  and sources of bugs, 341–344
  tests created for, 340
Power:
  concerns in circuit design, 231–235
  of microprocessors, 60
POWER architecture, 97

Index

Power bugs, 339, 374
Power grid, 252
Power leads, 310
Power on self test (POST), 62, 374
PowerNow!, 61, 82–84
PowerPC, 97
  defined, 374
  processors for, 96
  special architecture modes of, 107
PowerPC AltiVec, 106
PPC 970 (G5), 96
PQFP (plastic quad flat pack), 373
PR (*see* Photoresist)
Prd (prediction signal), 195
Precision Architecture RISC (PA-RISC), 97
  defined, 372
  and Hewlett Packard, 97
  and PA-RISC MAX2, 106
  processors for, 96
  special architecture modes of, 107
Prediction:
  static, 150
  two-level, 151
Prediction signal (Prd), 195
Prefetch, 161
Prelayout circuit design, 235
Prepeg, 306, 374
Pre-silicon validation, 181–182, 322, 341
  defined, 374
  in HDL model, 178–182
Printed circuit assembly (PCA), 306, 374
Printed circuit boards (PCBs), 304, 374
Probe testing, wafer, 351
Process scaling, 17
Processor(s), 1–2 (*See also specific types, e.g.*: Athlon processors)
  frequency of, 138
  markets of, 74–75
  microarchitecture/fabrication of, 97
  overclocking of, 46–47
  repackaging/disabling, 78
  used in IBM PC, 56
Processor architecture, 72, 127–168
Processor bus, 44–47
Processor design, 79
Processor design team, 81
Processor performance, 141–142
Processor roadmaps, 74–78
Product cost, 85–91
Programming languages, high-level, 98, 173
Proliferation design, 374
Proliferations, 79
Przybylski, Steven, 118
PSMs (*see* Phase-shift masks)
PS/2, 56
PTH (plated through hole), 373
P-to-N-ratio, 212, 375
P-type metal oxide semiconductor (PMOS), 375
P-type semiconductors, 6, 375
P-type silicon, 4
Pumped bus, 40

Quad package, 310, 375
Quad-pumped bus, 40
Quartz, 280
Quartz crystal oscillator, 61
Quasi-ideal interconnect scaling, 29, 32
Quasi-ideal scaling, 26
Queue, load, 165

Race (mindelay violation), 223
RAM (*see* Random access memory)
Rambus DRAM (RDRAM), 50
RAMDAC (*see* Random access memory digital-to-analog converter)
Ramp, silicon, 74
Random access memory (RAM):
  CMOS RAM, 61
  defined, 375
  double data rate SDRAM, 50
  Dynamic random access memory, 47
  RAMDAC, 51, 375
  SDRAM, 48, 379
  Static random access memory, 48, 378
Random access memory digital-to-analog converter (RAMDAC), 51, 375

RAS (row access strobe), 48–49
RAS to CAS delay (T_{RCD}), 48
RAT (*see* Register alias table)
RAW (read-after-write)
    dependency, 153
RC delay, 25
RCC (*see* Resin coated copper foil)
RDRAM (Rambus DRAM), 50
Reactive ion etching (RIE), 288, 375
Read only memory (ROM):
    BIOS ROM, 62
    defined, 375
    and microcode, 155
    and silicon debugging, 348
Read-after-write (RAW)
    dependency, 153
Real Time Clock (RTC), 61
    defined, 375
    quartz crystal oscillator for, 61
Reduced instruction set computing
        (RISC), 97
    addressing modes for, 108
    CISC vs., 157
    and computations, 102
    defined, 375
    EPIC vs., 119–122
    operand structure of, 103
    Precision Architecture RISC, 97
    and RISC-I processors, 96
Register(s), 100
    64-bit, 105
    addresses vs., 111
    defined, 375
    doubling size of, 116
Register alias table (RAT), 153,
        165, 375
Register architectures, 101, 102
Register file(s), 63–64, 166
Register indirect addressing,
        107, 108
Register renaming, 152–154,
        165, 375
Register transfer language (*see* RTL)
Register transfer level (*see* RTL)
Removable media, 54
Renaming (*see* Register renaming)
Rent, E. F., 309
Rent's Rule, 309–310, 376

Reorder buffer (ROB), 157–159
    defined, 376
    of Pentium 4, 164
    uops in, 164
Repackage design, 79, 376
Replay, 160
Required windows, 224–225, 376
Resin coated copper (RCC) foil,
        316, 317
Resistance, thermal, 320–323, 380
Resource conflict, 376
Reticle, 376
Retirement (instructions), 167
Retrograde wells, 272
Return instruction(s), 112
    defined, 376
    and stack addressing, 113
RIE (*see* Reactive ion etching)
Ripple carry adder, 190
RISC (*see* Reduced instruction set
        computing)
RISC-I processors, 96
ROB (*see* Reorder buffer)
ROM (*see* Read only memory)
Row access strobe (RAS), 48–49
R-scaling, constant, 26
RTC (*see* Real Time Clock)
RTL (register transfer language,
        register transfer level), 73, 174,
        375 (*See also specific types, e.g.*:
        Verilog)
RTL model, 174, 176

S (shared) ownership state, 148
Safe mode (Windows), 62
Sales, bin frequency affecting, 46
SATA (*see* Serial advanced
        technology attachment)
Saturation region, 203, 376
Scalable Architecture (*see under*
        SPARC)
Scaling:
    constant field, 23
    constant *R*-scaling, 26
    future of, 33
    ideal, 26
    interconnects, 23–27
    microprocessors, 27–30

Scaling (*Cont.*):
    quasi-ideal, 26
    quasi-ideal interconnect, 29, 32
    of transistors (*see* Transistor scaling)
Scan, 334, 336, 337
Scan sequentials, 376
Schematic device count, 250
SCSI (*see* Small Computer System Interface)
Scuzzy (*see* Small Computer System Interface)
SDRAM (*see* Synchronous dynamic random access memory)
Self-test, built-in (*see* Built-in self-test)
Semiconductor(s), 4
    defined, 376
    N-type, 372
    P-type, 375
Semiconductor crystals, 3–4
Semiconductor design, 353
Semiconductor manufacturing, 263–301
    CMOS process flow example for, 289–299
    and etching, 286–288
    layering steps for, 268–279
    and photolithography, 279–286
    wafer fabrication in, 265–268
Sequential circuits, 195
Sequential logic, 191–196
Sequential processors, 129, 130
Sequentials, 216–220
    defined, 376
    scan, 376
SER (soft error rate), 230
Serial advanced technology attachment (SATA), 54, 376
Serial port, 57
Server(s), 74
    1U, 357
    2U, 357
    blade, 74
Server farms, 74
Server processors, 74, 75
Setup noise, 229–230, 376
Shallow trench isolation (STI), 289, 376
Shared (S) ownership state, 148

Shielding, 257
Shima, Masatoshi, 16
Shmoo plot, 345–346, 376
Shockley, William, 6–11
Shockley Semiconductor, 11
Sidewall spacers, 295, 376
Signaling, differential, 39
Silicide, 295, 376
Silicon:
    germanium vs., 10
    implanted with oxygen, 376
    local oxidation of, 370
    N-type silicon, 4
    polycrystalline silicon, 15
    P-type silicon, 4
    silicon on insulator, 31
    strained, 31, 32
Silicon bugs, 338
    defined, 376
    solutions to, 347
Silicon debug, 73, 332, 344–350, 376
Silicon dioxide, 12, 26–27, 274
Silicon implanted with oxygen (SIMOX), 31, 376
Silicon junction transistor, 10
Silicon nitride, 274
Silicon ramp, 74
Silicon test, 350–353
Silicon transistors, 10, 12
Silicon Valley, 11
Silicon-on-insulator MOSFETs, 31
Silicon-on-insulator (SOI) wafer, 31, 267
    bonded etched back silicon on insulator wafers, 359
    defined, 376
SIMD (*see* Single instruction multiple data)
SIMOX (*see* Silicon implanted with oxygen)
SIMOX (silicon implanted with oxygen), 31
Single instruction multiple data (SIMD), 105
    defined, 376
    instructions for, 105
    Streaming SIMD Extension, 105, 106

Single precision, 376
Sintering, 316
Skew, clock (*see* Clock skew)
Slot packaging, 68
Slurry, 278
Small Computer System Interface (SCSI, "scuzzy"), 54, 378
SMT (*see* Surface mount technology)
Snooping, 147
Socket(s), 58, 307, 312
SOD (*see* Spin-on dielectric)
Soft bake, 279
Soft error rate (SER), 230, 378
Soft errors, 230–231
Software, 98
Software bugs, 342, 349
Software interrupts, 113–114
SOG (*see* Spin-on glass)
SOI (*see* Silicon-on-insulator wafer)
SOI MOSFETs, 31
Soldering:
  defined, 378
  wave, 307
Solid-state amplifier, 6–11
Solid-state diffusion, 269, 378
Source (MOSFET), 200
Source terminal, 378
Southbridge chip(s), 41, 43
  buses specifically designed for, 53
  defined, 378
Spacers, sidewall, 295, 376
SPARC, 96, 97
SPARC VIS, 106
Sparks, Morgan, 9–10
Spatial locality, 143, 378
SPEC (Standard Performance Evaluation Corporation), 138, 141
SPEC89 benchmarks, 151
SpecFP2000, 112
SPECfp2000 benchmarks, 139, 140
SPECint92 benchmarks, 98–99
SPECint2000, 112, 138–139
Speedpaths, 338, 378
SpeedStep, 61, 233
Spin-on deposition, 273, 378
Spin-on dielectric (SOD), 273, 378
Spin-on glass (SOG), 273, 378

Spitfire (processor), 82–84
Sputter etching, 287, 288
Sputtering, 273, 274, 280, 281, 378
SRAM (*see* Static random access memory)
SSE (*see* Streaming SIMD Extension)
Stack, 100, 378
Stack addressing, 113
Stack architecture, 100, 101
Standard Performance Evaluation Corporation (SPEC), 138
Stanford University, 11, 96
Static prediction, 150
Static random access memory (SRAM), 48
  defined, 378
  high-speed, 67
Steppers, 280, 379
Stepping, 347
STI (*see* Shallow trench isolation)
Storage devices, 53–55
Store instruction, 379
Stored-response test, 351–352
Stores, 103
Straight pins, 312
Strained silicon, 31, 32
Streaming SIMD Extension (SSE), 105, 106
Structural level model, 174
Stuck-at fault model, 335, 379
Substrates, and interconnects, 313–318
Subthreshold leakage, 21, 234
  defined, 379
  threshold voltage affecting, 30
Subthreshold slope, 23, 234
Sun (company):
  architecture and processors used by, 96
  and SPARC, 97
Sunlight, 230
Super I/O chip, 43, 379
Superscalar processors, 132
  defined, 379
  IPC of, 133
  register files of, 166
Supply noise, 228–229, 379

Surface mount technology (SMT), 307, 308
  defined, 379
  as lead, 311
Swapping, hot, 57
Swift, Jonathan, 106
Synchronous buses, 39
Synchronous dynamic random access memory (SDRAM), 48–49
  defined, 379
  double data rate SDRAM, 50
  latencies of, 49
Synthesis, 177, 178
System-on-a-chip, 43, 44

Tag array, 145
Taken signal (Tk), 194
Tape drives, 54
Tapeout, 73, 260, 379
TB (terabyte), 379
T_{CL} (CAS latency), 48
T_{DELAY}, 19
Teal, Gordon, 9–11
Technology node, 28, 29
Temporal locality, 143, 379
Temporal shielding, 257
Terabyte (TB), 379
Terman, Frederick, 11
Test circuits, design for, 333–338
Test costs, 89
Testing, 331–354
  design for test circuits for, 333–338
  post-silicon validation for, 338–344
  and silicon debug, 344–350
  and silicon test, 350–353
  wafer probe, 382
  (*See also specific tests, e.g.:* Electrical test)
Texas Instruments (TI), 10, 12
Texture maps, 51
Thermal field oxide, 289
Thermal interface material (TIM), 322, 379
Thermal oxidation, 276–278, 379
Thermal resistance, 320–323, 380
Thoroughbred (processor), 82–84
Three-operand architectures, 102, 103

Threshold voltage(s) (V_T), 20
  affecting subthreshold leakage, 30
  defined, 380
  of MOSFETs, 202, 203
  multiple, 30
Through hole technology (THT), 307, 308
  defined, 380
  as lead, 311
THT (*see* Through hole technology)
Thunderbird (processor), 82–84
TI (*see* Texas Instruments)
Tie bar, 380
T_{ILD} (vertical spacing of wires), 25
TIM (*see* Thermal interface material)
Timing, circuit, 221–226
T_{INT} (wire thickness), 25
Tk (taken signal), 194
TLB (*see* Translation lookaside buffer)
T_{OX} (gate oxide), 19
$T_{OX\text{-}E}$ (electrical oxide thickness), 21
$T_{OX\text{-}P}$ (physical gate oxide thickness), 21
Trace cache, 157, 380
"Traitorous eight," 11
Transfer curve, 212, 380
Transistor(s), 3–10
  Bipolar Junction Transistors, 10, 200
  bulk metal oxide semiconductor field-effect, 31
  capacitance of, 19
  defined, 380
  field-effect, 7, 8, 13
  history of, 5–8
  junction, 8
  legging a, 248–250
  MOSFETs, 370–371 (*See also* Metal oxide semiconductor field-effect transistors)
  NMOS transistors, 61, 200–201
  PMOS transistors, 200–201
  point-contact, 8
  silicon, 10, 12
  silicon junction, 10
  sizing of, 18, 212–216
Transistor counts, 250

Transistor legging, 248–250, 380
  (*See also* Legging a transistor)
Transistor length, 380
Transistor radio, 10
Transistor scaling, 19–24, 68, 71–72
Transistor width, 380
Translation lookaside buffer (TLB), 111, 146, 380
Transmeta, 61, 233
Trap exceptions, 115, 380
Trap handler, 115
T_{RCD} (RAS to CAS delay), 48
Tree cache, 163
Triple gates, 32–33
Tungsten, 274, 295
Tunneling current, 22
Twin tub processes, 290
Two-level cache, 65
Two-level prediction, 151
Two-operand architectures, 102, 103
Two's complement, 104, 381

UBM (*see* Under-bump metallization)
UF (*see* Underfill)
Ultra-ATA, 54
UltraSPARC IV, 96
Ultraviolet (UV) light, 283
Ultraviolet (UV) lithography, 381
Unbuffered latch, 218
Under-bump metallization (UBM), 327, 381
Underfill (UF), 319, 381
Unified cache, 162
Unity gain points, 381
Universal Serial Bus (USB), 57
  defined, 381
  FireWire vs., 58
  Intel support for, 57
  USB 2.0, 58
University of Illinois, 10
Uops, 155, 164–167 (*See also* Microinstructions)
U.S. Department of Defense, 14
USB (*see* Universal Serial Bus)
USB 2.0, 58
UV (*see under* Ultraviolet)

Vacuum tube, 3, 381
Valence bands, 5
Valid windows, 223–224, 381
Validation, 74
  defined, 381
  post-silicon, 339–340, 374
  pre-silicon, 341, 374
Value processors, 75
Variations, 79
VAX architecture, 96, 102
  addressing modes for, 108
  creation of, 118
  defined, 381
  operand structure of, 103
  processors for, 96
VAX-11/780, 137
V_{dd} (voltage swing), 19
V_{dd}, 381
Verification, design, 180
Verilog, 173–174
  C vs., 174
  defined, 382
Very high-speed integrated circuit hardware description language (VHDL), 173, 382
Very long instruction word (VLIW), 119, 382
VHDL (*see* Very high-speed integrated circuit hardware description language)
Vias, 243, 306
  blind, 306
  buried, 306
  defined, 382
  microvias, 317, 371
  punch and fill of, 315
VID (*see* Voltage identification)
Video adapters (graphics cards), 51–53
$V_{i}L$ (voltage input low) unity, 227
Virtual address(es), 109–111
Virtual memory, 65, 111, 154
  data transfer instructions for, 108–111
  defined, 382
  effects of, on programs, 110
  and multitasking, 109
VLIW (*see* Very long instruction word)

Voltage(s):
  control of, 60–61
  multiple, 30
  and NMOS transistors, 217
  threshold, 20
Voltage identification (VID), 60–61, 382
Voltage input low (V_{iL}) unity, 227
Voltage regulator (VR), 60–61, 382
Voltage swing (V_{dd}), 19
VR (*see* Voltage regulator)
V_{ss}, 382
V_T (*see* Threshold voltage)

Wafer fabrication, 265–268
Wafer probe testing, 351, 382
Wafer yield, 88, 382
Wafers:
  BESOI, 359
  bonded and etched-back SOI, 267
  bulk, 266
  epi-wafer, 267, 364
  fabrication of, 265–268
  probe testing of, 351
  silicon-on-insulator, 267, 376
WAR (write-after-read) dependencies, 153
Wave soldering, 307
WAW (write-after-write) dependencies, 153
Web sites, 97
Well(s), 382
Well taps, 245
Wet etches, 287, 382
Wet oxidation, 276
Wettable (term), 382
Willamette (processor), 76, 77
Windows (Microsoft), 62

W_{INT} (wire width), 24
Wire(s):
  in buses, 38–39
  copper, 26
  heating of, 258
  layers of, 27
  length of, 25
  polysilicon, 248
  vertical spacing of, 25
Wire bond package, 314, 383
Wire capacitance, 25–27
Wire limited layout, 383
Wire spacing (W_{SP}), 24
Wire thickness (T_{INT}), 25
Wire width (W_{INT}), 24
Wires capacitance, 24
Workstations, 74
Write (instruction), 129
Write-after-read (WAR) dependencies, 153
Write-after-write (WAW) dependencies, 153
W_{SP} (wire spacing), 24

x86 architecture, 96
  64-bit address support by, 110
  creation of, 118
  defined, 383
  extensions for, 122–124
  memory architectures for, 101
  operand structure of, 102, 103
  operations for, 98–99
  processors for, 96
  use of, 119
XNOR gates, 183, 383
XOR gates, 183, 383

z-buffer, 51

CPSIA information can be obtained at www.ICGtesting.com
Printed in the USA
BVOW03*0125181214
379506BV00006B/12/P

9 780071 459518